# Tributes
## Volume 22

# Foundational Adventures
## Essays in Honour of Harvey M. Friedman

Volume 12
Dialectics, Dialogue and Argumentation. An Examination of Douglas Walton's Theories of Reasoning and Argument
Chris Reed and Christoher W. Tindale, eds.

Volume 13
Proofs, Categories and Computations. Essays in Honour of Grigori Mints
Solomon Feferman and Wilfried Sieg, eds.

Volume 14
Construction. Festschrift for Gerhard Heinzmann
Solomon Feferman and Wilfried Sieg, eds.

Volume 15
Hues of Philosophy. Essays in Memory of Ruth Manor
Anat Biletzki, ed.

Volume 16
Knowing, Reasoning and Acting. Essays in Honour of Hector J. Levesque.
Gerhard Lakemeyer and Sheila A. McIlraith, eds.

Volume 17
Logic without Frontiers. Festschrift for Walter Alexandre Carnielli on the occasion of his 60$^{th}$ birthday
Jean-Yves Beziau and Marcelo Esteban Coniglio, eds.

Volume 18
Insolubles and Consequences. Essays in Honour of Stephen Read.
Catarina Dutilh Novaes and Ole Thomassen Hjortland, eds.

Volume 19
From Quantification to Conversation. Festschrift for Robin Cooper on the occasion of his 65$^{th}$ birthday
Staffan Larsson and Lars Borin, eds

Volume 20
The Goals of Cognition. Essays in Honour of Cristiano Castelfranchi
Fabio Paglieri, Luca Tummolini, Rino Falcone and Maria Miceli, eds.

Volume 21
From Knowledge Representation to Argumentation in AI, Law and Policy Making. A Festschrift in Honour of Trevor Bench-Capon on the Occasion of his 60$^{th}$ Birthday
Katie Atkinson, Henry Prakken and Adam Wyner, eds.

Volume 22
Foundational Adventures. Essays in Honour of Harvey M. Friedman
Neil Tennant, ed.


Tributes Series Editor
Dov Gabbay                                               dov.gabbay@kcl.ac.uk

# Foundational Adventures
## Essays in Honour of Harvey M. Friedman

edited by
# Neil Tennant

© Individual author and College Publications 2014. All rights reserved.

ISBN 978-1-84890-117-9

College Publications
Scientific Director: Dov Gabbay
Managing Director: Jane Spurr

http://www.collegepublications.co.uk

Originally published online by Templeton Press
http://www.foundationaladventures.com

Original cover illustration: Courtesy of The Ohio State University
Cover design by Laraine Welch

Printed by Lightning Source, Milton Keynes, UK

---

All rights reserved. No part of this publication may be reproduced, stored in a retrieval system or transmitted in any form, or by any means, electronic, mechanical, photocopying, recording or otherwise without prior permission, in writing, from the publisher.

# CONTENTS

| | |
|---|---|
| Foundational Adventures: Editorial<br>**Neil Tennant** | 1 |
| Why the Theory R is Special<br>**Albert Visser** | 7 |
| Finding the Phase Transition for Friedman's Long Finite Sequences<br>**Andreas Weiermann and Martijn Baartse** | 25 |
| Software Verification with Towers of Abstraction<br>**Bruce W. Weide** | 41 |
| Absolute Infinite – A Bridge between Mathematics and Theology?<br>**Christian Tapp** | 77 |
| Analytic Cut in Modal Logic: the System B<br>**Grigori Mints** | 91 |
| Disguising induction: Proofs of the Pigeonhold Principle for Trees<br>**Jeffry L. Hirst** | 113 |
| Friedman and the Axiomatization of Kripke's Theory of Truth<br>**John P. Burgess** | 125 |
| Xeno Semantics for Ascending and Descending Truth<br>**Kevin Scharp** | 149 |
| Pragmatic Platonism<br>**Martin Davis** | 169 |

$\omega$-Models and Well-ordering Principles  179
**Michael Rathjen**

Ordering Free Products in Reverse Mathematics  213
**Reed Solomon**

Towards $NP - P$ via Proof Complexity and Search  243
**Samuel R. Buss**

Equivalence Relations in Set Theory, Computation Theory, 267
Model Theory and Complexity Theory
**Sy-David Friedman**

An Application of Proof-theory in Answer Set Programming 285
**V. W. Marek and J. B. Remmel**

# Foundational Adventures: Editorial
NEIL TENNANT

This volume is a partial Proceedings of the extraordinary conference *Foundational Adventures*, in honor of Harvey M. Friedman on the occasion of his 60th birthday. The conference was held at The Blackwell Inn at The Ohio State University, May 14–17, 2009. It was made possible by a generous grant from the John Templeton Foundation, supplemented by a grant for travel support from the National Science Foundation. The editor and organizer would like to take this opportunity to express his deep appreciation for the enthusiastic and unstinting support extended by OSU's President Gordon E. Gee, and Professor Joan Leitzel, the Interim Dean of the Federation of Colleges of Arts and Sciences; and by Dr. Hyung Choi of the Templeton Foundation.

The conference celebrated Harvey's distinguished, ongoing, and still blossoming researches into the Foundations of Mathematics and related areas. A disciplinarily diverse and highly distinguished list of invited speakers, from the USA and from abroad, bore witness to the remarkable fecundity and profundity of Harvey's seminal contributions. A full account of the conference is available online at p. 3 of the Philosophy Department's annual newsletter:

http://philosophy.osu.edu/files/philosophy/logos2009-10.pdf

The collection of papers here assembled provides a rich sample of the stimuli that Harvey's ideas have provided to mathematicians, computer scientists, music theorists, logicians and philosophers. Rather than attempt to summarize them in advance for the reader, I shall enter a few anecdotes of a more personal kind about the ways in which Harvey, his interests and his work have meant a great deal to me. This editorial provides me an opportunity to pay tribute to a man I am proud and privileged to know as a colleague and friend, and who has enormously enriched my own intellectual life for over twenty years.

I first met Harvey when I was a PhD student at Cambridge. He visited to give a mini-series of three seminars on the failure of Beth's Theorem on implicit definability for certain infinitary languages. That *tour de force* eventually appeared as [Friedman(1973)], Harvey's first publication in model theory. His results proved directly relevant several years later to work of my own on the problem of supervenience and reductionism, a topic in the philosophy of science that cannot be adequately discussed in innocence of results such as Harvey's.

My next encounter with work of Harvey's that bore directly on interests of my own was more indirect. While I was at the Australian National University, I was pursuing a project to develop proof-search strategies based on constraints afforded by normalization results in proof theory. I was outside the ANU mainstream in this regard, where the Automated Reasoning Project was focused more on the kinds of relevance logics studied in the Anderson-Belnap tradition. They were seeking to develop matrix methods to quickly knock out sequents unprovable in such logics. According to one of the ARP's founding submissions by the late Robert (Bob) Meyer, their aim was to program an automated reasoner that might be able to prove *Fermat's Last Theorem*(!)—if only they could first settle a supposedly minor, vexatious matter: the truth of an outstanding conjecture about Peano Arithmetic based on the Anderson-Belnap relevance logic $R$. This conjecture was called 'Gamma for $R^\#$', basically to the effect that *modus ponens* is an admissible rule in that system. After fruitless years with the conjecture unsettled, Bob took the problem to Columbus, Ohio, to ask Harvey to look at it. Harvey did, and a few days later gave Bob the devastating news that Gamma for $R^\#$ is false. One can read the brilliantly marshalled reasons why this is so in [Friedman and Meyer(1992)].

Shortly after I moved from the ANU to Ohio State, Harvey graciously attended all the sessions of an Interdisciplinary Research Seminar on the Mechanization of Inference, organized through the Center for Cognitive Science. After learning of my interest in proof-search methods for constructive relevant logics, he slipped me a piece of scrap paper on which he had written a subtly intercalated cascade of biconditionals involving 21 propositional variables, with a note to the effect that if $n = 21$ didn't work, then $n = 22$ would. The sense of 'work', by the way, was something like this: either your computer will crash upon using up all its available memory, or it will run until the Big Crunch, in vain search for a proof that an expert human logician can find quite easily, as soon as s/he twigs on to the underlying pattern by means of which the $n$-variable formula is constructed.

More recently Harvey has displayed once again his astonishing ability to listen to someone with a problem that affords enough philosophical or methodological interest, and technical challenge, to pique his interest, and then solve it in a relative flash. In connection with work on rational belief revision, I wanted to know whether any mathematical theory might afford infinitely many distinct justifications for any of its theorems. Put another way: could there be a theorem that is logically implied by infinitely many pairwise logically inequivalent conjunctions of axioms? Within a couple of days Harvey had the answer: *No*. For any extant mathematical theory, every theorem is implied by at most finitely many pairwise logically inequivalent conjunctions of axioms. Moreover, for Peano Arithmetic there is a bound on this number of distinct selections of axioms: *nineteen*. I am honored to be able to present Harvey's proof of this result in Chapter 10 of [Tennant(2012)].

It is widely recognized that Harvey is endowed with formidable natural talent—after all, hard work alone does not secure one a PhD from MIT and a Professorship at Stanford by the age of eighteen. But it should also be put on record that Harvey is utterly and strenuously devoted to the Life of the Mind, in a fiercely uncompromising and independent way. I do not refer here only to his night-owl habits, and his phone calls at all hours, which are legion within the profession. I refer also to the way Harvey has championed the pursuit of a certain kind of intellectual career: a life devoted to the very deepest problems, on which publishable progress might not be made within the usual timeframes of a scientific career. Harvey has set an example which only a gifted few might dare to follow; and for his intellectual courage we should all be grateful. His is a rare combination of prodigious natural gifts with unflagging industry.

Harvey's favorite acronym is g.i.i. (general intellectual interest). His renowned deductive powers are laser-like, traversing enormous deductive distances in a flash. The metaphor, however, is even more appropriate when it comes to the concentrated precision of focus that Harvey maintains when seeking foundational results that will be of fundamental intellectual importance. His (meta)theorems *have to have* g.i.i. in order to be worth pursuing. Harvey is impatient with that brand of overspecialized footling that one unfortunately encounters all too often as the stock-in-trade of professional logicians and mathematicians—the kind of research that Lord Eric Ashby once called 'creeping along the frontiers of knowledge with a hand lens'.

Imagine Frege were alive today, but unaware of how his own foundational researches had come to enjoy such celebrated status among philosophers and logicians in the second half of the 20th century. If he were to be apprised of the things Harvey has chosen to work on, because of their technical depth and philosophical importance, he might voice once again his fear that mathematicians would think *metaphysica sunt, non leguntur!*, and philosophers would think *mathematica sunt, non leguntur!* ([Frege(1893; reprinted 1962)], p. xii). The warm gratitude and respect paid to Harvey's work on the occasion of our conference in his honor give one the hope that those days are receding. No longer will fundamental importance be judged only from within the confines of a single discipline, with the untoward result that genuinely interdisciplinary achievements earn only the occasional cautious nod from scholars who are hostage to a single disciplinary framework and who carry a very compact intellectual toolbox (inside which, admittedly, are some very sharp tools).

Harvey is a thinker with the Big Picture constantly in mind, and, indeed, in constant search for a deeper understanding of how certain things might lie Beyond the Horizon. He is this century's true heir to Kurt Gödel, who initiated the metamathematical study of the essential incompleteness of consistent and sufficiently rich formal mathematical systems. When Gödel hit upon his idea of a certain brilliantly constructed formal sentence $G$ 'saying of itself', via arithmetical coding of its syntax, *I am unprovable*, the system shown to be unable to prove $G$ was relatively weak (Peano Arithmetic) and the unprovable (but true) sentence $G$ was formidably complex when written out fully in ordinary mathematical notation. Mathematicians could look at $G$, shrug their shoulders, say 'Not my cup of tea, I'm afraid!', and return to their everyday work of proving results that struck them, by contrast, as simple, natural, intuitively accessible, and elegant.

The Friedman program set itself the aim of uncovering much more profoundly unsettling Gödel-phenomena: independence results that would grab the practising mathematician by the scruff of the neck, so that she would at last *have* to take notice, and begin to question the adequacy of the set-theoretic foundations she had tacitly assumed were available for her discipline. Harvey has achieved this (as of this time of writing) by simultaneously

(i) identifying unprovable $\Pi_1^0$-sentences $G$ that are simple, intuitively accessible and elegant; and

(ii) strengthening *both* the formal system (to ZFC, say) that is unable to prove such a sentence $G$, *and* the formal system (to ZFC plus certain large cardinals, say) to which one would have to resort in order to establish $G$ as true.

Results of this kind represent the full flowering of Harvey's seminal insights about logico-mathematical strength, on which the field of reverse mathematics is based. The 'core mathematician' with little or no exposure to mathematical logic and fully formalized systems might regard the original independent Gödel sentences (saying *I am unprovable*), and the consistency statements to which they are equivalent, as ideological excrescences—as having nothing to do, organically, with the practice of the core mathematical community. They might even think the same of the set theorist's postulation of ever-larger kinds of cardinal numbers. These infinities, so the complacent core-mathematician would contend, are internal to set theory conceived of as a very specialized branch of mathematics itself, and surely have nothing to do with the potential 'set-theoretical grounding' of mathematics as they know it and practice it (and intend to extend it). *'Bourbaki, yes; but all these higher infinities?... surely no!'* And didn't Harvey himself say, in [Friedman(1971)], that all known mathematical objects have their set-theoretic surrogates residing in $V_{\omega+\omega}$? Going yet further: they might hold the same view towards any extension of the language they use by adjoining an explicit truth predicate, and then allowing it to feature in new instances of axiom schemata to which they already subscribe, such as the Principle of Mathematical Induction. Surely, they might think, talk about truth won't be of much help in settling first-order matters concerning basic mathematical objects such as natural numbers, rational numbers, real numbers, finite-dimensional vectors of such objects, and so on?

Harvey's seminal insight was that, *modulo* weak enough base systems, these apparently irrelevant means of strengthening a formal system were respectively *provably equivalent* to whole hosts of standard theorems that the core mathematician prizes and loves. Different mathematical theorems which, on the face of it, are just about certain well-understood kinds of mathematical objects, have built into them, as it were, a certain level of 'consistency strength'. Moreover, the hierarchy of consistency strengths appears to be *linear*. Different schools of thought about the 'correct' or 'acceptable' foundations of mathematics then correspond to different cut-off points, as it were, on this linear scale of consistency strengths. The *coup de grâce* of this line of revelation is that certain seductively natural, short and elegant statements about tame structures of combinatorial objects have consistency strengths so high up on this scale that core mathematicians would have to scramble to swot up on the higher infinities in [Kanamori(1994)], all of which now loom as potentially *logically indispensable* for the settling of conjectures about what, beforehand, were believed to be utterly tame mathematical entities. Either that, or search for a completely different *kind* of foundation for mathematics altogether.

## BIBLIOGRAPHY

[Frege(1893; reprinted 1962)] Gottlob Frege. *Grundgesetze der Arithmetik. I. Band.* Georg Olms Verlagsbuchhandlung, Hildesheim, 1893; reprinted 1962.

[Friedman and Meyer(1992)] Harvey Friedman and Robert K. Meyer. Whither Relevant Arithmetic? *Journal of Symbolic Logic*, 7:824–831, 1992.

[Friedman(1971)] Harvey M. Friedman. Higher set theory and mathematical practice. *Annals of Mathematical Logic*, 2(3):325–357, 1971.

[Friedman(1973)] Harvey M. Friedman. Beth's Theorem in Cardinality Logics. *Israel Journal of Mathematics*, 14(2):205–212, 1973.

[Kanamori(1994)] Akihiro Kanamori. *The Higher Infinite. Large Cardinals in Set Theory from their Beginnings.* Perspectives in Mathematical Logic. Springer, Berlin, 1994.

[Tennant(2012)] Neil Tennant. *Changes of Mind*. Oxford University Press, Oxford, 2012.

# Why the Theory R is Special
ALBERT VISSER

ABSTRACT. In this paper we provide a 'coordinate-free' characterization of the Tarski-Mostowski-Robinson theory R by showing that R is the maximum of the global degrees of interpretability that are in the minimum local degree of interpretability. In more mundane terms, we show that a recursively enumerable theory is locally finite iff it can be globally interpreted R. This is the first non-trivial coordinate-free characterization of a canonical theory.

*Dedicated to Harvey Friedman on the occasion of his 60th birthday.*

## 1 Introduction

Wouldn't it be nice if we could characterize salient theories like Robinson's Arithmetic and Peano Arithmetic in a coordinate-free way, independent of particular choices of signature and axiomatization? A moment's reflection shows that such a characterization would only be possible modulo some notion of sameness of theories. For example, one could imagine that a certain RE (recursively enumerable) degree of interpretability was characterized by a first-order formula $A$ over the partial preorder of degrees of interpretability of RE theories. If our salient theory were in that degree, our formula $A$ would be the coordinate-free characterization we are looking for. The theory would be characterized modulo mutual interpretability.

Regrettably, we have very few examples of such characterizations and the ones we have are rather trivial. Here are three examples.

- The theory EQ of pure equality is the initial element of the category of direct interpretations (without parameters). (An interpretation is *direct* if it is identity preserving and unrelativized.) The theory EQ is thus characterized modulo synonymy (aka. definitional equivalence).

- Consider the lattice of degrees of local interpretability of theories, where we impose no restriction on the signature nor on the complexity of the theory. The intended notion of interpretation is many-dimensional interpretation with parameters. This structure is studied by Mycielski, Pudlák and Stern in [3]. They call these degrees: *chapters*. The maximum of the structure is the degree of inconsistent theories. There is also an element directly below this maximum: the maximal degree of consistent theories. The following two salient theories are in the degree: Th($\mathbb{N}$) and Th$_{\Pi_1^0}$($\mathbb{N}$) (the theory axiomatized by the true $\Pi_1^0$-sentences).[1]

- Consider the degree structure of degrees of one-dimensional interpretability with parameters. Let $\uparrow\mathbf{n}$, for $n = 1, 2, \ldots$, be EQ, the theory of pure equality, plus the sentence $\exists x_0, \ldots, x_{n-1} \bigwedge_{i<j<n} x_i \neq x_j$. Let $\uparrow\infty$ be the union of the $\uparrow\mathbf{n}$. Our degree structure yields a partial ordering of the following form: we first have $\omega + 1$ and above that something else. The theory $\uparrow\mathbf{n}$ is in the $(n-1)$th degree from below and $\uparrow\infty$ is in the $\omega$th degree. So each of these theories is characterized modulo mutual one-dimensional interpretability with parameters. (See Theorem 4 of this paper.)

In the present paper, we produce a less trivial example of a characterization. The structure in which the characterization is given is the double degree structure of local and global interpretability for RE theories. The notion of interpretation involved is: many-dimensional, piecewise interpretation with parameters.[2] We will show that the theory R is the maximum of the global degrees that are in the minimal local degree.[3] As we will see this means that an RE theory is locally finitely satisfiable iff it is globally interpretable in R.

The theory R was introduced by Tarski, Mostowski and Robinson in their book [7]. It is a very weak theory that is *essentially undecidable*. This means that every consistent RE extension of the theory is undecidable. It was observed by Cobham that one still has an essentially undecidable theory if one drops the axiom R6 (given below), obtaining the theory $\mathsf{R}_0$. See [8] and [2]. Cobham has shown that R has a stronger property than essential undecidability. Consider any RE theory $T$. Suppose we have translation $\alpha$ of the arithmetical language into the language of $T$. Suppose $T$ is consistent with $\mathsf{R}_0^\alpha$. Then, $T$ is undecidable.[4] For the proof of a closely related result, see Vaught's paper [8]. In fact one can show that, if $T$ is consistent with

---

[1] The result also holds if we restrict ourselves to one-dimensional interpretations and/or parameter-free interpretations.

[2] We will explain *piecewise* in the paper.

[3] We will prove a number of related results where we vary the notion of interpretation used in defining the degree structure.

[4] Cobham's proof remains unpublished, but, using the clues provided in [8], it is not hard to find a proof.

$\mathsf{R}_0^\alpha$, then there is a finitely axiomatized extension $A$ of $\mathsf{R}_0$ and a translation $\beta$, such that $T$ is consistent with $A^\beta$.

## 2 Theories and Interpretations

In this section, we fix some basic concepts and notations. The reader is advised to go over it lightly, returning just when a notation or notion is not clear.

We work with RE theories in one-sorted first order predicate logic of finite signature with identity. These theories have *officially* a relational signature. Unofficially, we use function symbols, but these can be eliminated using a well-known unwinding procedure.

Our most general notion of interpretation is *piecewise, many-dimensional, relative interpretation with parameters, where identity is not necessarily translated as identity*. We will first set up the machinery for the case of *many-dimensional, relative interpretation with parameters, where identity is not necessarily translated as identity*. Then, we will extend the framework to piecewiseness. Even if the basic idea behind the various kinds of interpretation is rather obvious, some care is needed to get the definitions right, mainly because some careful management of the use of variables is necessary.

### 2.1 Translations

To define an interpretation, we first need the notion of *translation*. We first define the notion of *many-dimensional translation with parameters*.

Let $\Sigma$ and $\Xi$ be finite signatures for first-order predicate logic. We fix a sequence containing all variables $u_0, u_1, \ldots$ for the signature $\Sigma$, and we fix three disjoint sequences of variables $v_0, v_1, \ldots$, and $w_0, w_1, \ldots$, and $z_0, z_1, \ldots$ for the signature $\Xi$.

A *relative translation* $\tau : \Sigma \to \Xi$ is given by a quintuple $\langle n, m, \pi, \delta, F \rangle$. Here $n$ and $m$ are natural numbers. The number $n$ is the *dimension* of the interpretations and the number $m$ gives us the number of parameters. The formula $\pi$ is $\Xi$-formula with free variables among $w_0, \ldots, w_{m-1}$. This formula gives a constraint on the possible parameters. The formula $\delta$ is a $\Xi$-formula with bound variables among the $z_0, z_1, \ldots$ and free variables among $v_0, \ldots, v_{n-1}, w_0, \ldots, w_{m-1}$. It defines the domain of the interpretation. The mapping $F$ associates to each relation-symbol $R$ of $\Sigma$ (including the binary symbol $=$) a $\Xi$-formula $F(R)$. Let the arity of $R$ be $k$. We demand that the bound variables of $F(R)$ are among the $z_i$, and that $F(R)$ has at most the variables $v_0, \ldots, v_{kn-1}, w_0, \ldots, w_{m-1}$ free. We will write $\vec{v}_i$ for the block of variables $v_{ni}, \ldots, v_{n(i+1)-1}$ and $\vec{w}$ for $w_0, \ldots, w_{m-1}$. So, $F(R)$ will have at most $\vec{v}_0, \ldots, \vec{v}_{k-1}, \vec{w}$ free. We translate $\Sigma$-formulas to $\Xi$-formulas as follows:

- $(R(u_{j_0}, \cdots, u_{j_{k-1}}))^\tau := F(R)(\vec{v}_{j_0}, \cdots, \vec{v}_{j_{k-1}}, \vec{w})$;
  the formula $F(R)(\vec{v}_{j_0}, \ldots)$ is the result of simultaneous substitution of $v_{nj_i+s}$ for $v_{ni+s}$, where $0 \le i < k$ and $0 \le s < n$;

- $(\cdot)^\tau$ commutes with the propositional connectives;

- $(\forall u_k\, A)^\tau := \forall \vec{v}_k\, (\delta(\vec{v}_k, \vec{w}) \to A^\tau)$;

- $(\exists u_k\, A)^\tau := \exists \vec{v}_k\, (\delta(\vec{v}_k, \vec{w}) \wedge A^\tau)$.

We have introduced translations with careful variable management, showing how this can be done. In practice we want to be sloppy. We use e.g. $x$ and $x_i$ as metavariables ranging over the $u_j$ (and we will overload the use of, e.g., $z$ in the same way). If $x$ stands for $u_i$, we write $\vec{x}$ for $\vec{v}_i$. If $\vec{x}$ stands for $x_0, \ldots, x_{k-1}$, we will use $\vec{\vec{x}}$ for $\vec{x}_0, \ldots, \vec{x}_{k-1}$. Etc.

Here are some convenient conventions and notations. Suppose $\tau$ is the translation $\langle n, m, \pi, \delta, F \rangle$.

- We write $\delta_\tau$ for $\delta$, etc.

- We write $R_\tau$ for $F_\tau(R)$.

- We write $\vec{\vec{x}} : \delta_{\vec{w}}$ for: $\delta(\vec{x}_0, \vec{w}) \wedge \ldots \wedge \delta(\vec{x}_{k-1}, \vec{w})$.

- We write $\forall \vec{\vec{x}}{:}\delta_{\vec{w}}\, A$ for: $\forall \vec{x}_0 \ldots \forall \vec{x}_{k-1}\, (\vec{\vec{x}}{:}\delta_{\vec{w}} \to A)$. Similarly for the existential case.

We can define the identity translation and composition of translations in the obvious way.

Consider $\tau = \langle n, m, \pi, \delta, F \rangle$. The translation $\tau$ is one-dimensional if $n = 1$. In this case we will write: $\tau = \langle m, \pi, \delta, F \rangle$. The translation $\tau$ is parameter-free if $m = 0$ and $\pi = \top$. In this case, we will write: $\tau = \langle n, \delta, F \rangle$. If $\tau$ is one-dimensional and parameter-free, we write: $\tau = \langle \delta, F \rangle$.

## 2.2 Interpretations and Interpretability

A translation $\tau$ supports a *relative interpretation* of a theory $U$ in a theory $V$, if, $V \vdash \exists \vec{w}\, \pi_\tau \vec{w}$ and, for all sentences $A$ of the language of $U$, we have $U \vdash A \Rightarrow V \vdash \forall \vec{w}\, (\pi_\tau \vec{w} \to A^\tau \vec{w})$. Thus, an interpretation has the form: $K = \langle U, \tau, V \rangle$.

We note that the above definition is equivalent to the following. A translation $\tau$ supports a *relative interpretation* of a theory $U$ in a theory $V$, if,

i. $V \vdash \exists \vec{w}\, \pi_\tau \vec{w}$

ii. $V \vdash \forall \vec{w}\, (\pi_\tau \vec{w} \to \exists \vec{v}\, \delta_\tau \vec{v} \vec{w})$

iii. for all axioms $A$ of the theory of identity, we have:
$U \vdash A \;\Rightarrow\; V \vdash \forall \vec{w}\,(\pi_\tau \vec{w} \to A^\tau \vec{w})$

iv. for all axioms $A$ of the the theory $U$, we have:
$U \vdash A \;\Rightarrow\; V \vdash \forall \vec{w}\,(\pi_\tau \vec{w} \to A^\tau \vec{w})$

The identity interpretation $\mathsf{ID}_U : U \triangleright U$ is defined as $\langle U, \mathsf{id}_\Sigma, U\rangle$, where $\Sigma$ is the signature of $U$ and $\mathsf{id}_\Sigma$ is the identity translation for $\Sigma$. The composition $K; M$ of $K : U \triangleleft V$ and $M : V \triangleleft W$ is $\langle U, (\tau_K; \tau_M), W\rangle$, where $\tau_K; \tau_M$ is the composition of the underlying translations of $K$ and $M$.

Par abus de langage, we write '$\delta_K$' for: $\delta_{\tau_K}$; '$P_K$' for: $P_{\tau_K}$; '$A^K$' for: $A^{\tau_K}$, etc. We define:

- We write $K : U \triangleleft V$ or $K : V \triangleright U$, for:
  $K$ is an interpretation of the form $\langle U, \tau, V\rangle$.

- $V \triangleright U :\Leftrightarrow U \triangleleft V :\Leftrightarrow \exists K\; K : U \triangleleft V$.
  We read $U \triangleleft V$ as: $U$ is interpretable in $V$. We read $V \triangleright U$ as: $V$ interprets $U$.

We say that *a theory $V$ locally interprets a theory $U$* if, for any finite subtheory $U_0$ of $U$, we have $V \triangleright U_0$. We write $V \triangleright_{\mathsf{loc}} U$ for: $V$ locally interprets $U$. It is easily see that both $\triangleleft$ and $\triangleleft_{\mathsf{loc}}$ are preorderings.

## 2.3 Piecewise Interpretability

The idea of piecewise interpretability is that we can build up the domain from a number of pieces that may or may not be of the same dimension and that may or may not overlap. The same object of the interpreting theory may occur in different roles posing as different objects of the interpreted theory.

A translation $\tau$ now has the form $\langle \ell, \nu, m, \pi, \delta, F\rangle$, where:

- $\ell$ is a natural number that stands for the set of pieces $0, \ldots, \ell-1$;

- $\nu$ is a function that assigns to each piece $j$ a dimension $\nu^j$;

- $m$ is again the number of parameters, and $\pi$ a constraint on the parameters;

- $\delta$ is a function from pieces $j$ to domains $\delta^j$ of dimension $\nu^j$;

- $F$ is now a function that sends a pair $P, f$ to an appropriate formula. Here $f$ assigns to each argument place of $P$ a piece.

To make it all work smoothly we again enumerate the variables of $\Sigma$ by $u_0, u_1, \ldots$ We fix $\ell + 2$ disjoint sequences of variables for the signature $\Xi$, to wit $v_0^0, v_1^0 \ldots$, and $\ldots$ and $v_0^{\ell-1}, v_1^{\ell-1}, \ldots$, and $w_0, w_1, \ldots$ and $z_0, z_1, \ldots$ We write $\vec{v}_i^j$ for $v_{\nu^j i}^j, \ldots, v_{\nu^j (i+1)-1}^j$ and $\vec{w}$ for $w_0, \ldots, w_{m-1}$. We demand:

- $\pi$ contains variables among $\vec{w}$.

- $\delta^j$ has bound variables among the $z_0, z_1, \ldots$ and free variables among $\vec{v}_0^j, \vec{w}$.

- Suppose $P$ has arity $k$. $F(P, f)$ has bound variables among $z_0, z_1, \ldots$ It has free variables among $\vec{v}_0^{f0}, \vec{v}_1^{f1}, \ldots, \vec{v}_{k-1}^{f(k-1)}, \vec{w}$.

We translate $\Sigma$-formulas to $\Xi$-formulas as follows. The basic form of translation is $A^{\tau,g}$, where $g$ is a function from the indices of the free variables of $A$ to pieces.

Before giving the somewhat unreadable atomic clause, we discuss it for a more concrete example. Suppose we have two pieces a and b. Let's say that a is 1-dimensional and b is 2-dimensional. Suppose we have no parameters.

We want to translate $P(u_3, u_1)$. How are we going to do it? Well, we need to know in what pieces $u_1$ and $u_3$ are supposed to be. The function $g$ is an oracle that tells us precisely that. Suppose that, according to $g$, $u_1$ is in piece a and $u_3$ is in piece b. We note that $g$ in combination with our formula $P(u_3, u_1)$ places b on the first argument place and a on the second argument place. Thus we need to consider $F(P, f)$, where $f(0) = $ b and $f(1) = $ b. Let $F(P, f)$ be $A(v_0^b, v_1^b, v_1^a)$. We note that $v_0^b, v_1^b$ is the block $\vec{v}_0^b$. We have to replace the variables in $A(v_0^b, v_1^b, v_1^a)$ by the ones corresponding to $u_3, u_1$. These are the blocks $v_3^b = v_6^b v_7^b$ and $\vec{v}_1^a = v_1^a$. Thus we find: $(P(u_3, u_1))^{\tau,g} = A(v_6^b, v_7^b, v_1^a)$.

- $(R(u_{j_0}, \cdots, u_{j_{k-1}}))^{\tau,g} := F(R, f)(\vec{v}_{j_0}^{g j_0}, \cdots, \vec{v}_{j_{k-1}}^{g j_{k-1}}, \vec{w})$,
  where $f(s) := g(j_s)$; the formula $F(R, f)(\vec{v}_{j_0}^{g j_0}, \ldots)$ is the result of simultaneous substitution of $v_{\nu^{fi} j_i + s}^{fi}$ for $v_{\nu^{fi} i + s}^{fi}$, where $0 \leq i < k$ and $0 \leq s \leq \nu^{fi} - 1$.

- $(A \wedge B)^{\tau,g} := A^{\tau,g \restriction \mathsf{fv}(A)} \wedge B^{\tau,g \restriction \mathsf{fv}(B)}$;
  similarly for the other propositional connectives;

- $(\forall u_k A)^{\tau,g} := \bigwedge_{j < \ell} \forall \vec{v}_k^j (\delta^j(\vec{v}_k^j, \vec{w}) \to A^{\tau,g[k:j]})$, where $g[k:j]$ is the result of setting $g$ at $k$ to $j$;

- $(\exists u_k A)^{\tau,g} := \bigvee_{j < \ell} \exists \vec{v}_k^j (\delta^j(\vec{v}_k^j, \vec{w}) \wedge A^{\tau,g[k:j]})$.

A translation $\tau$ is called *parameter-free* if $m_\tau = 0$ and $\pi = \top$, *one-dimensional* if $\nu_\tau j = 1$, for all $j < \ell_\tau$, *one-piece* iff $\ell_\tau = 1$. Similarly, for interpretations.

The rest of the development is the same as before.

EXAMPLE 1. We show that $\uparrow\mathbf{1} := \mathsf{EQ}$, the theory of pure identity interprets $\uparrow\ell := \mathsf{EQ} + \exists u_0, \ldots, u_{\ell-1} \bigwedge_{i<j<\ell} u_i \neq u_j$. We take $\tau$ with:

- $\ell_\tau := \ell$,

- $\nu_{\tau,j} = 1$,

- $m = 0$,

- $\pi_\tau := \top$,

- $\delta^{\tau,j} := (v_0^j = v_0^j)$,

- $F_\tau(=, ij) := (v_0^i = v_1^i)$, if $i = j$, and $F_\tau(=, ij) := \bot$, otherwise.

We can easily see that this construction does indeed yield the desired interpretation. It follows that $\uparrow\mathbf{1} \triangleright_{\mathsf{loc}} \uparrow\infty$, where $\uparrow\infty$ is the union of the $\uparrow\ell$.

If the target theory $V$ proves that we have at least two elements, we can always replace a piecewise interpretation by a many-dimensional one. Here is how this works. Suppose there are $\ell$ pieces and that $k$ is the maximum of the dimensions of the pieces. Without loss of generality we may assume that our pieces are represented by numbers $j < \ell$. Our new one-piece interpretation will be $k+\ell$-dimensional. We represent an object $v_0^j, \ldots v_{\nu^j-1}^j$ of $\delta^j$ as $v_0, \ldots v_{k+\ell-1}$. Here we demand that $\delta^j(v_0, \ldots, v_{\nu^j-1})$. The variables $v_{\nu^j}, \ldots, v_{k-1}$ are unconstrained: they serve as padding to get the right length. The variables $v_k, \ldots v_{k+\ell-1}$ satisfy $v_k = \ldots = v_{k+j} \neq v_{m+j+1} = \ldots = v_{k+\ell-1}$. They serve to keep the sequences from different pieces disjoint. From each sequence we can read off from which piece it comes. To find out whether $v_0, \ldots, v_{k+\ell-1}$ is equal to $v_{k+\ell}, \ldots, v_{2(k+\ell)-1}$, we first extract $j_0$ and $j_1$ from the end-strings and then check whether we have:

$$F(=, j_0 j_1)(v_0, \ldots, v_{\nu^{j_0}-1}, v_{k+\ell}, \ldots, v_{k+\ell+\nu^{j_1}-1}).$$

The treatment of the other predicates is similar. Note that our interpretation does not add new parameters. The connection between the old interpretation and the new one is quite close: there is a $V$-definable isomorphism between the old interpretation and the new one.

## 3  Double Degree Structures

The double degree structure $\mathcal{D}_{ijk}$ consists of the RE theories plus two partial preorders $\lhd$ and $\lhd_{\mathsf{loc}}$. Here the interpretations considered are (possibly) many-dimensional if $i = 1$ and 1-dimensional if $i = 0$, (possibly) piecewise if $j = 1$ and one-piece if $j = 0$, (possibly) with parameters if $k = 1$ and parameter-free if $k = 0$. (So the $i, j, k$ are booleans saying whether or not a certain feature is 'turned on'.)

Note that $\lhd$ is a subordering of $\lhd_{\mathsf{loc}}$.

The basic degree structures we work with are given as pairs of partial preorders. However, when convenient, we will flexibly switch to talk about degrees where the induced equivalence relations are divided out. Note that if we think of the degree structures in this last way, there is a projection functor $\pi$ mapping the degrees of global interpretability to the degrees of local interpretablity that is part of the double degree structure. We will use $[T]$ for the global degree of $T$ and $[T]_{\mathsf{loc}}$ for the local degree of $T$.

Let the lowest degree of global interpretability be $\bot$ and let the lowest degree of local interpretability be $\bot_{\mathsf{loc}}$. So, $\bot = [\mathsf{EQ}]$ and $\bot_{\mathsf{loc}} = [\mathsf{EQ}]_{\mathsf{loc}}$.

We provide a convenient characterization of $\bot$ and $\bot_{\mathsf{loc}}$ in $\mathcal{D}_{ijk}$ with $j = 1$. So only the presence of piecewise interpretations is essential for the result. A theory $T$ is *finitely satisfiable* if $T$ has a finite model. A theory $T$ is *locally finitely satisfiable* iff every finitely axiomatized subtheory of $T$ has a finite model.

THEOREM 2. *We work in $\mathcal{D}_{ijk}$ with $j = 1$. (i) The RE theory $T$ is in $\bot$ iff $T$ is finitely satisfiable. (ii) The RE theory $T$ is in $\bot_{\mathsf{loc}}$ iff $T$ is locally finitely satisfiable.*

**Proof.** Suppose $T$ is in $\bot$. So, for some $K$, we have: $K : \mathsf{EQ} \rhd T$. Let $\mathcal{M}$ be any finite model of $\mathsf{EQ}$. Clearly, the interpretation $K$ gives a finite 'internal' model $K(\mathcal{M})$ of $\mathcal{M}$.

Conversely, suppose $T$ is finitely satisfiable. Let $\mathcal{M}$ be a finite model of $T$. We may assume that the elements of $\mathcal{M}$ are $0, \ldots, m-1$. We now construct an interpretation that 'describes' $\mathcal{M}$ as follows. We take $\tau$ with:

- $\ell := m$,
- $\nu_j = 1$,
- $\mathfrak{m} = 0$,
- $\pi := \top$,
- $\delta^j := (v_0^j = v_0^j)$,

- $F(P, j_0 \ldots j_{k-1}) := \top$, if $P_{\mathcal{M}}(j_0, \ldots, j_{k-1})$, and $F(P, j_0 \ldots j_{k-1}) := \bot$, otherwise.

Part (ii) is immediate from (i). □

What happens when we do not have piecewise interpretations. Let us first consider the cases of $\mathcal{D}_{100}$ and $\mathcal{D}_{101}$. We have the following theorem.

THEOREM 3. *Consider $\mathcal{D}_{100}$ or $\mathcal{D}_{101}$.*

*i. Both $\perp\!\!\!\perp$ and $\perp\!\!\!\perp_{\mathsf{loc}}$ consist of precisely the theories that have a one-element model.*

*ii. The degree $\perp\!\!\!\perp$ has a unique immediate successor $\perp\!\!\!\perp^+$, that consists of the finitely satisfiable theories for which every model has at least two elements.*

*iii. The degree $\perp\!\!\!\perp_{\mathsf{loc}}$ has an a unique immediate successor $\perp\!\!\!\perp_{\mathsf{loc}}^+$, that consists of the locally finitely satisfiable theories, for which every model has at least two elements.*

Note that it follows that the (locally) finitely satisfiable theories can be characterized in terms of the degree structure.

**Proof.** Let $\uparrow\!\mathbf{1} := \mathsf{EQ}$ and $\uparrow\!\mathbf{2} := (\mathsf{EQ} + \exists x, y\ x \neq y)$.

*Ad (i):* It is easy to see that the elements of $\perp\!\!\!\perp$ are precisely those with a one-element model. Consider any $T$ such that $\mathsf{EQ} \vartriangleright_{\mathsf{loc}} T$. It follows that every finite subtheory of $T$ has a one-element model. Since the signature is finite, there are only finitely many such one-element models. So, there must be one model that satisfies arbitrarily large finite subtheories of $T$ and, hence, $T$ itself.

*Ad (ii):* We take $\perp\!\!\!\perp^+ := [\uparrow\!\mathbf{2}]$. We clearly have $\uparrow\!\mathbf{2} \not\trianglerighteq \uparrow\!\mathbf{1}$.

Suppose $T$ is finitely satisfiable and every model of $T$ has at least two elements. It is immediate that $T \vdash \uparrow\!\mathbf{2}$, and hence $T \vartriangleright \uparrow\!\mathbf{2}$. By the considerations at the end of Subsection 2.3, we can simulate piecewise interpretations by one piece interpretations (without introducing new parameters) as soon as we have at least two elements available in the interpreting theory. Thus, $\uparrow\!\mathbf{2} \vartriangleright T$. So if $T$ is finitely satisfiable, then $T \equiv \uparrow\!\mathbf{2}$.

For the converse, suppose $T \equiv \uparrow\!\mathbf{2}$. It is easy to see that $T$ must be finitely satisfiable. Moreover $T \vartriangleright \uparrow\!\mathbf{2}$, implies $T \vdash \uparrow\!\mathbf{2}$, hence every model of $T$ has at least two elements.

Finally, consider any RE theory $W$. If $W$ has a one-element model then $W \equiv \uparrow\!\mathbf{1}$, if not then $W \vartriangleright \uparrow\!\mathbf{2}$.

*Ad (iii):* This is immediate from (ii). □

Finally, we consider $\mathcal{D}_{001}$. Thus we will leave the case of $\mathcal{D}_{000}$ open. We remind the reader that, for $n > 0$,

$$\uparrow\mathbf{n} := (\mathsf{EQ} + \exists x_0, \ldots, x_{n-1} \bigwedge_{i<j<n} x_i \neq x_j),$$

and $\uparrow\infty$ is the union of the $\uparrow\mathbf{n}$.

THEOREM 4. *In $\mathcal{D}_{001}$ the situation is as follows.*

i. *A theory $T$ is finitely satisfiable iff $T \equiv_{(\text{loc})} \uparrow\mathbf{n}$, for some $n$. Here $n$ is the minimal size of a model satisfying $T$.*

ii. $\uparrow\mathbf{1} \lneq_{(\text{loc})} \uparrow\mathbf{2} \ldots \lneq_{(\text{loc})} \uparrow\infty.$

iii. *For any $T$, $T \equiv_{(\text{loc})} \uparrow\mathbf{n}$ for some $n$, or $T \rhd_{(\text{loc})} \uparrow\infty$.*

iv. *A theory $T$ is locally finitely satisfiable iff $T \lhd_{\text{loc}} \uparrow\infty$.*

v. $T \equiv_{\text{loc}} \uparrow\infty$ *iff $T$ is locally finitely satisfiable and has only infinite models, i.o.w. iff $T$ is locally finitely satisfiable but not finitely satisfiable.*

Note that the degrees $[\uparrow\mathbf{n}]$, $[\uparrow\infty]$, $[\uparrow\mathbf{n}]_{\text{loc}}$ and $[\uparrow\infty]_{\text{loc}}$ are all determined in terms of the degree structure.

**Proof.** We skip the easy proof. Note that we need the parameters to 'describe' a model of $n$ or less than $n$ elements in $\uparrow\mathbf{n}$. □

The theory R, due to Tarski, Mostowski and Robinson ([7]) is locally finitely satisfiable and not finitely satisfiable. The main result of our paper is that, for any locally finitely satisfiable theory $T$ is interpretable in R via a one-dimensional, one-piece, parameter-free, identity preserving interpretation. As a direct consequence of the above results, we find:

- In $\mathcal{D}_{ijk}$ with $j = 1$, R is in the $\lhd$-maximum of $\bot_{\text{loc}}$,
- In $\mathcal{D}_{100}$ and $\mathcal{D}_{101}$, R is in the $\lhd$-maximum of $\bot^+_{\text{loc}}$.
- In $\mathcal{D}_{001}$, R is in the $\lhd$-maximum of $[\uparrow\infty]_{\text{loc}}$.

In all cases the degree [R] is first order definable in the double degree structure.

REMARK 5. We note that in $\mathcal{D}_{i1k}$, the degree $[\uparrow\infty]$ is the immediate successor of $\bot$. In $\mathcal{D}_{100}$ and $\mathcal{D}_{101}$, the degree $[\uparrow\infty]$ is the immediate successor of $\bot^+$. Finally, in $\mathcal{D}_{001}$, the degree $[\uparrow\infty]$ is the supremum of the degrees $[\uparrow\mathbf{n}]$.

## 4 The Theories R, Q⁻ and Q

In this section, we briefly introduce three theories of number theory R, Q⁻ and Q. The theories Q⁻ and Q will each play a role in the proof of our main theorem. We consider the signature with constant and function symbols 0, S, + and ·. We define $\underline{0} := 0$, $\underline{n+1} := \mathsf{S}\underline{n}$, and $x \leq y :\leftrightarrow \exists z\, z + x = y$. We consider the following axioms.

R1. $\vdash \underline{m} + \underline{n} = \underline{m+n}$

R2. $\vdash \underline{m} \cdot \underline{n} = \underline{m \cdot n}$

R3. $\vdash \underline{m} \neq \underline{n}$, for $m \neq n$

R4. $\vdash x \leq \underline{n} \to \bigvee_{i \leq n} x = \underline{i}$

R5. $\vdash x \leq \underline{n} \vee \underline{n} \leq x$

R6. $\vdash\ \leq$ is a linear ordering

R7. $\vdash x < y \to \mathsf{S}x \leq y$

R8. $\vdash x \leq y \to \exists z {\leq} y\, z + x = y$

The theory $\mathsf{R}_0$ is axiomatized by R1,2,3,4. The theory R is axiomatized by R1,2,3,4,5. The theory R* is axiomatized by R1,2,3,4,5,7,8.

It was observed by Cobham that one still has an essentially undecidable theory if one drops the axiom R5 (given above), obtaining the theory $\mathsf{R}_0$. See [8] and [2]. In fact R is interpretable in $\mathsf{R}_0$. In [10], it is shown that R* is interpretable in $\mathsf{R}_0$, and hence in R.

Robinson's Arithmetic Q was introduced in [7]. Using Solovay's method of shortening cuts (see [5]), one can show that Q interprets seemingly much stronger theories like $I\Delta_0 + \Omega_1$. See [4] and [1]. Here are the axioms of Q.

Q1. $\vdash \mathsf{S}x = \mathsf{S}y \to x = y$

Q2. $\vdash 0 \neq \mathsf{S}x$

Q3. $\vdash x = 0 \vee \exists y\, x = \mathsf{S}y$

Q4. $\vdash x + 0 = x$

Q5. $\vdash x + \mathsf{S}y = \mathsf{S}(x+y)$

Q6. $\vdash x \times 0 = 0$

Q7. $\vdash x \times \mathsf{S}y = x \times y + x$

The theory $Q^-$ is due to Andrzej Grzegorczyk. It is a prima facie weakening of Q in which addition and multiplication are partial. Thus, instead of $+$ and $\times$ we have ternary relation symbols A and M. We define $\leq$ by $x \leq y :\leftrightarrow \exists z\, \mathsf{A}zxy$. The theory is axiomatized as follows.

G1. $\vdash \mathsf{S}x = \mathsf{S}y \to x = y$,

G2. $\vdash 0 \neq \mathsf{S}x$,

G3. $\vdash x = 0 \vee \exists y\, x = \mathsf{S}y$,

G4. $\vdash (\mathsf{A}xyz_0 \wedge \mathsf{A}xyz_1) \to z_0 = z_1$

G5. $\vdash \mathsf{A}x0x$

G6. $\vdash \mathsf{A}xyz \to \mathsf{A}x(\mathsf{S}y)(\mathsf{S}z)$

G7. $\vdash (\mathsf{M}xyz_0 \wedge \mathsf{M}xyz_1) \to z_0 = z_1$

G8. $\vdash \mathsf{M}x00$

G9. $\vdash (\mathsf{M}xyz \wedge \mathsf{A}zxw) \to \mathsf{M}x(\mathsf{S}y)w$

The theory Q is interpretable in $Q^-$ by a result of Vítězlav Švejdar. See [6].

## 5  R is Top

In this section, we prove our main result.

THEOREM 6. *For any locally finitely satisfiable RE theory $T$, we have $\mathsf{R} \triangleright T$, via a one-dimensional, one-piece, parameter-free interpretation.*

The proof has the following structure. We construct a certain class of 'numbers' Good in R using a predicate $\alpha$ that describes the axiom set of $T$. Either Good provides a certain uniquely determined element $g^\star$ or not. If it does we can use $g^\star$ to uniquely specify an internally finite model $z$ of a non-standardly finite part of $T$. Satisfaction in this model provides the desired interpretation of $T$. If $g^\star$ does not exist, this shows that Good is so rich that we can construct an interpretation of $Q^-$ in Good. Hence, by Švejdar's result we have in interpretation of Q in Good. But as soon as we have Q, we have a whole range of well known techniques available to construct an interpretation of $T$. We will exhibit two such ways.

Suppose $T$ is a locally finitely satisfiable RE theory. Let the axiom set of $T$ be given by the $\Sigma_1^0$-formula $\alpha$. We may assume that $\alpha y$ is of the form $\exists x\, \alpha_0 xy$, where $\alpha_0$ is $\Delta_0$. We can always arrange that $\alpha_0 mn$ implies $n < m$.

We need to speak about finite models and satisfaction in the context of number theory. Here we sketch one way how to do it. We code finite structures for the signature of $T$ by numbers. We use the iterations of Cantor Pairing $(\cdot,\cdot)$ to implement finite sequences $(\cdot,\cdot,\ldots)$ of fixed standard length. We use Ackermann coding to represent finite sets. This means that we define $x \in y$ as $\exists u, v \leq y\ (y = u \cdot 2^{x+1} + v \wedge 2^x \leq v < 2^{x+1})$. (Remember that the graph of exponentiation is $\Delta_0$-definable.) Suppose the signature of $T$ is $\Sigma := \langle P, Q, \ldots \rangle$. We code a finite structure for $\Sigma$ as a sequence $n := \langle m, p, q, \ldots \rangle$, where $m$ stands for domain $\{0, \ldots, m-1\}$, and where $p$ codes, in Ackermann style, a finite set of sequences with length the arity of $P$, etc. Just to make the argument run smoothly we stipulate that 0 is a code for the one-element model in which all atomic statements made with predicate symbols from the signature are false. We also assume that 0 is not a Gödel number of a formula.

We can find a $\Sigma^0_1$-formula $z, f \models a$ meaning: the model $z$ and the assignment $f$ satisfy the formula $a$. For any formula $A$ we let $A^y$ be the result of bounding all quantifiers in $A$ by $y$. We define $\mathsf{comm}(u)$ as the conjunction of the following statements.

i. For all models $z < u$, the assignments $f$ for $z$ on variables $w'$, with $w' < u$, are closed under random reset for the domain of $z$. In other words, for a variable $w < u$, and $a < (z)_0$ and $f$ an assignment for $z$ on variables $w' < u$, there exists an assignment $f[w:a]$ such that, for all $w', b < u$, with $w'$ a variable and $b < (z)_0$, we have:

$$f[w:a](w') = b \leftrightarrow ((w' \neq w \wedge fw' = b) \vee (w' = w \wedge b = a)).$$

ii. All formulas $a < u$, have a unique analysis into terms and formulas below $u$.

iii. We have the commutation conditions for formulas $a$ and models $z$ below $u$.

We find that, if $n \geq 2^{2^{c \cdot m^2}}$, then $\mathsf{comm}^n(m)$, where $c$ is a sufficiently large standard number.

Let $\mathsf{good}(x)$ be a predicate coding the conjunction of (a) and (b).

a. There is a largest $y \leq x$, such that there is a minimal $z \leq x$, such that $\mathsf{comm}^x(\max(z, y))$, and, for any $a \leq y$, if $a \in \alpha^y$, then $z \models^x a$.

b. For all $u, v, w \leq x$, we have: $\mathsf{S}u \neq 0$, $\mathsf{S}u = \mathsf{S}v \to u = v$, $u = 0$ or $\exists q \leq x\ \mathsf{S}q = u$, $u + 0 = u$, $u + \mathsf{S}v = \mathsf{S}(u+v)$, $u \times 0 = 0$, $u \times \mathsf{S}v = u \times v + u$.

We work in $\mathsf{R}^\star$. Let $\mathsf{Good}$ be the virtual class of all $g$ such that, for all $x \leq g$, $\mathsf{good}(x)$. Clearly, for every standard $n$, we can show that $\underline{n}$ is in $\mathsf{Good}$. We note that Good is downwards closed w.r.t. $\leq$. Either Good is closed under successor or it is not.

We consider first the case that Good is not closed under successor. Consider $g$ and $g'$ such that $g, g' \in$ Good, $Sg \notin$ Good, $Sg' \notin$ Good. Suppose $g < g'$. Then by axiom R8, $Sg \leq g'$. So, by the downwards closure of Good, $Sg \in$ Good. Quod non. So, $g \not< g'$. Similarly, $g' \not< g$. Since, by R7, $\leq$ is linear, we find $g = g'$. So there is a unique element $g^*$ in Good, such that $g^*$ does not have a successor in Good.

Since $g^*$ is good, we can find a largest $y^* \leq g^*$, such that there is a minimal $z^* \leq g^*$, such that $\mathsf{comm}^{g^*}(\max(z^*, y^*))$, and, for any $a \leq y^*$, if $a \in \alpha^{y^*}$, then $z^* \models^{g^*} a$.

Since $T$, is locally finite, for every standard $n$, we can show that $\underline{n} < y^*$. It follows that for any standard axiom $A$ of $T$, we have $z^* \models^{g^*} \ulcorner A \urcorner$.

To simplify inessentially, let's assume that $T$ has just one binary predicate symbol $P$. We define a (parameter-free, one-dimensional, one-piece, identity-preserving) translation $\nu$ as follows:

- $\delta_\nu(v) :\leftrightarrow v < (z^*)_0$,

- $P_\nu(v_0, v_1) :\leftrightarrow z^*, f \models^{g^*} \ulcorner P(v_0, v_1) \urcorner$, where $f$ codes the assignment on just $\ulcorner v_0 \urcorner$ and $\ulcorner v_1 \urcorner$ such that $f \ulcorner v_0 \urcorner = v_0$ and $f \ulcorner v_1 \urcorner = v_1$. This assignment exists since $\mathsf{comm}^{g^*}(\max(z^*, y^*))$. We need only two resets to obtain $f$.

We can now prove, by external induction, that for any standard $A$, $z^* \models^{g^*} \ulcorner A \urcorner$ iff $A^\nu$, since we have the commutation conditions for $\models^{g^*}$ for formulas below $y^*$. Hence, for any axiom $A$ of $T$, we have $A^\nu$.

Note that $\nu$ is parameter-free since the starred elements are definable. Also $\nu$ is identity preserving. Thus, we have shown that R* plus 'Good is not closed under successor' interprets $T$.

We turn to the case where Good is closed under successor. In other words, we work in R* plus 'Good is closed under successor'. Clearly 0 is in Good. Consider the following (parameter-free, one-dimensional, one-piece) translation $\gamma$ for the language of Q$^-$.

- $\delta_\gamma(x) :\leftrightarrow \mathsf{Good}(x)$,

- $0_\gamma := 0$,

- $S_\gamma x := Sx$,

- $A_\gamma xyz :\leftrightarrow x, y, z \in \mathsf{Good} \wedge x + y = z$,

- $M_\gamma xyz :\leftrightarrow x, y, z \in \mathsf{Good} \wedge x \times y = z$.

We evidently find that R* plus 'Good is closed under successor' proves $(Q^-)^\gamma$. Thus we have shown that R* plus 'Good is closed under successor' interprets $Q^-$. As mentioned earlier, Vítězlav Švejdar shows that $Q^-$ interprets Q. See [6]. So, we have R* plus 'Good is closed under successor' interprets Q.

If we can show that $Q \triangleright T$, we have both R* plus 'Good is not closed under successor' interprets $T$, and R* plus 'Good is closed under successor' interprets $T$. Then, we can form a disjunctive interpretation to show that $R^* \triangleright T$.

Finally, we prove that Q interprets $T$. There are two ways to do it. Here is the first. We know that Q interprets a convenient theory like Buss' $S_2^1$ on a definable cut. (See e.g. [1].) So it is sufficient to show that $S_2^1$ interprets $T$. As before, we can write down a satisfaction predicate for finite models. We can find definable cut $J$ (downwards closed w.r.t. $\leq$, closed under S, +, × and $\omega_1$, or equivalently #.) such that $S_2^1$ proves that we have the commutation conditions for formulas in $J$. The crux here is that the witnesses for satisfaction need not be in $J$.[5] Let's write $\alpha \restriction y$ for the set of all axioms witnessed below $y$. Using this, we can find a definable cut $J^*$, such that:

$$S_2^1 \vdash \forall y, z \in J^* \ (z \models \alpha \restriction y \to \forall p, a \in J^* \ (\text{proof}_{\alpha \restriction y}(p, a) \to z \models a))$$

From this it follows that, for every $n$, $S_2^1 \vdash \text{con}^{J^*}(\alpha \restriction n)$. In more sloppy notation: $S_2^1 \vdash \text{con}^{J^*}(T \restriction n)$. We can now use the Henkin-Feferman argument to build an interpretation of $T$ in $S_2^1$. (See e.g. [9].) In terms of that paper, we have shown: $Q \triangleright \mho(T)$.) This interpretation is one-dimensional, one-piece and parameter free. Moreover, by [11], Corollary 6.1, we find that we can make the interpretation identity-preserving.

REMARK 7. Since the axiomatization of R is reasonably simple, we can do more. One can show that $S_2^1 \vdash \text{con}^{J^*}(R)$.

We can use this insight to produce an example of a theory $U$ that is not locally finitely satisfiable but such that Q still interprets Q+con($U$). Reason in $S_2^1$. We either have $\text{con}^{J^*}(Q)$ or $\text{incon}^{J^*}(Q)$. In the first case, we find, by Pudlák's version of the second incompleteness theorem, that $\text{con}^{J^*}(Q + \text{incon}(Q))$. So, a fortiori, we have $\text{con}^{J^*}(R+\text{incon}(Q))$. In the second case, we have $\text{incon}^{J^*}(Q)$. We find, by $\exists \Sigma_1^b$-completeness, that $\text{con}^{J^*}(R + \text{incon}(Q))$. So in both cases we have $\text{con}^{J^*}(R + \text{incon}(Q))$. We may conclude that Q interprets $Q + \text{con}(R + \text{incon}(Q))$. Evidently, $R + \text{incon}(Q)$ is not locally finitely satisfiable.

---

[5] In fact, the argument is closely analogous with the construction of a $\Sigma_1^0$-truth predicate that works for formulas in a cut $J$.

We turn to the second proof of the interpretability of $T$ in $\mathsf{Q}$. By a result of Wilkie, we know that $\mathsf{Q}$ interprets $\mathrm{I}\Delta_0 + \Omega_1$ on a definable cut. So, it is sufficient to show the interpretability of $T$ in $\mathrm{I}\Delta_0 + \Omega_1$. Let $J$ be a definable cut such that we have $\mathrm{I}\Delta_0 + \Omega_1 \vdash \forall x \in J\; 2^{2^{cx^2}} \downarrow$. As is well-known we have $(\mathrm{I}\Delta_0 + \Omega_1) \rhd (\mathrm{I}\Delta_0 + \Omega_1 + \mathsf{incon}^J(\mathsf{Q}))$. This interpretation uses the Henkin-Feferman construction and is one-dimensional, one-piece and parameter-free. We can arrange it to be also identity-preserving. So it is sufficient to show that $W := \mathrm{I}\Delta_0 + \Omega_1 + \mathsf{incon}^J(\mathsf{Q})$ interprets $T$.

We work in $W$. Let $p^\star$ be the smallest proof of $\mathsf{incon}(\mathsf{Q})$. Let $y^\star$ be the largest number below $p^\star$, such that there is a model $z$ below $p^\star$ that satisfies all axioms of $\alpha$ witnessed below $y^\star$. The number $y^\star$ exists since we can bound the existential quantifier in the satisfaction predicate by $2^{2^{c(p^\star)^2}}$. Now we take the smallest such model $z^\star$ for the given $y^\star$. Again we use the above bounding argument to show that $z^\star$ exists. We use $z^\star$ to construct the desired interpretation of $T$ in a way that is completely analogous to our earlier construction of $\gamma$.

Since all interpretations we used along the way are one-dimensional, one-piece, parameter-free and identity-preserving, their composition also has these desired properties.

## Acknowledgements

I thank the two anonymous referees for their remarks and suggestions.

## BIBLIOGRAPHY

[1] P. Hájek and P. Pudlák. *Metamathematics of First-Order Arithmetic*. Perspectives in Mathematical Logic. Springer, Berlin, 1991.
[2] J. P. Jones and J.C. Shepherdson. Variants of Robinson's essentially undecidable theory R. *Archiv für Mathematische Logik und Grundlagenforschung*, 23:61–64, 1983.
[3] J. Mycielski, P. Pudlák, and A.S. Stern. *A lattice of chapters of mathematics (interpretations between theorems)*, volume 426 of *Memoirs of the American Mathematical Society*. AMS, Providence, Rhode Island, 1990.
[4] E. Nelson. *Predicative arithmetic*. Princeton University Press, Princeton, 1986.
[5] R. M. Solovay. Interpretability in set theories. Unpublished letter to P. Hájek, ⟨http://www.cs.cas.cz/~hajek/RSolovayZFGB.pdf⟩, 1976.
[6] V. Švejdar. An interpretation of Robinson's Arithmetic in its Grzegorczyk's weaker variant. *Fundamenta Informaticae*, 81:347–354, 2007.
[7] A. Tarski, A. Mostowski, and R.M. Robinson. *Undecidable theories*. North–Holland, Amsterdam, 1953.
[8] R.L. Vaught. On a theorem of Cobham concerning undecidable theories. In E. Nagel, P. Suppes, and A. Tarski, editors, *Logic, Methodology and Philosophy of Science. Proceedings of the 1960 International Congress*, pages 14–25. Stanford University Press, Stanford, 1962.
[9] A. Visser. Can we make the Second Incompleteness Theorem coordinate free. *Journal of Logic and Computation*, 2009. doi: 10.1093/logcom/exp048.
[10] A. Visser. Cardinal arithmetic in the style of baron von Münchhausen. *Review of Symbolic Logic*, 2(3):570–589, 2009. doi: 10.1017/S1755020309090261.

[11] A. Visser. The predicative Frege hierarchy. *Annals of Pure and Applied Logic*, 160(2):129–153, 2009. doi: 10.1016/j.apal.2009.02.001.

# Finding the Phase Transition for Friedman's Long Finite Sequences

ANDREAS WEIERMANN AND MARTIJN BAARTSE

## 1 Introduction

In [4] Friedman defines a property $\mathcal{F}$ of sequences over $\{1, \cdots, k\}$, $k \in \mathbb{N}$ as follows.

DEFINITION 1. A sequence $x_1 \ldots x_n$ has property $\mathcal{F}$ if there are no $i < j \leq n/2$ such that the window $x_i \ldots x_{2i}$ is a subsequence of the window $x_j \ldots x_{2j}$. A sequence $a_1 \ldots a_n$ is a subsequence of the sequence $b_1 \ldots b_m$ if there exists a strictly increasing embedding function $E : \{1, \ldots n\} \to \{1, \ldots m\}$ such that $a_i = b_{E(i)}$ for all $i \in \{1, \ldots, n\}$.

He then continues to show that sequences with property $\mathcal{F}$ are finite, that the maximum length of a sequence over $\{1\}$ with property $\mathcal{F}$ is 3, that the maximum length of a sequence over $\{1, 2\}$ with property $\mathcal{F}$ is 11 and that the maximum length of a sequence over $\{1, 2, 3\}$ with property $\mathcal{F}$ is bigger than $A_{7198}(158386)$. Here, $A_{7198}$ is the 7198th branch of the Ackermann function. This grows so fast that the length function $L$, which maps a number $k$ to the maximum length that a sequence over $\{1, \ldots, k\}$ with property $\mathcal{F}$ can have, is not provably total in $I\Sigma_2$.

We will introduce a function parameter $f$ (for a sublinear function $f : \mathbb{N} \to \mathbb{N}$) in the property $\mathcal{F}$ and find out for which values of this parameter the length function is provably total in $I\Sigma_2$ and for which values it is not. This dramatic change of behaviour, from provability to non-provability, is called a phase transition. The intuitive picture of the phase transition is sketched in

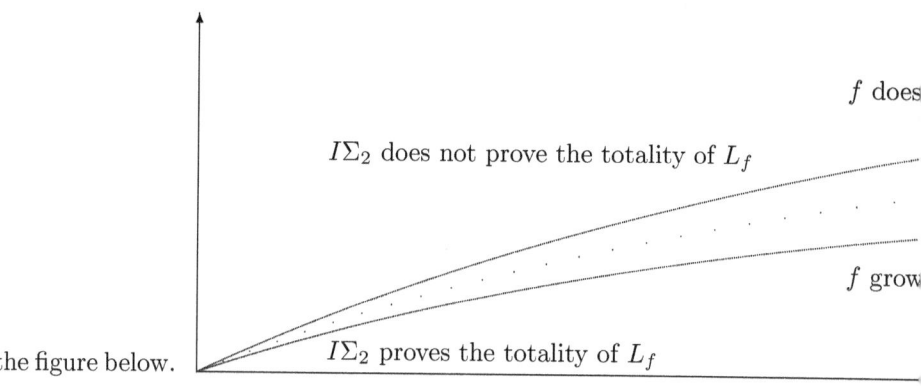

the figure below.

Some other papers on interesting phase transitions are [10, 11, 12].

The parameter function $f$ will determine the length of the windows in the following way.

DEFINITION 2. A sequence $x_1 \ldots x_n$ has property $\mathcal{F}_f$ if there are no $i < j$ with $i+f(i), j+f(j) \leq n$ such that the window $x_i \ldots x_{i+f(i)}$ is a subsequence of the window $x_j \ldots x_{j+f(j)}$.

Note that $\mathcal{F}$ is $\mathcal{F}_I$ where $I$ is the identity function. Let $L_f$ be the length function associated with $\mathcal{F}_f$. Experience has shown that it is usually easy to prove the provability part. So we will start by finding functions $f$ for which the totality of $L_f$ is easy to prove and then try to show unprovability of $L_f$ for other functions $f$.

## 2 Provability

To show provability of the totality of $L_f$ (in $I\Sigma_2$) we have to find a bound on the length that a sequence over $\{1, \ldots, k\}$ with property $\mathcal{F}_f$ can have. An easy way to do this would be to find for each $k$ an $l$ such that $|\{n | f(n) = l - 1\}| > k^l$. By the pigeon hole principle it would follow that for every sequence $x_1 \ldots x_m$ over $\{1, \ldots, k\}$ with $m \geq \max\{n | f(n) = l - 1\} + l - 1$ there are $i < j$ such that $x_i \ldots x_{i+f(i)} = x_j \ldots x_{j+f(j)}$. Hence $L_f(k) < \max\{n | f(n) = l - 1\} + l - 1$. The sort of functions that accomplish this are the logarithmic ones. However, if we take $f$ to be a logarithm with some fixed base then the argument above will not work when $k$ gets too large. Therefore we use a logarithmic function with a slowly decreasing factor in front. This factor will be of the form $1/(g^{-1}(n))$. We define inverse functions as follows.

DEFINITION 3. If $h : \mathbb{N} \to \mathbb{N}$ is any unbounded function then we define $h^{-1}$ to be the function $m \mapsto \min\{n|h(n) \geq m\}$.

So we set
$$f(n) = \lfloor \frac{1}{g^{-1}(n)} \log_2 n \rfloor$$
where $g : \mathbb{N} \to \mathbb{N}$ is an increasing function which tends to infinity. The base of 2 in the logarithm is not important. It could as well be any other number $> 1$. We will now formalize the argument above and see that it works if $I\Sigma_2$ proves the totality of $g$.

THEOREM 4. *If $g$ is provably total in $I\Sigma_2$ then the function $L_f$ is provably total in $I\Sigma_2$.*

**Proof.** We set $j = \max\{g(2\lceil \log_2 k \rceil), k^4\}$ and show that any sequence with length at least $4j + f(4j)$ cannot have property $\mathcal{F}_f$ because it will contain two windows that are the same. So suppose that we have a sequence $w$ over $\{1, \ldots, k\}$ of length at least $4j + f(4j)$. For $i \geq j$ we have
$$f(i) \leq \frac{\log_2 i}{2 \log_2 k} = \frac{1}{2} \log_k i$$
and in particular $f(4j) \leq 1 + \frac{1}{2} \log_k j$ (for $k \geq 4$). We now look at the windows of $w$ which start at a position $i$ for $j \leq i \leq 4j$. There are more than $3j$ such windows. These windows are sequences over $\{1, \cdots k\}$ with length at most $2 + \frac{1}{2} \log_k j$. The number of possible sequences over $\{1, \cdots k\}$ with length at most $2 + \frac{1}{2} \log_k j$ is limited by
$$\sum_{m=1}^{2+\lfloor \frac{1}{2} \log_k j \rfloor} k^m \leq 2k^{2+\lfloor \frac{1}{2} \log_k j \rfloor} \leq 2k^2 \sqrt{j}.$$
Using $k^4 \leq j$ we get $k^2 \sqrt{j} \leq j$, thus $2k^2 \sqrt{j} < 3j$ and we can conclude that indeed two windows must be the same which implies that $L_f(k) < 4j + f(4j)$. Since we assume that $g$ is provably total in $I\Sigma_2$ the function $k \mapsto 4j + f(4j)$ is provably total in $I\Sigma_2$ and this proves the theorem. □

## 3 Unprovability

In this section we will use the same functions $f$ (which depends on $g$) and show that if every function that is provably total in $I\Sigma_2$ is eventually dominated by $g$, then $L_f$ is not provably total in $I\Sigma_2$. So we have to show that the function $L_f$ grows very fast. We do this by using sequences over $\{1, \ldots, k\}$ with property $\mathcal{F}$ and constructing out of them sequences over

$\{1,\ldots,h(k)\}$ (where $h$ is some elementary function) with property $\mathcal{F}_f$ that have about the same length.

The whole construction consists of the combination of three constructions. The first construction will give a sequence with property $\mathcal{F}_\phi$ where $\phi$ is a log like function. The second construction will transform this sequence a little so that it has property $\mathcal{F}_\psi$ where $\psi$ is a small modification of $\phi$ such that $\psi$ is non decreasing and has small enough values at small arguments. The third construction then finally gives us a sequence with property $\mathcal{F}_f$. In these constructions we will view numbers as finite sequences. We will use the following coding functions that depend on the number $k$ which will stand for the cardinality of the set of elements of the input sequence. The function $N$ maps a sequence $(a_1,\ldots a_q)$ to $1 + \sum_{i=1}^{q}(a_i - 1)k^{i-1}$ (we will only use inputs in which for every $i$, $1 \leq a_i \leq k$). The function $p_j$ maps the number $N((a_1,\ldots a_q))$ to $a_j$, so $p_j$ maps $n$ to a number in $\{1,\ldots k\}$ that is equivalent to $\lfloor (n-1)/k^{j-1} \rfloor + 1$ modulo $k$.

### 3.1 Construction I

*Input*
The input of this construction is a sequence $x_1 x_2 \ldots x_n$ (with $n > 1$) over $\{1,\ldots,k\}$ with property $\mathcal{F}$

*Output*
The output of this construction is a sequence $y_1 y_2 \ldots y_m$ with $m = \beta(\lfloor n/2 \rfloor)$ ($\beta$ is defined below) over $\{1,\ldots,k^2 + 2\}$ with property $\mathcal{F}_\phi$. We will now define this function $\phi$ (the idea behind this definition is given in the informal description below). In this definition we need the following function. Let $\beta : \mathbb{N} \to \mathbb{N}$ be defined recursively by

$$\begin{aligned}\beta(0) &= 0 \\ \beta(b+1) &= \beta(b) + (b+3) \cdot k^{b+2} - b.\end{aligned}$$

We now define $\phi$ as follows.

$$\phi(i) = \begin{cases} \beta^{-1}(i) + 1 & \text{if } \beta^{-1}(i + \beta^{-1}(i) + 1) = \beta^{-1}(i) \\ \beta^{-1}(i) + 3 & \text{otherwise} \end{cases}$$

*Example*
If the input is 112222 ($k = 2$) then the output will be

$$\begin{matrix}1&1\\0&0\end{matrix}S\begin{matrix}1&1\\0&1\end{matrix}S\begin{matrix}1&1\\1&0\end{matrix}S\begin{matrix}1&1\\1&1\end{matrix}T\begin{matrix}2&2\\0&0\end{matrix}S\begin{matrix}1&2&2\\0&0&1\end{matrix}S\begin{matrix}1&2&2\\0&1&0\end{matrix}S\begin{matrix}1&2&2\\0&1&1\end{matrix}S\begin{matrix}1&2\\1&0\end{matrix}$$

$$\begin{matrix}1&2&2\\1&0&1\end{matrix}S\begin{matrix}1&2&2\\1&1&0\end{matrix}S\begin{matrix}1&2&2\\1&1&1\end{matrix}T\begin{matrix}2&2\\0&0\end{matrix}S\begin{matrix}2&2&2&2\\0&0&0&1\end{matrix}S\begin{matrix}2&2&2&2\\0&0&1&0\end{matrix}S\begin{matrix}2&2&2\\0&0&1\end{matrix}$$

$$\begin{matrix}2&2&2&2\\0&1&0&0\end{matrix}S\begin{matrix}2&2&2&2\\0&1&0&1\end{matrix}S\begin{matrix}2&2&2&2\\0&1&1&0\end{matrix}S\begin{matrix}2&2&2&2\\0&1&1&1\end{matrix}S\begin{matrix}2&2&2&2\\1&0&0&0\end{matrix}S\begin{matrix}2&2&2\\1&0&0\end{matrix}$$

$$\begin{matrix}2&2&2&2\\1&0&1&0\end{matrix}S\begin{matrix}2&2&2&2\\1&0&1&1\end{matrix}S\begin{matrix}2&2&2&2\\1&1&0&0\end{matrix}S\begin{matrix}2&2&2&2\\1&1&0&1\end{matrix}S\begin{matrix}2&2&2&2\\1&1&1&0\end{matrix}S\begin{matrix}2&2&2\\1&1&1\end{matrix}$$

where $\begin{matrix}a\\c\end{matrix}$ stand for $N((a, c+1))$, $S$ stands for $k^2+1$ and $T$ stands for $k^2+2$.

*Informal description*
The output in the example consists of three blocks because the input 112222 has three windows, namely 11, 122 and 2222. The end of each block is marked by a $T$ (standing for $k^2 + 2$). The $i$th block consists of repetitions of window $i$ of the input $(x_i \ldots x_{2i})$ on the first line and these repetitions are counted in base $k$ on the second line. So the $i$th block contains $k^{i+1}$ repetitions since the length of window $i$ of the input is $i + 1$. These repetitions are separated by an $S$ ($k^2 + 1$). The first repetition in each block is different. These repetitions only consist of the last two elements of the corresponding window of the input. This is because these two elements are new in the sense that the other elements in this window are also contained in the previous window. The function $\beta$ maps a number $i$ to the position of $i$th $T$. The function $\phi$ is defined in such a way that if a window of the output is contained in block $i$ then this window contains exactly one $S$ and a cyclic permutation of the corresponding window from the input. If a window of the output starts in block $i$ and ends in block $i+1$ then this window contains exactly one $S$ and one $T$ and a cyclic permutation of the window $x_{i+1} \ldots x_{2i+2}$ from the input.

*Formal description*
We will still use $S = k^2 + 1$ and $T = k^2 + 2$ here. Given the input $x_1 \ldots x_n$ over $\{1, \ldots k\}$ with property $\mathcal{F}$ we will define the output $y_1 \ldots y_m$ over $\{1, \ldots, k^2 + 2\}$ and prove that this sequence has property $\mathcal{F}_\phi$. Let $m = \beta(\lfloor n/2 \rfloor)$. To improve readability we will denote $\beta^{-1}(i)$ by $b$ in the

next definition.

$$y_i = \begin{cases} k^2 + 2 & \text{if } \beta(b) = i \\ k^2 + 1 & \text{if } \beta(b) \neq i \text{ and} \\ & i - \beta(b-1) + b - 1 \text{ is a multiple of } b+2 \\ N((x_{b+j-1}, c+1)) \cdot k & \text{if } \exists q \, (i - \beta(b-1) + b - 1 = q \cdot (b+2) + j \text{ and} \\ & 1 \leq j \leq b+1 \text{ and } 0 \leq c < k \\ & \text{and } \lfloor q/(k^{b+1-j}) \rfloor \equiv c \pmod{k}) \end{cases}$$

One can check that for every $i$ exactly one of the three conditions is the case and in case of the third condition the values of $j$, $q$ and $c$ are uniquely determined.

LEMMA 5. *Let $0 \leq i' \leq i$, $0 \leq j' \leq j$ and let $u_1 \ldots u_i$ and $v_1 \ldots v_j$ be sequences over $\{1, \ldots, l\}$ and let $w > l$. If $u_{i'+1} \ldots u_i w u_1 \ldots u_{i'}$ is a subsequence of $v_{j'+1} \ldots v_j w v_1 \ldots v_{j'}$ then $u_1 \ldots u_i$ is a subsequence of $v_1 \ldots v_j$.*

**Proof.** Since $w \notin \{1, \ldots, l\}$ it follows that it must be the case that $u_{i'+1} \ldots u_i$ is a subsequence of $v_{j'+1} \ldots v_j$ and $u_1 \ldots u_{i'}$ is a subsequence of $v_1 \ldots v_{j'}$. Hence $u_1 \ldots u_i$ is a subsequence of $v_1 \ldots v_j$. □

LEMMA 6. *If $u_1 \ldots u_i$ is a subsequence of $v_1 \ldots v_j$ then for any function $h$ the sequence $h(u_i) \ldots h(u_{i'})$ is a subsequence of $h(v_1) \ldots h(v_j)$ with the same embedding.*

**Proof.** Clear. □

LEMMA 7. *If $i < j$, $j + \phi(j) \leq m$, $\phi(i) = \beta^{-1}(i) + 1$, $\phi(j) = \beta^{-1}(j) + 1$, $\beta^{-1}(i) < \beta^{-1}(j)$ and $y_i \ldots y_{i+\phi(i)}$ is a subsequence of $y_j \ldots y_{j+\phi(j)}$ then $x_{\beta^{-1}(i)} \ldots x_{2\beta^{-1}(i)}$ is a subsequence of $x_{\beta^{-1}(j)} \ldots x_{2\beta^{-1}(j)}$.*

**Proof.** From the definition of $y_i$ it follows that both $y_i \ldots y_{i+\phi(i)}$ and $y_j \ldots y_{j+\phi(j)}$ contain exactly one element that is bigger than $k^2$ and since the one is a subsequence of the other these elements have to be equal. Let $h$ equal the identity on numbers $> k^2$ and $p_1$ on numbers $\leq k^2$. There exist $i'$, $j'$ such that

$$h(y_i) \ldots h(y_{i+\phi(i)}) = x_{i'+1} \ldots x_{2\beta^{-1}(i)} w x_{\beta^{-1}(i)} \ldots x_{i'}$$

and

$$h(y_j) \ldots h(y_{j+\phi(j)}) = x_{j'+1} \ldots x_{2\beta^{-1}(j)} w x_{\beta^{-1}(j)} \ldots x_{j'}$$

where $w > k^2$. By lemmas 6 and 5 we conclude that $x_{\beta^{-1}(i)} \ldots x_{2\beta^{-1}(i)}$ is a subsequence of $x_{\beta^{-1}(j)} \ldots x_{2\beta^{-1}(j)}$. □

**LEMMA 8.** *If $i < j$, $j + \phi(j) \le m$, $\phi(i) = \beta^{-1}(i) + 1$, $\phi(j) = \beta^{-1}(j) + 3$ and $y_i \ldots y_{i+\phi(i)}$ is a subsequence of $y_j \ldots y_{j+\phi(j)}$ then $x_{\beta^{-1}(i)} \ldots x_{2\beta^{-1}(i)}$ is a subsequence of $x_{\beta^{-1}(j)+1} \ldots x_{2\beta^{-1}(j)+2}$.*

**Proof.** Let $h$ be defined as in the proof of the previous lemma. There exist $i', j'$ such that $\beta^{-1}(i) - 1 \le i' \le 2\beta^{-1}(i)$, $\beta^{-1}(j) \le j' \le 2\beta^{-1}(j)$ and

$$h(y_i)\ldots h(y_{i+\phi(i)}) = x_{i'+1}\ldots x_{2\beta^{-1}(i)} w x_{\beta^{-1}(i)} \ldots x_{i'}$$

where $w = S$ or $w = T$ and

$$h(y_j)\ldots h(y_{j+\phi(j)}) = x_{j'+1}\ldots x_{2\beta^{-1}(j)} T x_{2\beta^{-1}(j)+1} x_{2\beta^{-1}(j)+2} S x_{\beta^{-1}(j)+1}\ldots x_{j'}.$$

By lemma 6 the first sequence is a subsequence of the second. In case $w = S$ this will still be the case if we delete the $T$ from the second sequence and then by lemma 5 we conclude that $x_{\beta^{-1}(i)}\ldots x_{2\beta^{-1}(i)}$ is a subsequence of $x_{\beta^{-1}(j)+1}\ldots x_{2\beta^{-1}(j)+2}$. In case $w = T$ the first sequence will still be a subsequence of the second if we delete the $S$ from the second sequence and then by lemma 5 we conclude that $x_{\beta^{-1}(i)}\ldots x_{2\beta^{-1}(i)}$ is a subsequence of $x_{\beta^{-1}(j)+1}\ldots x_{2\beta^{-1}(j)+2}$. □

**LEMMA 9.** *If $i < j$, $j + \phi(j) \le m$, $\phi(i) = \beta^{-1}(i) + 3$, $\phi(j) = \beta^{-1}(j) + 3$, $\beta^{-1}(i) + 1 < \beta^{-1}(j) + 1$ and $y_i \ldots y_{i+\phi(i)}$ is a subsequence of $y_j \ldots y_{j+\phi(j)}$ then $x_{\beta^{-1}(i)+1} \ldots x_{2\beta^{-1}(i)+2}$ is a subsequence of $x_{\beta^{-1}(j)+1} \ldots x_{2\beta^{-1}(j)+2}$.*

**Proof.** Let $h$ be as defined in the proof of the previous lemma. There exist $i', j'$ such that $\beta^{-1}(i) \le i' \le 2\beta^{-1}(i)$, $\beta^{-1}(j) \le j' \le 2\beta^{-1}(j)$ and

$$h(y_i)\ldots h(y_{i+\phi(i)}) = x_{i'+1}\ldots x_{2\beta^{-1}(i)} T x_{2\beta^{-1}(i)+1} x_{2\beta^{-1}(i)+2} S x_{\beta^{-1}(i)+1}\ldots x_{i'}.$$

and

$$h(y_j)\ldots h(y_{j+\phi(j)}) = x_{j'+1}\ldots x_{2\beta^{-1}(j)} T x_{2\beta^{-1}(j)+1} x_{2\beta^{-1}(j)+2} S x_{\beta^{-1}(j)+1}\ldots x_{j'}.$$

By lemma 6 the first sequence is a subsequence of the second. Two applications of lemma 5 now yield the result. □

**LEMMA 10.** *If $i < j$, $j + \phi(j) \le m$, $\phi(i) = \beta^{-1}(i) + 3$ and $\phi(j) = \beta^{-1}(j) + 1$ then it cannot be the case that $y_i \ldots y_{i+\phi(i)}$ is a subsequence of $y_j \ldots y_{j+\phi(j)}$.*

**Proof.** The sequence $y_i \ldots y_{i+\phi(i)}$ contains an $S$ and a $T$ while the sequence $y_j \ldots y_{j+\phi(j)}$ does not contain an $S$ or does not contain a $T$. □

**LEMMA 11.** *If $i < j$, $j + \phi(j) \le m$, $\phi(i) = \beta^{-1}(i) + 1$, $\phi(j) = \beta^{-1}(j) + 1$ and $\beta^{-1}(i) = \beta^{-1}(j)$ then it cannot be the case that $y_i \ldots y_{i+\phi(i)}$ is a subsequence of $y_j \ldots y_{j+\phi(j)}$.*

**Proof.** Since the sequences $y_i \ldots y_{i+\phi(i)}$ and $y_j \ldots y_{j+\phi(j)}$ have the same length, all we have to do is show that they are not equal. Let $h$ equal the identity on numbers $> k^2$ and $p_2$ on numbers $\leq k^2$. If these sequences are equal then there exists a number $c$ such that $0 \leq c \leq \beta^{-1}(i) + 1$ and

$$h(y_i) \ldots h(y_{i+\phi(i)}) = h(y_j) \ldots h(y_{j+\phi(j)}) = a_{c+1} \ldots a_{\beta^{-1}(i)+1} w a_1 \ldots a_c$$

where $w > k^2$. Let $d$ be the number $q$ that is used in the definition of $y_i$ if $y_i \leq k^2$ and let $d$ be one less than the $q$ used in the definition of $y_{i+1}$ otherwise. So $q = d$ is used in the definition of $y_i \ldots y_{i+\beta^{-1}(i)-c}$ and $q = d+1$ is used in the definition of $y_{i+\beta^{-1}(i)-c+1} \ldots y_{i+\phi(i)}$. The number $d$ is uniquely determined by $a_1 \ldots a_{\beta^{-1}(i)+1}$. This implies that $i = j$ which contradicts the assumption $i < j$. □

**LEMMA 12.** *If $i < j$, $j + \phi(j) \leq m$, $\phi(i) = \beta^{-1}(i) + 3$, $\phi(j) = \beta^{-1}(j) + 3$, $\beta^{-1}(i) + 1 = \beta^{-1}(j) + 1$ then it cannot be the case that $y_i \ldots y_{i+\phi(i)}$ is a subsequence of $y_j \ldots y_{j+\phi(j)}$.*

**Proof.** Since the length of $y_i \ldots y_{i+\phi(i)}$ is the same as the length of $y_j \ldots y_{j+\phi(j)}$ it suffices to show that they are not equal. This is clear since both sequences must contain exactly one $T$. This is the $T$ at position $\beta(\beta^{-1}(i))$ ($= \beta(\beta^{-1}(j))$) and since $i < j$ there cannot exist a $c$ such that $y_{i+c} = y_{j+c} = T$. □

**LEMMA 13.** *There are no $i < j$ such that $y_i \ldots y_{i+\phi(i)}$ is a subsequence of $y_j \ldots y_{j+\phi(j)}$.*

**Proof.** For every $i < j$ the conditions of one of the lemmas 7 - 12 are satisfied and it either follows directly that $y_i \ldots y_{i+\phi(i)}$ is not a subsequence of $y_j \ldots y_{j+\phi(j)}$ or the assumption that $y_i \ldots y_{i+\phi(i)}$ is a subsequence of $y_j \ldots y_{j+\phi(j)}$ implies that there are $i' < j' \leq \lfloor n/2 \rfloor$ such that $x_{i'} \ldots x_{2i'}$ is a subsequence of $x_{j'} \ldots x_{2j'}$ which contradicts the assumption that the sequence $x_1 \ldots x_n$ has property $\mathcal{F}$. □

## 3.2 Construction II

*Input*

The input of this construction are functions $h, h'$ and a sequence $x_1 \ldots x_n$ over $\{1, \ldots, k\}$ with property $\mathcal{F}_h$. The functions $h, h'$ must satisfy the conditions $\forall i(i + 1 + h(i+1) \geq i + h(i))$, $\forall i(i + 1 + h'(i+1) \geq i + h'(i))$ and $\forall i(0 \leq h(i) - h'(i) \leq 1)$.

*Output*

The output of this construction is a sequence $y_1 \ldots y_m$ with $m = \max\{i +$

$h'(i)|\exists j \ (i+h'(i) < j + h(j) \leq n)\}$ over $\{1, \ldots, 2k^2\}$ with property $\mathcal{F}_{h'}$.

*example*
If
$$h(i) = \begin{cases} 3 & \text{if } i \text{ is a multiple of 3} \\ 2 & \text{otherwise} \end{cases},$$
$h'(i) = 2$ and $x_1 \ldots x_n = 1122333$ then the output is

$$\begin{array}{cccccc} 1 & 1 & 2 & 2 & 3 & 3 \\ 1 & 2 & 2 & 3 & 3 & 3 \\ 0 & 0 & 1 & 0 & 0 & 1 \end{array}$$

where $\begin{array}{c} a \\ b \\ c \end{array}$ stands for $N((a, b, c+1))$.

*informal description*
The function $h'$ is almost the same as $h$. The only difference is that for some arguments the value of $h'$ is one less. To ensure that for $i < j$ with $h'(i) = h(i) - 1$ it is not the case that $y_i \ldots y_{i+h'(i)}$ is a subsequence of $y_j \ldots y_{j+h'(j)}$ we write the next element on the second line and indicate the difference between $h(i)$ and $h'(i)$ at position $i$ on the third line.

*formal description*
The sequence $y_1 \ldots y_m$ over $\{1, \ldots, 2k^2\}$ is defined as follows.
$$y_i = N((x_i, x_{i+1}, h(i) - h'(i) + 1)).$$
We show that this sequence has property $h'$. Assume that we have $i < j$ such that $j + h'(j) \leq m$.

**LEMMA 14.** *If $h'(i) = h(i)$ and $y_i \ldots y_{i+h'(i)}$ is a subsequence of $y_j \ldots y_{j+h'(j)}$ then $x_i \ldots x_{i+h(i)}$ is a subsequence of $x_j \ldots x_{j+h(j)}$.*

**Proof.** Applying lemma 6 to the function $p_1$ and the sequences $y_i \ldots y_{i+h'(i)}$ and $y_j \ldots y_{j+h'(j)}$ yields the result. □

**LEMMA 15.** *If $h'(i) = h(i) - 1$ and $y_i \ldots y_{i+h'(i)}$ is a subsequence of $y_j \ldots y_{j+h'(j)}$ then there exists $j' \geq j$ such that $x_i \ldots x_{i+h(i)}$ is a subsequence of $x_{j'} \ldots x_{j'+h(j')}$.*

**Proof.** Let $E$ be an embedding from $y_i \ldots y_{i+h'(i)}$ into $y_j \ldots y_{j+h'(j)}$. Since $p_3(y_i) = 2$ it must also be the case that $p_3(y_{E(i)}) = 2$. From the definition of $y$ we see that this implies that $h'(E(i)) = h(E(i)) - 1$ and from the conditions on $h$ and $h'$ and the definition of $m$ it follows that there exists $j'$ such that $j \leq j' \leq E(i)$ and $j + h'(j) < j' + h(j') \leq n$. Let $E'$ : $\{i,\ldots,i+h(i)\} \to \{j'',\ldots,j''+h(j')\}$ be an extension of $E$ with $E'(i+h(i)) = E(i+h(i)-1)+1$. We claim that $E'$ is an embedding from $x_i \ldots x_{i+h(i)}$ into $x_{j'} \ldots x_{j'+h(j')}$. For $i \leq i' < i + h(i)$ we have $y_{i'} = y_{E'(i')}$ and thus $p_1(y_{i'}) = p_1(y_{E'(i')})$ which can be rewritten as $x_{i'} = x_{E'(i')}$ and we also have $y_{i+h(i)-1} = y_{E'(i+h(i)-1)}$ and thus $p_2(y_{i+h(i)-1}) = p_2(y_{E'(i+h(i)-1)})$ which can be rewritten as $x_{i+h(i)} = x_{E'(i+h(i)-1)+1} = x_{E'(i+h(i))}$ □

The above two lemmas show that if the conditions on the input are satisfied then the output sequence does indeed have the property $\mathcal{F}_{h'}$.

### 3.3 Construction III

*Input* The input of this construction is a non decreasing function $h$, a function $f$ and a sequence $x_1 \ldots x_n$ over $\{1,\ldots,k\}$ with property $\mathcal{F}_h$. The function $f$ must satisfy the following conditions.

$$\text{For all } j, \quad |\{i \in \{j,\ldots,j+f(j)-1\}|f(i) \neq f(i+1)\}| \leq 2 \quad (1)$$

and

$$\text{If } f(j) > f(j+1)+1 \text{ then for every } i \text{ in the interval}$$
$$\{j+1+f(j+1)+1\ldots j+f(j)\} \text{ it is the case that } i+f(i) \geq j+f(j) \quad (2)$$

*Output* The output of this construction is a sequence $y_1 \ldots y_m$ over $\{1,\ldots 3k^3 + s\}$ with property $\mathcal{F}_f$. The numbers $m$ and $s$ depend on $n$ and the functions $h$ and $f$

*Example*
If $h(i) = \lfloor\sqrt{i}\rfloor$, $x_1 \ldots x_{15} = 113223333111122$ and $f$ is given by the table

| $i$    | 1 | 2 | 3 | 4 | 5 | 6 | 7 | 8 | 9 | 10 | 11 | 12 | 13 |
|--------|---|---|---|---|---|---|---|---|---|----|----|----|----|
| $f(i)$ | 2 | 2 | 1 | 1 | 1 | 3 | 3 | 3 | 3 | 2  | 2  | 3  | 3  |

then the output sequence is

```
2 2 3 3 - - - - - 3 3 3 3
- - 1 1 3 2 - - - - - - -
- - - - - 3 1 1 1 1 2 2 -
1 1 2 2 2 3 3 3 3 1 1 2 2
```

where $\begin{smallmatrix}a\\b\\c\\d\end{smallmatrix}$ stands for $N((a,b,c,d))$. For clarity, the positions that are not important (i.e. are not used in the proof that the output has property $\mathcal{F}_f$) are marked with $-$.

*Informal description*
The part of $x_1 \ldots x_{15}$ that consists of windows of length 2 is $x_1 \ldots x_{3+h(3)} = 1132$.
The part of $x_1 \ldots x_{15}$ that consists of windows of length 3 is $x_4 \ldots x_{8+h(4)} = 2233331$.
The part of $x_1 \ldots x_{15}$ that consists of windows of length 4 is $x_9 \ldots x_{15} = 3111122$.
We have $f(1) = f(2) = 2$ so we start by using a beginning of $x_4 \ldots x_{4+h(4)}$ and putting that on the first line. On the fourth line we write a 1 so we know that we are using the first line at these positions. We then have $f(3) = f(4) = f(5) = 1$ so we now use a beginning of $x_1 \ldots x_{3+h(3)}$ and put that on the second line. On the fourth line we write a 2 so we know that we are using the second line at these positions. Now $f(6) = f(7) = f(8) = f(9) = 3$ and we do the same thing again. Then $f(10) = f(11) = 2$ so we use a piece from $x_4 \ldots x_{8+h(8)}$. Since we already used the beginning at two positions earlier we will now start at $x_6$ and jump to the first line again. The fact that we use the same piece again on the first line is a coincidence. In case there are arguments $i$ such that $f(i+1) < f(i) - 1$ we will use a new number each time that this happens and put it on the positions $i+1+f(i+1)+1, \ldots, i+f(i)$. This will enable us to show that if the output does not have property $\mathcal{F}_f$ then the input cannot have property $\mathcal{F}_h$.

*Formal description*
We need the following three functions in the definition of $y_i$.
$\ell : \mathbb{N} \to \{1, 2, 3\}$ with $\ell(i) \equiv |\{j < i | f(j) \neq f(j+1)\}| + 1 \pmod{3}$. This function tells us which coordinate is currently used.
$c(i) = |\{j < i | f(j) = f(i)\}|$. This function gives the number of sequences of length $f(i)$ that are already used.
$e(a, i) = \max(\{1\} \cup \{j \leq i | \ell(j) = a\})$. This function gives the latest position at which coordinate $a$ was active.
Let $m$ be maximal such that for all $i \leq m$ we have that

$$h(h^{-1}(f(i)) + c(i)) = f(i) \tag{3}$$

and

$$n \geq h^{-1}(f(i)) + c(i) + f(i) \tag{4}$$

(for each length we must have enough windows to use). We will first define a sequence $z_1 \ldots z_m$ which we will then modify into a sequence $y_1 \ldots y_m$.

$$z_i = N((\begin{array}{l} x_{h^{-1}(f(e(1,i)))+c(e(1,i))+i-e(1,i)}, \\ x_{h^{-1}(f(e(2,i)))+c(e(2,i))+i-e(2,i)}, \\ x_{h^{-1}(f(e(3,i)))+c(e(3,i))+i-e(3,i)}, \\ \ell(i))).\end{array}$$

Let $z^0 = z$ and construct $z^q$ (of the same length) out of $z^{q-1}$ by letting $j$ be the $q$th element in the set $\{r | r + f(r) > r + 1 + f(r+1)\}$ and setting $z_i^q = z_i^{q-1}$ if $i \notin \{1+j+f(j+1)+1 \ldots j+f(j)\}$ and $z_i^q = 3k^3 + q$ otherwise. Let $s$ be the least number such that $z^s = z^{s+1}$ and set $y = z^s$.

**LEMMA 16.** *If $i < j$ and $y_i \ldots y_{i+f(i)}$ is a subsequence of $y_j \ldots y_{j+f(j)}$ then $y_i \ldots y_{i+f(i)} = z_i \ldots z_{i+f(i)}$.*

**Proof.** We will show that $z_i^s \ldots z_{i+f(i)}^s$ does not contain a $3k^3 + s$. This implies that $z_i^s \ldots z_{i+f(i)}^s = z_i^{s-1} \ldots z_{i+f(i)}^{s-1}$ and thus $z_i^{s-1} \ldots z_{i+f(i)}^{s-1}$ is a subsequence of $z_j^{s-1} \ldots z_{j+f(j)}^{s-1}$. By repetition of the argument we conclude that $z_i^s \ldots z_{i+f(i)}^s = z_i^0 \ldots z_{i+f(i)}^0$ which proves the lemma. We now show that $z_i^s \ldots z_{i+f(i)}^s$ does not contain a $3k^3 + s$. Suppose for a contradiction that it did. Let $j'$ be the number such that $\{j'+1+f(j'+1)+1\ldots j'+f(j')\}$ is exactly the set of indices of $3k^3+s$ elements in $z^s$. If $z_i^s = 3k^3 + s$ then by (2) we have that $i + f(i) \geq j' + f(j')$ and $j + f(j) \geq j' + f(j')$. Since $i < j$ this means that $z_i^s \ldots z_{i+f(i)}^s$ contains more $3k^3 + s$ elements than $z_j^s \ldots z_{j+f(j)}^s$ contradicting the assumption that $z_i^s \ldots z_{i+f(i)}^s$ is a subsequence of $z_j^s \ldots z_{j+f(j)}^s$. If $z_i^s \neq 3k^3 + s$ then let $a$ be the least number such that $z_{i+a}^s = 3k^3 + s$ and let $b$ be the least number such that $z_{j+b}^s = 3k^3 + s$. The assumption that $z_i^s \ldots z_{i+f(i)}^s$ is a subsequence of $z_j^s \ldots z_{j+f(j)}^s$ implies that $a \leq b$, but $i < j$ yields $b < a$ and we have the desired contradiction. $\square$

**LEMMA 17.** *If $i < j$, $i+f(i), j+f(j) \leq m$ and $y_i \ldots y_{i+f(i)}$ is a subsequence of $y_j \ldots y_{j+f(j)}$ then there exist $i', j'$ such that $i' < j'$ and $x_{i'} \ldots x_{i'+h(i')}$ is a subsequence of $x_{j'} \ldots x_{j'+h(j')}$.*

**Proof.** Let $E$ be an embedding from $y_i \ldots y_{i+f(i)}$ into $y_j \ldots y_{j+f(j)}$. Let $a = \min\{b+f(b) | j \leq b \leq j+f(j)\}$. From the construction of $y$ we see that every element in $y_{a+1} \ldots y_{j+f(j)}$ is $> 3k^3$. By lemma 16 it follows that $E$ is an embedding from $y_i \ldots y_{i+f(i)}$ into $y_j \ldots y_a$ and it also follows that $E$ is an embedding from $z_i \ldots z_{i+f(i)}$ into $z_j \ldots z_a$. By definition of $a$ we have that $E(i) + f(E(i)) \geq a$ and thus $E$ is also an embedding from

$p_{p_4(z_i)}(z_i)\ldots p_{p_4(z_i)}(z_{i+f(i)})$ into $p_{p_4(z_i)}(z_{E(i)})\ldots p_{p_4(z_i)}(z_{E(i)+f(E(i))})$. By the definition of $z_i$ and the conditions (1), (3) and (4) we have for $q, r$ with $q + f(q) \leq m$ and $0 \leq r \leq f(q)$

$$p_{p_4(z_q)}(z_{q+r}) = x_{h^{-1}(f(q))+c(q)+r}$$

and thus

$$p_{p_4(z_i)}(z_i)\ldots p_{p_4(z_i)}(z_{i+f(i)}) =$$
$$x_{h^{-1}(f(i))+c(i)} \cdots x_{h^{-1}(f(i))+c(i)+h(h^{-1}(f(i))+c(i))}$$

and

$$p_{p_4(z_i)}(z_{E(i)})\ldots p_{p_4(z_i)}(z_{E(i)+f(E(i))}) =$$
$$x_{h^{-1}(f(E(i)))+c(E(i))} \cdots x_{h^{-1}(f(E(i)))+c(E(i))+h(h^{-1}(f(E(i)))+c(E(i)))}.$$

Set $i' = h^{-1}(f(i)) + c(i)$ and $j' = h^{-1}(f(E(i))) + c(E(i))$. Since $h$ is non decreasing equation (3) yields $i' < j'$ in case $f(i) < f(E(i))$. In case $f(i) = f(E(i))$ we have $c(i) < c(E(i))$ since $i < E(i)$ and thus in this case we also have $i' < j'$. □

## 3.4 Putting the constructions together

We will now be able to show that $L_f$ grows about as fast as $L$ or as $g$ depending on which grows slower.

THEOREM 18. *Let $f(i) = \lfloor \frac{1}{g^{-1}(i)} \log_2 i \rfloor$ with $g$ a strictly increasing function satisfying the condition $g(1) \geq 2$, $g(i+1) \geq g(i)^4$. Then the following holds.*

$$L_f(3 \cdot 2^{45}(k^2+2)^{48} + \lfloor \log_2 k \rfloor) \geq \min\{g(\lfloor \log_2 k \rfloor), \lfloor L(k)/2 \rfloor - 1\}.$$

**Proof.** Let $w = x_1\ldots x_n$ be a sequence over $\{1,\ldots,k\}$ with property $\mathcal{F}$ and $n = L(k)$. Applying the first construction to it we obtain a sequence $w' = x'_1\ldots x'_{n'}$ over $\{1,\ldots k^2+2\}$ with property $\mathcal{F}_\phi$ and $n' = \beta(\lfloor n/2 \rfloor)$. Then, using the second construction four times, we can get a sequence $w'' = x''_1\ldots x''_{n''}$ over $\{1,\ldots 2^{15}(k^2+2)^{16}\}$ with property $\mathcal{F}_{\beta^{-1}-1}$ and $n'' = n' - 4$. Finally we want to apply the third construction to $w''$ to produce a sequence $w''' = x'''_1\ldots x'''_{n'''}$ over $\{1,\ldots,3\cdot 2^{45}(k^2+2)^{48} + \lfloor \log_2 k \rfloor\}$ with property $\mathcal{F}_f$ and $n''' \geq \min\{g(\lfloor \log_2 k \rfloor), \lfloor L(k)/2 \rfloor - 1\}$.

We verify that conditions (1) and (2) are met. If $f(i) \neq f(i+1)$ then $\lfloor \log_2 i \rfloor \neq \lfloor \log_2(i+1) \rfloor$ or $g^{-1}(i) \neq g^{-1}(i+1)$. The interval $\{j,\ldots,j+f(j)-1\}$ is contained in $\{j,\ldots,j+\lfloor \log_2 j \rfloor - 1\}$ and by the condition on

$g$ it is clear that both possibilities can occur at most once in this interval. Hence (1) is satisfied.

It is clear that on an interval $\{g(i)+1,\ldots,g(i+1)\}$ the function $f$ is non decreasing. By the condition on $g$ it follows that for all $i$, $f(g(i+1)+1) > f(g(i))$. Hence, if $i > j$ then $f(i)+1 \geq f(j)/2$. So if $i \in \{j+1+f(j+1)+1,\ldots,j+f(j)\}$ then $i+f(i) \geq j+f(j+1)+1+f(i)+1 \geq j+f(j)/2+f(j)/2 = j+f(j)$ and thus (2) is satisfied.

We will now show that $n''' \geq \min\{g(\lfloor \log_2 k \rfloor), \lfloor L(k)/2 \rfloor - 1\}$. So we have to show that (3) and (4) hold for $i \leq \min\{g(\lfloor \log_2 k \rfloor), \lfloor L(k)/2 \rfloor - 1\}$. Since $f(i) < i$ we have for $i \leq \lfloor L(k)/2 - 1 \rfloor$

$$|\{j|j+h(j) \leq n'' \text{ and } h(j) = f(i)\}| = (f(i)+3)k^{f(i)+2} - f(i).$$

Since for all $i$, $f(g(i+1)+1) > f(g(i))$ it follows that for a fixed $l$ there can be at most two values of $g^{-1}(j)$ such that $f(j) = l$. If we let $d = \max\{g^{-1}(j)|f(j) = l\}$ then $|\{j|f(j) = l\}| \leq 2 \cdot 2^{l \cdot d}$. Hence

$$\begin{aligned}|\{j|j \leq g(\lfloor \log_2 k \rfloor) \text{ and } f(j) = f(i)\}| &\leq 2 \cdot 2^{f(i) \cdot \log_2 k} \\ &= 2 \cdot k^{f(i)} \\ &< (f(i)+3)k^{f(i)+2} - f(i) \\ &= |\{j|j+h(j) \leq n'' \text{ and } h(j) = f(i)\}|\end{aligned}$$

and (3) follows.

If $i \leq \lfloor L(k)/2 - 1 \rfloor$ then by (3) $h^{-1}(f(i)) + c(i) < h^{-1}(f(i)+1) \leq h^{-1}(\lfloor L(k)/2 - 1 \rfloor) = \beta(\lfloor L(k)/2 - 1 \rfloor) + 1$. Since $f(i) \leq \lfloor L(k)/2 \rfloor \leq \beta(\lfloor L(k)/2 \rfloor) - 4 - \beta(\lfloor L(k)/2 - 1) - 1$, (4) follows. □

## 4 The phase transition

Using the Hardy functions we can now nicely describe the phase transition

DEFINITION 19. The Hardy functions are defined by ordinal recursion as follows

$$\begin{aligned}H_0(x) &= x \\ H_{\alpha+1}(x) &= H_\alpha(x+1) \\ H_\lambda(x) &= H_{\lambda[x]}(x)\end{aligned}$$

where $\lambda$ is a limit ordinal and $\lambda[x]$ is the $x$th term from the standard fundamental sequence for $\lambda$ (cf., for example, [3, 5] for a definition).

It is a known fact (see, for example, [1, 3, 5] for a proof) that $I\Sigma_2$ proves the totality of $H_\alpha$ if and only if $\alpha < \omega^{\omega^\omega}$. The next theorem now follows from theorem 4 and 18.

THEOREM 20. *Let*

$$f_\alpha(n) = \lfloor \frac{1}{H_\alpha^{-1}(n)} \log_2 n \rfloor$$

$I\Sigma_2$ *proves the totality of* $L_{f_\alpha}$ *if and only if* $\alpha < \omega^{\omega^\omega}$.

**Note added in print:** The second author has obtained a similar classification of the $PA$-unprovability threshold regarding long finite sequences of natural numbers fulfilling non embeddability conditions in terms of Friedman's gap condition (See, for example, [6, 7] for a definition of Friedman's gap condition).

**Acknowledgement:** We thank Florian Pelupessy for his helpful comments.

## BIBLIOGRAPHY

[1] W. Burr: Verschiedene Charakterisierungen der $I\Sigma_{n+1}$-beweisbar rekursiven Funktionen, Thesis Münster 1996.
[2] D. de Jongh and R. Parikh: Well-partial orderings and hierarchies. Nederl. Akad. Wetensch. Proc. Ser. A 80=Indag. Math. 39 (1977), no. 3, 195–207.
[3] M. Fairtlough and S. Wainer: Hierarchies of provably recursive functions. Handbook of proof theory, 149–207, Stud. Logic Found. Math., 137, North-Holland, Amsterdam, 1998.
[4] H. Friedman: Long finite sequences. J. Combin. Theory Ser. A 95 (2001), no. 1, 102–144.
[5] H. Friedman and M. Sheard: Elementary descent recursion and proof theory. Ann. Pure Appl. Logic 71 (1995), no. 1, 1–45.
[6] K. Schütte and S. G. Simpson: Ein in der reinen Zahlentheorie unbeweisbarer Satz über endliche Folgen von natürlichen Zahlen. Arch. Math. Logik Grundlag. 25 (1985), no. 1-2, 7589.
[7] S. G. Simpson: Ordinal Numbers and the Hilbert Basis Theorem, The Journal of Symbolic Logic, Vol. 53, No. 3. (Sept., 1988), 961–974.
[8] S. G. Simpson: Nonprovability of certain combinatorial properties of finite trees. Harvey Friedman's research on the foundations of mathematics, 87–117, Stud. Logic Found. Math., 117, North-Holland, Amsterdam, 1985.
[9] R. Smith: The consistency strengths of some finite forms of the Higman and Kruskal theorems. Harvey Friedman's research on the foundations of mathematics, 119136, Stud. Logic Found. Math., 117, North-Holland, Amsterdam, 1985.
[10] A. Weiermann: An application of graphical enumeration to PA. J. Symbolic Logic 68 (2003), no. 1, 5–16.
[11] A. Weiermann: Phase transition thresholds for some Friedman-style independence results. Math. Log. Q. 53 (2007), no. 1, 4–18.
[12] A. Weiermann: Phase transitions for Gödel incompleteness. Ann. Pure Appl. Logic 157 (2009), no. 2-3, 281–296.

# Software Verification with Towers of Abstractions

BRUCE W. WEIDE

ABSTRACT. Reasoning about software system behavior is much like reasoning about physical system behavior—except in one critical respect. Empirical observation combined with the required disclaimer that "past experience is no guarantee of future performance" is inherently part of reasoning about whether a physical system will behave as hoped: whether the bridge will stand, the airplane will fly, or the hardware will properly execute programs. On the other hand, in principle one might hope to decide by purely mathematical reasoning and formal symbolic proof whether a software system will always behave as intended, *assuming* the physical computer running it will properly execute programs. As a practical matter, though, it has been far from clear until recently that proofs of correctness of software behavior can be produced consistently except for trivial software. There is now evidence suggesting that it might soon be routinely possible to verify—fully automatically—the correct behavior of significant pieces of software. An essential feature for software verification to scale up in this way is the use of an approach that modularizes proofs of software correctness by focusing attention on no more than two adjacent levels in a "tower of abstractions". An example illustrates how software verification proceeds when the specifications of desired behavior and the programs to be executed are written in a language that explicitly incorporates such towers of abstractions.

## 1 Introduction

Who hasn't used a computer program that crashed at an inopportune moment? Wouldn't it be great if that would never happen again? Many people believe this is a pipe dream. They argue that it is literally impossible to rid non-trivial software of all errors, and that even if reliable and defect-free software were possible in principle it would be too expensive to produce in practice. After all, it has been argued [10] that "[s]oftware entities are more complex for their size than perhaps any other human construct". If other engineers cannot guarantee that their artifacts will always work, then why should we expect software engineers to be able to make such guarantees for software?

Those who claim bug-free software is impossible draw two important conclusions from a single hidden assumption: that the way industrial-strength software is developed today is essentially the *only way* it could be done. They routinely assert, based on this assumption, that:

- Software must be tested in order that one should gain confidence in its correctness; but of course, there is no such thing as enough testing to show that software is entirely bug-free.

- The cost of reasoning carefully about all possible software behaviors must grow combinatorially (exponentially or probably worse) in the size of the software, e.g., the number of lines of code, and hence is infeasible for large programs.

## 1.1 On Testing

The shortcomings of testing—not just for software but across all of engineering—are well known. One of the pioneers of computing, Edsger W. Dijkstra, famously observed [15] that "[p]rogram testing can be used to show the presence of bugs, but never to show their absence!" The proper conclusion from this quite correct observation is that no amount of testing can hope to show software is bug-free. The improper (but popular and oft-repeated) conclusion is that it is impossible to show that software is bug-free.

The relevant question is whether testing is the only possible means of establishing software correctness. Rather than looking for software errors in the manner a civil engineer looks for weak spots that might become potholes in pavement, is it possible to reason about the behavior of a program in the manner of a mathematician proving a theorem? The theorem to be proved says, in effect, "for all permissible values of the inputs, this program produces correct outputs." A kind of mathematical reasoning like this is actually what many software professionals are taught to do while *developing* a program, though the process is rarely explained to them this way.[1] The practical difficulty is that humans are asked to carry out this reasoning only informally and in their heads. Humans make mistakes, especially when trying to process massive amounts of information such as the possible states of and paths through program code. Therefore, as a practical matter most large programs have some defects. Moreover, as noted by critics of mathematical approaches to establishing software correctness, residual

---

[1] Other software professionals subscribe to "test-driven design" which, as its name implies, simply assumes testing will be used to decide whether software might be correct.

errors in software professionals' reasoning inevitably escape detection now, and would continue to do so even if they were subjected to the usual social processes that historically have governed "proof" in the mathematics literature [37].

Might it be possible, though, to *formalize* the reasoning software professionals perform while developing programs, thereby permitting the use of computerized symbolic reasoning tools to catch these admittedly unavoidable human errors before software is released?

## 1.2 On The Combinatorial Explosion

Suppose it were possible, i.e., that we could to turn the establishment of the correct behavior of software into a mathematical rather than an empirical matter. There would remain a huge practical problem: real industrial-strength software consists of thousands or millions of lines of code. For a variety of technical reasons inherent in how most software is currently written, there are inevitably "unanticipated interactions between structural components" [57] in today's software systems. Surely, keeping track of all of them would be infeasible even for automated tools.

The relevant question then becomes whether the current languages and methods for software development are the only viable approaches to creating large software systems. Rather than designing software systems that are merely *structurally* modular (which certainly has already been achieved), might it be possible to design them to be *behaviorally* modular? This basic idea is hardly foreign to software professionals. Indeed, most would agree they try to achieve this, and some would argue they actually do. For otherwise how could mere humans possibly understand the large software systems they create? These massive and complex creations sometimes fail—but to be honest, the best of them also work correctly nearly all the time and do amazingly sophisticated things. Here is a particularly cogent explanation of the idea that underlies how software professionals try to develop software in order to maintain intellectual control over it. In this passage, "module" is taken to be "procedure" [43]:

> We reason separately about the correctness of a procedure's implementation and about parts of the program that call the procedure. To prove the correctness of a procedure definition, we show that the procedure's body satisfies its specification. When reasoning about invocations of a procedure, we use only the specification.

If the behavioral correctness of each software module truly could be reasoned about in isolation, then the cost of reasoning about an entire system would need to grow only linearly in the number of lines of code [57]. The problem today is that even those who think they know how to achieve behavioral modularity generally cannot *always* achieve it when working within the confines of state-of-the-practice software development methods and programming languages. And even if they were able to achieve it, they generally would be unable to make a concise convincing argument that they have done so, which means their dramatically simplified reasoning about the overall software system behavior as a result of assuming behavioral modularity might not be sound.

While the combinatorial explosion involved in reasoning about overall system behavior has not been tamed by present methods of software development, then, at least it has been caged. The challenge it poses might not be nearly as imposing as originally imagined. Might we just need to tweak programming languages and software design techniques to conquer the residual threats to modular reasoning about behavior?

## 1.3 Prospects for Practical Automated Software Verification

This article explores the thesis that automated **software verification**—mechanized fully formalized reasoning about software behavior relative to some formalized specification of correct behavior—might be not only possible but on the verge of becoming practical. What can make this claim meaningful and believable?

First, one needs a language in which to write formal specifications of the "permissible inputs" and "correct outputs" of a program—to describe *what* the software is supposed to do, i.e., its desired behavior. Fortunately, the language of mathematics and logic is available as the foundation for describing desired behavior. Unfortunately, it is currently unrealistic to expect software engineers to write not only ordinary programming codes in a formal language, but also to write specifications in a (related but distinct) formal language. The problem lies in education, though; it is not a problem in principle. Most current software professionals have not been taught to write formal specifications using mathematical and symbolic logic notations and have not practiced this skill. Yet those who are trained and experienced at doing so find it no more difficult (and usually far less so) than writing code to meet those specifications. In fact, the more complex the software, the more likely it seems that a formal specification of behavior is more concise and simpler to write and understand than code to achieve it. So, while the challenge of writing formal specifications cannot be ignored, it is far from fatal to the vision of software verification, given that software engineers evidently are able to write very complex program code in equally formal languages. Moreover, without precise and unambiguous specifications written in some language—either informal but adequately rigorous, or formal—even testing has no hope of providing confidence that software is correct. Without such specifications, there is simply no firm standard against which to judge correctness either for testing or for verification. Finally, experience shows that the very process of formalizing specifications leads to better software designs and results in mistakes being caught earlier in the design process [44]. The vision of software verification does not exacerbate the problems of capturing informal software requirements into rigorous specifications. If anything, it helps solve them by offering well-understood conventional notations from standard mathematics in which to express those requirements as formal specifications.

Second, one needs a formal statement of the proposed algorithm (code) that transforms inputs into outputs. Fortunately, there are many computer programming languages that seem to be available as the basis for this task—to describe *how* to achieve the desired behavior. Unfortunately, programming languages in widespread use for developing industrial-strength software are not the right ones for this job. One needs a way to combine the above symbolic entities—formal specifications of behavior and code purported to achieve it—into a set of mathematical statements called **verification conditions** (VCs) whose total size grows slowly in the total size of the above entities. There are now specification-plus-programming languages that make this possible. Although they are not the languages currently used for industrial-strength software or taught in most classrooms, they are not different in kind but rather in a few essential details.

Finally, one needs a way to mechanically prove each of the VCs in order to complete the proof-of-correctness of the software. Fortunately, there has been considerable progress on this front in the past 10-15 years, partly as a result of improvements in automated theorem-proving technology [21, 22] and partly as a result of the availability of faster computers. Unfortunately, these automated provers themselves are very complex pieces of software that lead to a bootstrapping problem: who proves the correctness of the provers, and can we trust them if they are not themselves verified? It turns out one really need not trust these most complex links in the verification tool chain. Rather than worrying about possibly faulty automated theorem-proving code to *discover* a proof, one can insist that an automated theorem-prover that purports to have discovered a proof emit a proof certificate to substantiate that claim. A proof certificate is essentially a step-by-step proof in symbolic logic that can be automatically *checked* by a separate tool. Checking a putative proof is far easier than discovering a proof in the first place, so much so that verifying such a proof-checker once seems quite feasible—far easier than tasks already claimed to be accomplished, such as verifying (with a one-time herculean effort) a compiler [39, 42] or an operating system kernel [34, 59].

One key point in the above paragraph should give reason for pause: how have faster computers (faster by a mere factor of perhaps 100) been able to make a serious difference in the ability to find proofs of mathematical statements, a problem considered computationally intractable under the best of circumstances? The situation is analogous to that facing chess-playing programs until fairly recently. It was long considered impossible that a computer could defeat a human chess champion, but chess-playing programs underwent some improvements and the computers running them became faster, even as the difficulty of playing championship chess *remained essentially fixed*. Similarly over the years, the mathematical statements that one needs to prove in order to establish software correctness have not grown fundamentally more difficult, yet the tools to prove them have advanced and the computers that run those tools have become substantially more powerful.

The conclusion is that automated software verification no longer can be dismissed as a real possibility for practical use in the near future.

## 1.4 Basics of Software Verification: An Example

Before moving on to the details of the above thesis, here is an example that illustrates the basic idea of software verification—for a trivial program, yes, but it is intended as an illustration of the idea. Imagine a simple imperative programming language with a programming type **Natural**, whose values are considered to be non-negative mathematical integers (initially 0) without upper bounds. There are already operations to **Increment**, **Decrement**, and compare the values of **Natural** variables, but not to add them. Figure 1 shows how one might write a new (recursive) programming operation **Add** to fill this gap. It is intended to compute the sum of the values of the parameters **n** and **m** and place the result in **n**, without changing the value of **m**. Assume there are formal symbolic specifications written using the language of mathematics to describe the behavior of **Natural** variables in the program, the behavior of the existing programming operations, and the intended behavior of **Add**—all of which will be elaborated later in this article for a more interesting example.

Software verification for such code consists of two steps:

1. Combine the above specifications and code to produce a number of putative theorems, the verification conditions, whose validity entails the correctness of the program.

2. Prove each of the VCs.

```
procedure Add (updates n: Natural, restores m: Natural)
    if IsPositive (m) then
        Decrement (m)
        Add (n, m)
        Increment (n)
        Increment (m)
    end if
end Add
```

Figure 1: Code for **Add** Operation

Prove
$m_0 - 1 < m_0$

Given
1. $0 \leq n_0$
2. $0 \leq m_0$
3. $0 \leq 0$
4. $0 < m_0$
5. $0 \leq m_0 - 1$

(a) VC #1

Prove
$(m_0 - 1) + 1 = m_0$

Given
1. $0 \leq n_0$
2. $0 \leq m_0$
3. $0 \leq 0$
4. $0 \leq m_0 - 1$
5. $0 \leq n_0 + (m_0 - 1)$
6. $0 \leq n_0 + (m_0 - 1) + 1$
7. $0 \leq (m_0 - 1) + 1$
8. $0 < m_0$

(b) VC #2

Figure 2

**Prove**

$n_0 + (m_0 - 1) + 1 = n_0 + (m_0 - 1) + 1$

**Given**
1. $0 \leq n_0$
2. $0 \leq m_0$
3. $0 \leq 0$
4. $0 \leq m_0 - 1$
5. $0 \leq n_0 + (m_0 - 1)$
6. $0 \leq n_0 + (m_0 - 1) + 1$
7. $0 \leq (m_0 - 1) + 1$
8. $0 < m_0$

(a) VC #3

**Prove**

$n_0 = n_0 + m_0$

**Given**
1. $0 \leq n_0$
2. $0 \leq m_0$
3. $0 \leq 0$
4. $m_0 \leq 0$

(b) VC #4

Figure 3

Figures 2 and 3 show screen-shots of the results of step 1 of software verification using our group's software verification tools. These are the result of rote syntactic translation of the code of Figure 1 and the relevant specifications into four VCs by following a set of sound and relatively complete proof rules for the specification-and-programming language used, followed by routine simplification by substitution of equals for equals. Each VC has the form of a universal statement about the variables appearing in it, which in this case are all mathematical integers. The proofs of these four VCs are left to the reader—who might notice that in some important respects they do not resemble theorems that any self-respecting mathematician would write. While VCs are technically just "little theorems" as far as a human or automated prover is concerned, the techniques needed to prove them might be somewhat different than those that would be applied to prove deep mathematical theorems posited by professional mathematicians—or even homework exercises in a textbook. A typical VC includes a number of assumptions that are not at all relevant to the conclusion to be proved and, in fact, might even overwhelm a human with so many useless details that it is difficult to separate the wheat from the chaff. It is therefore important to remember that VCs like these are expected to be proved by automated theorem-provers in step 2 and generally are not meant to be seen by humans.

The remainder of this article:

- reviews the history of software verification by identifying important milestones, both positive and negative;

- elaborates on why software verification—not to mention fully automated software verification—has not been widely believed to be workable (e.g., see the implicit and explicit criticisms above);
- discusses in some detail why these criticisms are unfounded (e.g., see the brief responses above), including some arguments based on the conceptual differences between drawing conclusions about physical systems and about software systems; and
- demonstrates via a more interesting example what is required to make software verification scalable to non-trivial software.

There is no review of contemporary related work. This is a concession to brevity, particularly as many research groups around the world have tackled Tony Hoare's **verifying compiler grand challenge** [26, 27]. The focus instead is more tutorial in nature with respect to our group's approach to software verification, in order to match the needs of the intended audience: mathematicians, logicians, philosophers, and computer scientists for whom software verification might otherwise be viewed as a somewhat mysterious undertaking. Some of the discussion involves arguments that seem closely related to ones that arise in the philosophy of science, but there are no references to such related work, either, and no digression into secondary objections that might be raised from that direction.

The reader is assumed to have some basic grounding in computer programming, mathematics, science, and logic—and an open mind.

## 2 Historical and Philosophical Perspective

Before examining software verification in more technical detail, it is worth reviewing its short history as well as its connections with concepts that are somewhat more familiar to most scientists, engineers, and mathematicians.

### 2.1 Sources of Difficulty and Hope

The idea of formally, symbolically, and automatically proving the correctness of a program relative to a full behavioral specification can be traced to Floyd [18], Hoare [24] and King [31] only about 40 years ago. Just a few years later, a paper critical of this vision [13] produced a profoundly chilling effect on both community and financial support for software verification research, at least in the U.S. That paper argued against the practical prospects for verifying software, as follows: "[T]he theorems that arise in trying to *prove* real programs are not simple... [T]otally automated *proof* systems are out of the question..." The italics are in the original: *prove/proof* in italics is defined by the authors as formal proof, in contrast to proof by fallible social processes. Interestingly, the claimed complexity of proving programs correct is neither as superficially compelling nor as true as the obvious objection that software verification researchers still encounter far more often: "Software correctness is undecidable." This observation presumably is intended to lead the reader/listener to the (misguided) conclusion that, therefore, software verification is a practical impossibility and any vision of it ought to be dismissed out of hand.

It is interesting to revisit [13] now in light of recent successes of automated verification of hardware and software, including among many others [30, 41, 4, 50, 60, 61, 42, 40, 59]. Whatever doubts one might have had about the feasibility or potential impact of automated verification a third of a century ago, the dream of automatically verified software can hardly be dismissed as "out of the question" today. Indeed, evidence now suggests that, contrary to the claim of [13], VCs are generally quite simple mathematically. In most situations, software verification is an enormous bookkeeping task involving VCs that modern theorem-provers often—yet still not always—can prove mechanically. Practical difficulties in verifying correct software seem to have little to do with VCs being fundamentally difficult to prove [32, 61] and nothing to do with undecidability. An important conclusion from recent software verification research is that automatically proving valid VCs that are *obvious* to mathematicians, and devising specifications that lead to such VCs also being obvious to automated provers, present the real challenges.

Why should it be the case that VCs for correct programs are "obvious"? In short, VCs capture exactly the properties that a software developer would have to be convinced are true in order to write correct code in the first place. Software professionals, typically being at best amateur mathematicians, generally do not write code that depends on new theorems that would be publishable in the mathematics literature. The formal reasoning underlying software verification merely captures the informal reasoning of the software developer and carefully checks it. Most of the time, the developer's reasoning checks out; occasionally it does not.

As might be expected, then, careful study of *failed* attempts to prove VCs resulting from typical code (e.g., [32, 55]) has been enlightening. Empirical observation to date suggests that:

- When software fails to always behave correctly when it is executed—hence software verification leads to the inability to prove a VC—it is usually traceable to mistakes arising from the fact that humans are unable to keep completely straight in their heads the overwhelming number of bookkeeping details that impact the correctness of the code they write. It is plausible that automated software verification should be feasible and helpful in rooting out human error in software development precisely because keeping track of an enormous number of details is the sort of thing computers can do well.

- When software always behaves correctly when it is executed—yet software verification leads to the inability to prove a VC—it is usually traceable to human failure to adequately *justify* correctness [33, 61] with appropriate annotations in the code, not to insidious underlying properties such as undecidability or even inherent mathematical complexity. The unproven VC is true, is obvious to a mathematician, and yet is not sufficiently obvious to the automated prover, so its proof is not discharged mechanically. Annotations (e.g., invariants) to justify correctness might be too weak or wrong, or the mathematics on which those justifications are based might contain insufficient developments and/or results (e.g., the prover does not know about certain established theorems from the underlying mathematical domains). Other times, specifications, annotations, and supporting mathematics might be technically adequate but not engineered to match the capabilities of automated provers. In yet other cases, automated provers might simply lack the fine-tuning necessary to reach obvious conclusions.

In other words, the practical barriers to automated software verification are quite a bit less daunting than the ones that typically have been advanced by critics as insurmountable.

## 2.2 Mathematical Models: Physical System Behavior *vs* Software System Behavior

At its heart, software verification deals with a fundamental and somewhat philosophical question that arises across the physical sciences and engineering:

> Is a proposed mathematical model of some behavior a valid "cover story" for that phenomenon, in the sense that it permits one to analyze and predict relevant features of the phenomenon?

In physical science, a mathematical model can be a useful cover story even when it contributes nothing to causal understanding and helps explain no causal connections, but instead merely predicts certain observable features. A mathematical model of higher-level phenomena (say, the behaviors of measured properties of electrical circuits) might be related by purported explanatory or causal links to putative mathematical models of underlying lower-level phenomena (say, electron mobility and charge flux). On the other hand, that mathematical model might be treated simply as a law.

Consider Ohm's Law: $V = IR$. Here, $V$ is the voltage across two terminals of a component in an electrical circuit, $I$ is the current flowing through that component, and $R$ is the resistance of the component. Without making any reductionist commitment to an explanation of what voltage and current are or how they arise from other physical phenomena, one might be willing to assume that resistance—whatever that is—is a constant for a given circuit component, and conclude from Ohm's Law that $V$ and $I$ for that component are proportional to one another. This is a mathematical conclusion drawn entirely from the mathematical model. That model describes no causal physical relationships to explain the origins of the measured voltage, current, or resistance, but it is consistent with empirical observations of electrical circuits and (experience has shown) can be used to predict certain features of previously unexamined cases. The scientific "truth" of Ohm's Law is based on this utility rather than on its ability to help explain the underlying physical causes of the measured quantities involved in it.

A law of this sort is, of course, not something that either a god or a legislature has decreed. It is a no more and no less than a mathematical model of the physical world that allows one to predict some measurable behaviors of the physical world. It abstracts away details that are not needed in order to make these predictions.

In (a Popperian view of) science, someone postulates such a mathematical model of a phenomenon that is consistent with observations already made. Then the broader science community searches for counter-examples to the claim that the proposed mathematical model is a valid cover story for the phenomenon. Finding no counter-examples after a long and presumably careful search, the science community tends to accept as a matter of social process that the mathematical model is a valid cover story for the behavior of interest. Confidence grows in the validity of the mathematical model in this sense so long as only corroborating evidence is found; if certain observations seem anomalous, then sometimes the model can be adjusted and thereby saved from the scrap heap of history.

If mathematical relationships between models at distinct levels can be postulated and checked, so much the better. Then, such mathematical models have the potential for leading to explanation and understanding of causal connections between physical phenomena at different levels rather than merely for making predictions at a single level [11]. But even when no such connection is made, a mathematical model can be useful in practice for its predictive power alone.

The above view pertains to the role of mathematical modeling in nearly all of science and engineering. The exception is software engineering, where there is a crucial difference: a mathematical model of software-system behavior may be specified entirely without regard to natural phenomena. A software system can be treated separately from the computer on which it runs, i.e., as a purely symbolic entity whose meaning (an abstract dynamic behavior) is determined by the semantics of the programming language in which it is written. Its desired behavior is similarly a purely symbolic entity whose meaning is determined by the semantics of the specification language, which is largely that of mathematics. A piece of software thus treated is, in principle, a closed system whose behavior can be completely analyzed and predicted in its own terms: the description of a mathematical model of software behavior is essentially the software itself.

Of course, proving that a software system in this sense is "correct" most certainly does not prove that, when executed on a physical computer rather than on the conceptual computational model underlying the programming language semantics, the software actually behaves as advertised. The leverage one gains in treating a software system purely conceptually is one of separation of concerns and potentially assignment of blame for defects. If software that has been proved correct does not execute correctly on a real computer, then the defect lies outside the lines of code that have been verified. It might be in the compiler, which could have generated improper code for execution by not matching the semantics of the programming language source code; or in the operating system, which could have loaded the executable code generated by the compiler into the wrong place in memory; or in the computer itself, which could have been designed or implemented or manufactured incorrectly relative to its own specification. Fortunately, it recently has been shown possible to prove (using interactive proof methods, i.e., not fully automatically) that system software such as a compiler [39, 42] or an operating system [34, 59] is correct in the same sense of not containing lines of code that are in error when executed on an idealized non-physical computer. Computer hardware designers have been using related methods to verify that symbolic representations of computer operation in their designs faithfully capture specified behavior. In other words, in principle, the entire software milieu (extending even to symbolic representations of computer hardware designs) is subject to the same sort of formal verification processes as those described in this article. Once the software verification vision is fully realized, defects in computer-based systems will have been isolated to the physical realizations of computers themselves and the physical systems they interact with.

The observation that a software system can be treated as a purely symbolic entity and reasoned about as such is behind the notion popularly called **virtual reality (VR)**. VR is a misnomer: nothing about the phenomena explained by the software engineer's mathematical models need be even remotely "real" or "realistic". A software engineer is free to design a mathematical model of truly imaginary and/or entirely abstract behavior that would never be observable in the natural physical world, and then to try to create the desired behavior in a software system by building it on top of existing lower-level software components that have their own mathematical models of their own behaviors. Because a software professional not only *designs* the desired mathematical model of new behavior but also *designs* one or more realizations of such behavior—rather than relying on nature to have done that—understanding and explaining the precise causal connections between the specified higher-level behavior and the stipulated lower-level behaviors are vital aspects of a software engineer's job.

## 2.3 Towers of Abstractions

It is interesting to note what happens when students in an early computer science course are asked to think about what might be meant by "levels" in the discussion above. Consider any popular video game, e.g., Nintendo®Wii bowling. Are there bowling balls and bowling pins inside the computer? Of course not! What is inside the computer, then, that makes this bowling-like behavior? The answers typically offered are enlightening. Vectors? Numbers? Bits? None of these "exists" inside the computer in the usual sense that one could directly observe them as physical objects or entities. Instead, each of these *concepts* is nothing but a figment of a software professional's imagination. Each is but a mathematical model that can be presented to a software-system user and/or to one learning computer science, as though it did exist inside the computer, in such a compelling and logically irrefutable way that one is hard-pressed *not* to believe that each one enjoys a physical reality embodied in the hardware. We can easily forget that someone—again, a person, not a god—designed in gory detail exactly how to **represent** "higher-level" concepts in terms of "lower-level" concepts already in place: bowling balls and pins using vectors; vectors using numbers; numbers using bits; bits using ... voltages, right?

This is what one learns in a digital electronics course. Surely voltages physically exist inside the computer, don't they? Actually, there are no voltages inside the computer, either, in the sense that someone has ever directly observed a physical "voltage object" in a computer or anywhere else. Voltage is an abstract mathematical quantity that can be measured by what looks at first like a magic device and that can be related to other measurable quantities by, e.g., Ohm's Law. What really exists inside the computer that makes these voltages is ... electrons, right? No one has ever directly observed one of these, either. If one could see an electron, would it look more like Pluto orbiting the sun or more like a wave on water? Probably not much like either; after all, these are metaphors arising from similarities between mathematical models that help predict certain aspects of the behavior of the theoretical objects physicists call electrons and the mathematical models they use for predicting certain behaviors of the solar system or of water waves. Maybe what's *really* inside the computer are even smaller entities that make up the electrons.

The above scenario illustrates a **tower of abstractions**. Strictly speaking, "tower" is too restrictive a metaphorical term selected more for the image it evokes (e.g., Figure 4) than for technical merit. In fact, each new abstraction might be built "on top of" *multiple* abstractions from "lower" layers.[2]

Figure 4: A Famous Tower

Do software professionals need to think about the answers to the physics questions listed above? Only to pass physics exams. What physically exists inside the computer is of little consequence to software engineers. All that

---

[2]The idea that direction in a tower of abstractions should be viewed as up/down (though only approximately in the case of Figure 4) is also arbitrary and metaphorical, but the senses of "higher" and "lower" are easily remembered: consider the term "underlying" applied to explanation and causality among physical phenomena.

matters is that the bits we *imagine* to exist inside it behave as bits are supposed to behave; similarly, the numbers, vectors, etc., that we *imagine* to exist inside it. Metaphysical questions, such as whether there might be an infinite regress at play here, are also of little consequence to software engineers. All that matters right now is the layer of immediate interest and (sometimes) the one immediately below it. The layers above and below these are simply out of the picture when reasoning about the behavior—hence the behavioral correctness—of a particular software component.

Still, it is interesting to think for another moment about the point where we relate bits to voltages as we move across those two adjacent levels in this tower of abstractions. For voltages and below, the mathematical models have been developed *a posteriori* to be consistent with the observed behavior of natural physical systems and (for the surviving models) to predict the behavior of those systems. For bits and above, the mathematical models have been developed *a priori* to create new synthetic or artificial behaviors that need not adhere to any particular laws other than those of the mathematical theories used in these mathematical models—and the laws of logic. Wii bowling pins act very much like real bowling pins, but there is no reason whatsoever they could not have been designed by software professionals to exhibit behavior that could never be seen in a real bowling alley. Some video games, in fact, are popular precisely because they do not depict a virtual reality that mimics the natural world in which we live.

In summary, then, mathematical modeling of software-system behavior is different from mathematical modeling of natural physical-system behavior in two key ways:

- The mathematical models of software-system behavior are *not limited* to describing behavior that might be observed in the natural physical world.

- The connections between mathematical models that describe behavior at two neighboring levels of a software system's tower of abstractions are always critically important: an explanation that *relates* two such levels in the software-models region of a tower of abstractions is never optional, as it might be in the physical-models region of a tower of abstractions.

## 2.4 Testing *vs* Verification for Towers of Abstractions

The remainder of this article deals with a special version of the earlier question in the context of towers of abstractions:

> Is a proposed mathematical model of software component/system behavior a valid "cover story" for that phenomenon, in the sense

that it permits one to analyze and predict relevant features of the phenomenon?

This question can be approached in two distinct ways. One can search for counter-examples to the claim that the higher-level model is a valid cover story, as when using a science paradigm, by observing the behavior of software as it executes. This is known, in software engineering as in the rest of science and engineering, as testing. Alternatively, one has—for software—the option of trying to prove mathematically the claim that the higher-level mathematical model provides a valid cover story for the software's behavior, using a mathematics paradigm. This approach is known in software engineering as verification. While the term might be used elsewhere with different meanings, there is no comparable notion to software verification in natural science and engineering but only in mathematics and logic. For example, there is no way to prove mathematically that a bridge being designed by a civil engineer will not fall down. One can hope to prove that a mathematical statement of such a property holds for a particular mathematical model of the bridge, but this *proves* nothing about the physical bridge itself.

Yet in principle there might be a way to prove that software will not crash or give wrong answers, because the mathematical models involved in software behavioral specifications at two adjacent levels in a tower of abstractions are purely symbolic entities, as are the computer programs that bring together lower-level mathematical models to create behaviors consistent with a higher-level one. Mathematics and logic can directly address such questions. If it ever becomes practical to prove that a higher-level mathematical model provides a valid cover story for software-system behavior implemented by carefully relating the new mathematical model to existing lower-level mathematical models in a tower of abstractions, it inevitably will become a professional responsibility for software engineers to carry out such verification—at first for the most critical software systems, eventually for all commercial or otherwise important software.

A significant question, then, is whether automated software verification that accounts for *connections between levels in a tower of abstractions* rather than working only at *a single level of abstraction within a tower* (as in the simple example of Section 1.4) will ever become practical. The next section explains why we believe it will, first with an overview of how our research group's current software verification tools do the job, and then by using another (still reasonably simple) example as an illustration of proof-of-concept.

## 3 Automated Software Verification in RESOLVE

Our group's research in automated software verification directly addresses the verifying compiler grand challenge [26, 27]. This challenge effectively has been reissued from the similar vision mentioned earlier, and first expressed over 40 years ago [18, 24, 31]. Briefly paraphrased, the challenge is:

> Given a mathematical model specifying the desired behavior of new software, and like-kind specifications of existing software that can be combined to achieve that behavior, prove automatically that given program code correctly combines the latter components/systems to implement the desired behavior.

A verifying compiler should, of course, respect the crucial role of towers of abstractions. For programs written in the specification-and-programming language being compiled, it should be able to work for code that operates within a single level of a tower. It also should be able to work for code that bridges two adjacent levels in a tower—indeed, it must support this in order that a new level in the tower should be constructed correctly in the first place. If it can do both, then in principle there are no limits to the complexity of software with which it can deal. If one has the mathematical theories to model any desired software behavior, then the verifying compiler should be able to produce VCs (with content involving those theories) whose validity is tantamount to correctness of the software that is claimed to exhibit that behavior. Afterward, it should be able to prove those VCs.

### 3.1 The Vision

The long-term vision guiding our group's research is that of a future in which no production software is considered properly engineered unless it has been fully specified and automatically verified as satisfying these specifications. The verifying compiler grand challenge, when met, will offer significant benefits. Of course, it will not imply that verified application software will always operate perfectly: other software with which it communicates might not work properly, the hardware on which it executes might not work properly, etc. However, full behavioral specification plus modular verification that software meets its specification will imply a clear separation of concerns. For verified software, residual errors will be limited to whether the specification captures the requirements and to whether the supporting software and underlying hardware behave as advertised. Questions about the correctness of verified software components relative to their specifications will be effectively moot.

The approach used in our RESOLVE framework for software verification [53, 51] begins with respect for modular reasoning. We use the term **soft-**

ware component because nearly all software is intended to be used as a piece (or module) of another, larger, software system—which very likely is itself a software component in the same sense. Scalability of reasoning about software behavior to large software systems demands that each software component be verifiable in isolation, so its proof of correctness need not be revisited, revised, and/or redone every time the component is used in the context of a larger software system. This is the essence of modular reasoning. A much weaker notion is modular construction, in which syntactic modularity is ensured but semantic modularity is not: all the plugs and sockets "fit together", but there is no way to reason about whether an entire assemblage behaves as expected without flattening it out into its atomic pieces. Most programming languages support modular construction of software but not modular reasoning about its behavior. RESOLVE is unique among modern imperative programming languages in that it has been designed from the beginning to support modularity in both these senses.

Modular verification of a RESOLVE software component is a two-step process. First, we combine the specifications describing the component's intended behavior, the specifications of the components used in a proposed realization of that behavior, and the code that combines the latter to implement the former, generating as output a set of verification conditions (VCs, as introduced earlier). The VCs are assertions in the formal language of mathematical logic with variables, functions, and predicates ranging over the mathematical theories appearing in the specifications. The validity of these VCs implies the correctness of the code to be verified. So, the second step is to prove that the VCs are valid. The first step is done by a tool called a **verification-condition generator**, which is comparable in complexity to a normal compiler except that it generates VCs rather than executable code. The second step is done by a separate tool, an automated theorem-prover, as discussed earlier.[3] We have built our own special-purpose theorem-prover called SplitDecision [1], but we also use Isabelle/HOL [46], Z3 [14], and Dafny/Z3 [40], which have larger and more diverse user communities and are more fully developed tools. Fortunately, VCs often tend to be rather easy to prove when compared to the sorts of theorems a sophisticated tool like Isabelle can prove interactively with a professional mathematician or logician at the helm (e.g., [2]). Rather than being mathematically deep, typical VCs are generally bookkeeping jungles in which conclusions are to be proved from many (mostly irrelevant) assumptions, as seen in Figures 2 and 3. The reason a practical verifying compiler is envisioned as fully automated is that computers are better at bookkeeping tasks like this, while people are better at creative tasks. The division of labor between humans

---

[3] Figure 5 also shows a proof-checker tool for reasons raised in Section 1.3.

and computers should acknowledge this. Hence, for us, a verifying compiler should require no input from a software professional that looks like a "proof command".

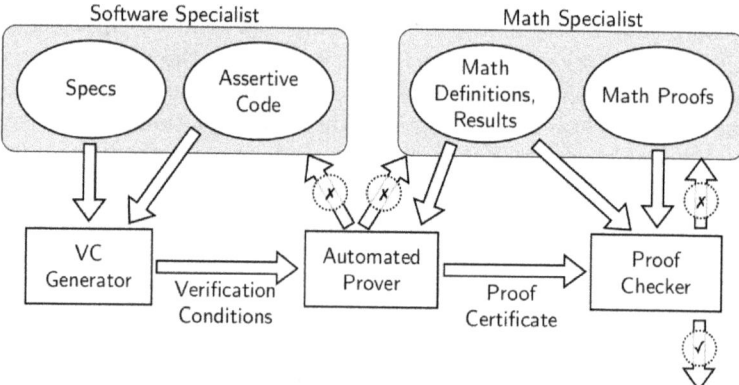

Figure 5: System Overview of a RESOLVE Verification System

A number of research groups recently have achieved considerable progress toward addressing various aspects of the verifying compiler grand challenge, e.g., [4, 8, 6, 5, 28, 48, 7, 9, 17, 38, 60, 12, 61, 29, 40]. Meanwhile, others have achieved impressive results in verifying—albeit at great expense in terms of human time spent with interactive theorem proving tools—significant pieces of potentially important software systems such as compilers [39, 42] and operating system kernels [34, 59]. We do not survey such tools or results, but instead concentrate on specific aspects of software verification using RESOLVE to illustrate how towers of abstractions play a crucial role in the process.

## 3.2 Proof of Correctness of Data Representation

As noted earlier, a mathematical model of higher-level phenomena in the physical world might be treated simply as a law, or it might be related to putative mathematical models of underlying lower-level phenomena. Both situations also arise in software verification. The former is the case when a new software component is an enhancement, or extension, of an existing component: it provides incremental additional functionality within a single level in a tower of abstractions. Here, mathematical modeling is homogeneous in the sense that the mathematical vocabulary involved in specifying the incremental behavior is the same as that used in specifying the behavior of the existing component it enhances. The key technical feature of this subproblem is that the new software component introduces no new **types** (which we consider to be what are traditionally called abstract data types, or ADTs) and hence does not entail a change of mathematical theories between the specifications of an existing software component and the desired behavior of the new component. One assumes that the mathematical model of the component being enhanced is a valid cover story for its behavior, and must prove that a mathematical model involving the same mathematical theories is a valid cover story for the behavior of program code that adds a piece of new functionality to the underlying component. We have previously published a variety of verification results on this subproblem [54, 52, 58, 51].

The rest of this article outlines the RESOLVE approach to the other situation: the new component introduces a new ADT with a mathematical model that (in general) involves different mathematical theories from those used in the mathematical models of the behaviors of the underlying software components. Here, mathematical modeling is heterogeneous in the sense that the mathematical vocabulary involved in specifying the incremental behavior might not be the same as that used in specifying the behavior of the existing components. This problem is known in computer science as **proof of correctness of data representation** [25]. It is analogous to the situation faced by physical scientists and engineers when connecting two related levels in a tower of abstractions: a mathematical model proposed as a cover story for some behavioral phenomena must be related to lower-level mathematical models of underlying phenomena that are claimed to be responsible for producing the higher-level behavior to be explained.

The purported connection between a higher level and a lower level in a tower of abstractions involves some sort of "bridge law" that relates them. Focusing on how one is to interpret values of variables of a new higher-level ADT from its data representation in terms of the values of lower-level ADTs, Figure 6 uses the terminology of computer science for this situation: the **abstract state** (or abstract value) of a higher-level variable is related by an **abstraction relation** to the **concrete state**, which consists of a collection of one or more variables of lower-level ADTs.

When the mathematical models of neighboring levels in the tower involve different mathematical theories, it can be difficult to reason fully automatically about VCs that make connections between them. It is one thing to prove VCs that involve only, say, linear arithmetic over the integers (as in Figures 2 and 3). It is one thing to prove VCs that involve only, say, finite sets of objects. It is one thing to prove VCs that involve only, say, finite strings of objects. It is another thing entirely to prove VCs that involve all three of these theories. There are techniques to combine decision procedures for individual theories [45, 3, 36] under certain technical conditions.

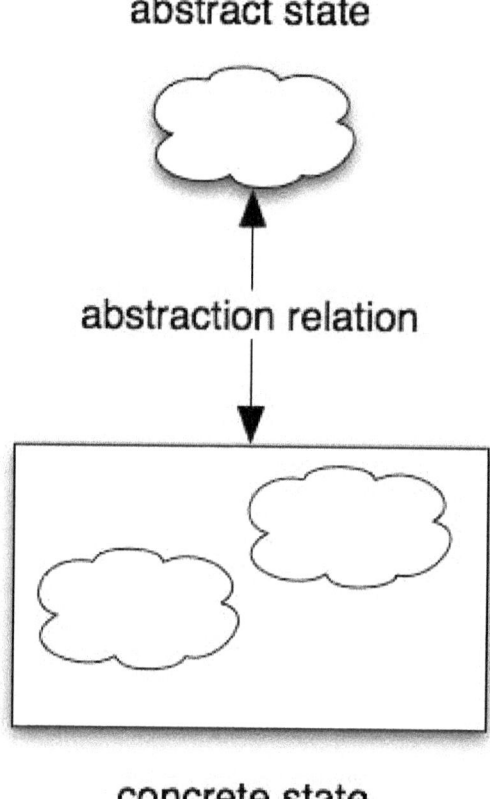

Figure 6: How a Higher-Level ADT is Related to Lower-Level ADTs

However, VCs involving lengths of strings, cardinalities of sets, mappings between strings and sets, etc., stir together all the above theories in ways that defy immediate mechanical combination of individual decision procedures (where they exist). Moreover, some of the individual theories used in software specifications are, of course, undecidable. Yet inevitably such theories will be needed in software specifications. It is simply unacceptable to limit expressiveness of specifications by insisting that all descriptions of behavior involving integers should use only linear arithmetic, on the grounds that this would make verification easier.

The approach used by Harvey Friedman [19, 20] as part of our group's software verification research has been to identify interesting fragments of theories involving integers, sets, strings, as well as partial and total functions and relations. These are the theories that commonly arise in RESOLVE software specifications. What mathematical predicates and functions are involved in actual VCs? Where quantifiers appear, which quantifier structures arise in actual VCs? How many variables do actual VCs have? Can one devise decision procedures for theory fragments that satisfy technical restrictions suggested by these empirical observations? Such questions lead to plenty of challenges for logicians. The focus turns away from the classical question of whether a full theory is decidable—because it simply isn't decidable if the theory is generally useful for modeling full functional behavior of software. Instead, the focus turns to whether the restricted forms of actual VCs make it decidable: structural restrictions on the sentences to be proved, size or complexity restrictions on those sentences, etc.

Even within our own group, there is no consensus that new decision procedures for such theory fragments are the best way to go in order to improve automated theorem-provers for software verification. Would it be better to seek powerful but necessarily heuristic automated general-purpose proof techniques that work well on actual VCs? No one knows yet. It is worthwhile to explore both approaches.

### 3.3 Example: A Set ADT Represented Using a Queue ADT

We now consider another relatively simple example. The earlier example (Figure 1) showed an enhancement to add new functionality to an existing ADT. This one illustrates how proofs of correctness of data representation arise and are carried out in the RESOLVE framework. RESOLVE has the standard imperative programming control constructs and basic constructs supporting component-based software design. The first major difference between RESOLVE and a typical industrial language like Java is that, in RESOLVE, we program over arbitrary mathematical spaces: the type of each variable in the program is an ADT, also called its **program type**, and each program type has an associated mathematical model that determines the variable's **mathematical type** for purposes of reasoning about the values the variable might have during program execution.[4] Software engineers barely realize they are following exactly this approach when they program with a built-in **value type** in a language like Java. The value of a variable of program type `int` in Java, say, is treated for reasoning purposes as if it were a mathematical integer between some bounds, not as a block of 32 bits in computer memory—and certainly not as a vector of 32 voltages or as a particular configuration of electrons inside a memory chip. The reason, of course, is that it would be far more difficult to reason about the behavior of programs that use `int`s if their (incidental) data representation in terms of bits were not hidden behind the cover story that `int`s *behave* like mathematical integers whose values are constrained to lie within certain bounds. RESOLVE applies this approach to all variables of all program types, so *every* variable ranges over the values of its program type's mathematical model. This distinguishes it from languages like Java, where a variable of a programmer-defined type—a so-called **reference type**—is thought of as ranging over references to, or addresses of, memory locations where data representations are stored during program execution.

The required modularity of reasoning for scalable software verification is actually quite difficult to achieve because of a number of common programming practices that routinely appear in today's industrial software by virtue of programming languages with features that are too permissive. The interested reader may consult [56] for details about a chief impediment to modularity: interference via aliased references. A language with reference types inevitably permits two variables to have the same (reference) value, which is known as **aliasing**. The most popular approach to reasoning about references in current verification systems is separation logic [49, 47]. Separation logic has been shown viable for dealing with references that are aliased *within* a component, but not for dealing with references that are aliased *across component boundaries*. A recent paper focusing on reasoning about such cross-boundary interference [16] concludes: "Fundamentally, because it concerns aliased pointers, freedom from interference is extremely difficult to protect against using programming language restrictions, and too expensive to protect against with runtime checking. It is better to say that *if* there is no interference *then* refinement reasoning is sound, rather than to say that it is unconditionally sound." Suffice to say that, despite protestations

---

[4]The mathematical model, or mathematical type, associated with each program type justifies the term "abstract data type". Rather than being based on how a program type is represented, reasoning is raised up one level in the tower of abstractions to something that is presumably easier to reason about than the underlying data representation would be.

that it is "difficult to protect against" such interference, the RESOLVE language nevertheless does just that. In summary, the RESOLVE approach to verification is modular because the RESOLVE language has been carefully designed to support modular reasoning about software component behavior in this respect (as well as others).

The specification of the example higher-level software component to be realized and verified is shown in Figure 7. This component provides client programmers with a new ADT, Set, and operations by which to manipulate Set variables. The mathematical model of a Set is a finite mathematical set of objects of some type Item (whose own mathematical model is undetermined at this point, but fixed, and is known by the name Item in mathematical contexts as well). So, for example, the **initialization ensures** clause means that the value of any newly declared programming variable of type Set is initially the empty set. Operations are specified in the generally accepted software design style known as design-by-contract: a **requires** clause introduces a precondition that must be true in the client (calling) program in order that the operation call is legal, and an **ensures** clause introduces a postcondition that the component implementer guarantees to be true in the client program upon return from the call. In a postcondition, a variable decorated with a prefix # denotes the value of that variable at the time of a call to the operation, where the undecorated version denotes its value upon return from the operation. So, for example, the Add operation has two parameters: a Set s (whose values $\#s$ and $s$ are considered to be finite mathematical sets of Items) and an Item x (whose values $\#x$ and $x$ are considered to be whatever the specification for the actual type Item declares its mathematical model to be). In code that uses this component, the Add operation may be called whenever $x \notin s$. If this is true at the time of the call, then when the call returns, it will be the case that $s = \#s \bigcup \{\#x\}$.

The code to be verified uses a single Queue to represent a Set. The specification for the existing lower-level component, QueueTemplate, is shown in Figure 8; the code that uses it to implement the higher-level SetTemplate specification is shown (split into two parts for typesetting purposes) in Figures 9 and 10. It can be seen that the specification in Figure 8 is similar in kind to that in Figure 7—but the theory involved in this specification is not the mathematics of finite sets but rather that of finite strings. Intuitively, strings record order and allow multiplicity (repetition), while sets do not. This suggests designing the obvious relationship between the two levels: the value of a Set variable at the higher level is represented by having a Queue containing exactly one copy of each of the elements of the Set variable's value, in arbitrary order.

```
contract SetTemplate (type Item)

    uses UnboundedIntegerFacility

    math subtype SET_MODEL is finite set of Item

    type Set is modeled by SET_MODEL
        exemplar s
        initialization ensures
            s = { }

    procedure Add (updates s: Set, clears x: Item)
        requires
            x is not in s
        ensures
            s = #s union {#x}

    procedure Remove (updates s: Set, restores x: Item,
                      replaces xCopy: Item)
        requires
            x is in s
        ensures
            s = #s \ {x}  and  xCopy = x

    procedure RemoveAny (updates s: Set, replaces x: Item)
        requires
            s /= { }
        ensures
            x is in #s  and  s = #s \ {x}

    function Contains (restores s: Set, restores x: Item): control
        ensures
            Contains = (x is in s)

    function IsEmpty (restores s: Set): control
        ensures
            IsEmpty = (s = { })

    function Size (restores s: Set): Integer
        ensures
            Size = |s|

end SetTemplate
```

Figure 7: SetTemplate Specification

```
contract QueueTemplate (type Item)

    uses UnboundedIntegerFacility

    math subtype QUEUE_MODEL is string of Item

    type Queue is modeled by QUEUE_MODEL
        exemplar q
        initialization ensures
            q = < >

    procedure Enqueue (updates q: Queue, clears x: Item)
        ensures
            q = #q * <#x>

    procedure Dequeue (updates q: Queue, replaces x: Item)
        requires
            q /= < >
        ensures
            #q = <x> * q

    function IsEmpty (restores q: Queue): control
        ensures
            IsEmpty = (q = < >)

    function Length (restores q: Queue): Integer
        ensures
            Length = |q|

end QueueTemplate
```

Figure 8: QueueTemplate Specification

```
realization QueueRealization (
      function AreEqual (restores i: Item, restores j: Item): cont
          ensures
              AreEqual = (i = j)
 ) implements SetTemplate

   uses QueueTemplate
   uses Concatenate for QueueTemplate
   uses IsPositive for UnboundedIntegerFacility

   facility QueueFacility is QueueTemplate (Item)
       enhanced by Concatenate

   type representation for Set is (
           items: Queue
       )
       exemplar s
       convention
           |s.items| = |elements (s.items)|
       correspondence function
           elements (s.items)
   end Set

   local procedure SplitAndExtract (updates q1: Queue,
           restores x: Item, replaces q2: Queue, replaces y: Item
       )
       requires
           |q1| > 0
       ensures
           #q1 = q2 * <y> * q1  and
           (if x is in elements (#q1) then y = x)

       Clear (q2)
       Dequeue (q1, y)
       loop
           maintains
               q2 * <y> * q1 = #q2 * <#y> * #q1  and
               x is not in elements (q2)  and  x = #x
           decreases |q1|
       while not IsEmpty (q1) and not AreEqual (x, y) do
           Enqueue (q2, y)
           Dequeue (q1, y)
       end loop
   end SplitAndExtract

   procedure Add (updates s: Set, clears x: Item)
       Enqueue (s.items, x)
   end Add
```

```
    procedure Remove (updates s: Set, restores x: Item,
                      replaces xCopy: Item)
        variable tmp: Queue
        SplitAndExtract (s.items, x, tmp, xCopy)
        Concatenate (s.items, tmp)
    end Remove

    procedure RemoveAny (updates s: Set, replaces x: Item)
        Dequeue (s.items, x)
    end RemoveAny

    function Contains (restores s: Set, restores x: Item): control
        variable qLength: Integer
        qLength := Length (s.items)
        if IsPositive (qLength) then
            variable y: Item
            variable tmp: Queue
            SplitAndExtract (s.items, x, tmp, y)
            Contains := AreEqual (x, y)
            Enqueue (s.items, y)
            Concatenate (s.items, tmp)
        end if
    end Contains

    function IsEmpty (restores s: Set): control
        IsEmpty := IsEmpty (s.items)
    end IsEmpty

    function Size (restores s: Set): Integer
        Size := Length (s.items)
    end Size

end QueueRealization
```

Figure 10: QueueTemplate Specification, part 2

The function **elements** maps a string to the set of its entries, thereby effectively removing information about the ordering and multiplicity of string entries. In addition, the (overloaded) set cardinality operator $|\bullet|$ and string length operator $|\bullet|$ are both integer-valued functions. This leads to a situation where most of the VCs arising from the code in Figures 9 and 10 simultaneously involve finite sets, strings, and integers, and mathematical functions and predicates from these theories—unions of finite sets, concatenation * of strings, set $\{\bullet\}$ and string $<\bullet>$ constructors from entries, etc. The design-by-contract approach with explicit mathematical modeling, in which programming variables' values range over their stated mathematical models, underlies VC generation. No further details are provided here because the present focus is on the nature of the VCs themselves rather than on how they are generated or why the rules for generating them are sound and relatively complete; the reader interested in such details is referred to [35, 23, 52, 51].

The code for QueueRealization is similar to the code one might write in any modern object-based imperative programming language. The primary difference is that the programmer of this code has had to write a few extra lines—of mathematics, not ordinary programming code. The two most important of these for proof of correctness of data representation are the **correspondence function** clause, which introduces the **abstraction function**; and the **convention** clause, which introduces the **representation invariant**. The former clause is the interpretation mapping that says precisely how the value of the higher-level Set variable, as described in the SetTemplate specification this code purports to implement, can be obtained from the lower-level Queue that represents it. The latter clause characterizes the domain of this mapping.[5] In addition, the programmer has had to write a contract for the local operation SplitAndExtract that is used as a "helper" operation within QueueRealization, as well as a loop invariant for the loop in that operation (the **maintains** clause) and a justification that the loop terminates (the **decreases** clause).

For the code of Figures 9 and 10, the RESOLVE VC generator emits 37 VCs, as summarized in Table 1. All 37 are proved automatically by SplitDecision, which is the least sophisticated of the back-end automated theorem-provers we use. SplitDecision does the entire job in a total of just over one second.

A couple of the most difficult VCs, where difficulty is based on proof time by SplitDecision, are shown in the screen-shots of Figures 11 and 12.

---

[5]The interpretation mapping in RESOLVE is currently limited to being a function to simplify certain aspects of the verification tools, though it is known that in general it might need to be a relation.

Table 1: VC Summary for `QueueRealization`

| Source of VCs | Number of VCs |
|---|---|
| type representation for `Set` (initialization) | 2 |
| `SplitAndExtract` | 11 |
| `Add` | 2 |
| `Remove` | 7 |
| `RemoveAny` | 4 |
| `Contains` | 9 |
| `IsEmpty` | 2 |
| `Size` | 0 |
| Total for `QueueRealization` | 37 |

**Verification Condition #2 (state index: 3, ensures clause)**

Prove

$\text{elements}(s.\text{items}_2 \circ \text{tmp}_2) = \text{elements}(\text{tmp}_2 \circ \langle x_0 \rangle \circ s.\text{items}_2) \setminus \{x_0\}$

Given
1. $x_0 \in \text{elements}(\text{tmp}_2 \circ \langle x_0 \rangle \circ s.\text{items}_2)$
2. $|\text{tmp}_2 \circ \langle x_0 \rangle \circ s.\text{items}_2| = |\text{elements}(\text{tmp}_2 \circ \langle x_0 \rangle \circ s.\text{items}_2)|$

Figure 11: A relatively difficult VC arising from `QueueRealization`

**Verification Condition #9 (state index: 10, convention)**

Prove

$|s.\text{items}_6 \circ \langle y_6 \rangle \circ \text{tmp}_6| = |\text{elements}(s.\text{items}_6 \circ \langle y_6 \rangle \circ \text{tmp}_6)|$

Given
1. $|\text{tmp}_6 \circ \langle y_6 \rangle \circ s.\text{items}_6| = |\text{elements}(\text{tmp}_6 \circ \langle y_6 \rangle \circ s.\text{items}_6)|$
2. $\text{is\_initial}(y_4)$
3. $x_0 \notin \text{elements}(\text{tmp}_6 \circ \langle y_6 \rangle \circ s.\text{items}_6)$
4. $\text{is\_initial}(y_8)$
5. $0 < |\text{tmp}_6 \circ \langle y_6 \rangle \circ s.\text{items}_6|$

Figure 12: Another relatively difficult VC arising from `QueueRealization`

The VC in Figure 11, which comes from the code for **Remove**, takes Split-Decision about 100 milliseconds to prove. It can be seen that this VC is not entirely trivial. Intuitively, its truth rests on interpreting Given #2 as a statement that there are no duplicate entries in a particular string. With this observation, an informal argument that the VC is true can be made by someone who knows what all the symbols[6] mean. The formal symbolic proof carried out by an automated prover requires not only Given #2 but also some knowledge relating strings and sets, e.g., a lemma about commutativity and associativity of string concatenation in the argument of **elements**, and/or expansion of elements of a concatenation of strings into a union of sets, etc.

The VC in Figure 12 comes from the code for **Contains**, and also takes SplitDecision about 100 milliseconds to prove. Its human proof is left to the reader.

The point of showing these VCs is that the human who developed this code was able to write it correctly without relying on any deep mathematical properties of sets, strings, or integers. It is little wonder, then, that the VCs needed to establish that the code is correct—fully formally and symbolically—do not demand deep mathematical insights for their proofs. We have done similar but more systematic analyses of hundreds of other VCs arising from the verifications of software components that are similarly representative of the kind of code written daily by thousands of software professionals [32, 55]. While not all are proved automatically, even the most difficult we have encountered are rather easily proved by humans in a matter of a few minutes of thinking, much of which involves wading through a sea of assumptions for the few (often just one or two) that are relevant to proving the conclusion.

## 4 Conclusion

Proofs of correctness of routine software inevitably will involve routine mathematical reasoning. Of course, not all software is routine in this sense. Software intended to be used, say, in support of a proof of the four-color conjecture might well be based on some deep mathematical results. Yet even here it is not likely that such results will appear spontaneously in the form of particularly tricky VCs. Instead, there will be some VCs whose proofs will rely on these deep mathematical results being proved off-line, with human assistance, and then invoked as lemmas in the proofs of VCs arising from that code. Such proofs are likely to be based on a reasonably small set

---

[6]In this human-readable view of VCs displayed by the RESOLVE tools, string concatenation is written using o, where the same operator appears as ∗ in specifications typed in by the software developer; set difference is denoted in both places by \.

of mathematical lemmas about the mathematical theories involved in the software specifications. We now have strong empirical evidence of this [55], but the case can be argued directly as follows. Mathematics that is difficult for—let alone unknown to—the software developer is highly unlikely to be needed to prove VCs arising from that programmer's code.

There is, of course, considerable bookkeeping involved in proving some VCs. We have seen VCs with dozens of assumptions, all but a few of which are irrelevant to the proof of the conclusion (e.g., all but Given #1 in Figure 12). These few happen to be the ones the developer really needed to think about when writing the code. We therefore conjecture that automated theorem-provers that are able to order assumptions into decreasing likelihood of relevance to the proof, and to assume that if there are many then most of them are irrelevant, will rarely be distracted and/or run out of time or memory while carrying out proofs arising from software verification. This, combined with Friedman's research program of looking for decision procedures for theories that are restricted in ways consistent with actual VCs observed in practice, suggests that the logic problems to be addressed to make automated software verification practical are not exactly the same as those traditionally studied by logicians. To paraphrase Friedman, there seems to be quite a bit of fundamental logic "red meat" lurking in this realm of applied logic. If progress can be made on such questions, then a practical verifying compiler will not be far behind.

## 5 Acknowledgments

I am delighted to acknowledge the special contributions of Bruce Adcock, Jeremy Avigad, Derek Bronish, Paolo Bucci, Harvey M. Friedman, Wayne Heym, Jason Kirschenbaum, Bill Ogden, Murali Sitaraman, Hampton Smith, Aditi Tagore, Diego Zaccai, and the anonymous referees. This material is based upon work supported by the National Science Foundation under Grants No. DMS-0701260, CCF-0811737, ECCS-0931669, and DUE-0942542. Any opinions, findings, conclusions, or recommendations expressed here are those of the author and do not necessarily reflect the views of the National Science Foundation.

## BIBLIOGRAPHY

[1] B. Adcock. *Working Towards the Verified Software Process.* PhD thesis, Computer Science and Engineering, The Ohio State University, 2010.
[2] J. Avigad, K. Donnelly, D. Gray, and P. Raff. A formally verified proof of the prime number theorem. *ACM Trans. Comput. Logic*, 9, December 2007.
[3] J. Avigad and H. M. Friedman. Combining decision procedures for the reals. *Logical Methods in Computer Science*, 2:1–42, 2006.
[4] J. Barnes. *High Integrity Software: The SPARK Approach to Safety and Security.* Addison-Wesley Longman, 2003.

[5] M. Barnett, B.-Y. Chang, R. DeLine, B. Jacobs, and K. Leino. Boogie: a modular reusable verifier for object-oriented programs. In *Formal Methods for Components and Objects*, volume 4111 of *Lecture Notes in Computer Science*, pages 364–387. Springer, 2006.

[6] M. Barnett, K. R. M. Leino, and W. Schulte. The Spec# programming system: An overview. In *Construction and Analysis of Safe, Secure, and Interoperable Smart Devices*, volume 3362 of *Lecture Notes in Computer Science*. Springer, 2005.

[7] B. Beckert, R. Hähnle, and P. H. Schmitt. *Verification of Object-Oriented Software: The KeY Approach*, volume 4334 of *Lecture Notes in Computer Science*. Springer, 2007.

[8] Y. Bertot and P. Casteran. *Interactive Theorem Proving and Program Development*. Springer, 2004.

[9] A. Bradley and Z. Manna. *The Calculus of Computation: Decision Procedures with Applications to Verification*. Springer, 2007.

[10] F. P. Brooks, Jr. *The Mythical Man-Month (Anniversary Edition)*. Addison-Wesley Longman, 1995.

[11] B. Chandrasekaran and J. R. Josephson. Function in device representation. *Engineering with Computers*, 16:162–177, 2000.

[12] E. Cohen, M. Dahlweid, M. A. Hillebrand, D. Leinenbach, M. Moskal, T. Santen, W. Schulte, and S. Tobies. VCC: A practical system for verifying concurrent C. In *Proceedings of 22nd International Conference on Theorem Proving in Higher Order Logics*, volume 5674 of *Lecture Notes in Computer Science*, pages 23–42. Springer, 2009.

[13] R. A. DeMillo, R. J. Lipton, and A. J. Perlis. Social processes and proofs of theorems and programs. *Commun. ACM*, 22:271–280, May 1979.

[14] L. DeMoura and N. Bjørner. Z3: an efficient SMT solver. In *Proceedings of the Theory and Practice of Software, 14th International Conference on Tools and Algorithms for the Construction and Analysis of Systems*, volume 4963 of *Lecture Notes in Computer Science*, pages 337–340. Springer, 2008.

[15] E. W. Dijkstra. Notes on structured programming. Technical Report T.H.-Report 70-WK-03 (EWD249), Technological University Eindhoven, The Netherlands, April 1970.

[16] I. Filipovic, P. O'Hearn, N. Torp-Smith, and H. Yang. Blaming the client: on data refinement in the presence of pointers. *Formal Aspects of Computing*, 22:547–583, September 2010.

[17] J. C. Filliâtre and C. Marché. The Why/Krakatoa/Caduceus platform for deductive program verification. In *Proceedings of the 19th International Conference on Computer Aided Verification*, volume 4590 of *Lecture Notes in Computer Science*, pages 173–177. Springer, 2007.

[18] R. Floyd. Assigning meanings to programs. In *Proc. of Symposium on Applied Mathematics*, volume 19, pages 19–32. AMS, 1967.

[19] H. M. Friedman. Decidability and undecidability involving string replacement. Technical Report OSU-CISRC-8/09-TR43, Computer Science and Engineering, The Ohio State University, 2009.

[20] H. M. Friedman. Deciding statements about strings with applications to program verification. Technical Report OSU-CISRC-8/09-TR42, Computer Science and Engineering, The Ohio State University, 2009.

[21] J. Harrison. Formal proof – theory and practice. *Notices of the American Mathematical Society*, 55:1395–1406, 2008.

[22] J. Harrison. *Handbook of Practical Logic and Automated Reasoning*. Cambridge University Press, 1st edition, 2009.

[23] W. D. Heym. *Computer Program Verification: Improvements for Human Reasoning*. PhD thesis, Computer and Information Science, The Ohio State University, Columbus, OH, 1995.

[24] C. A. R. Hoare. An axiomatic basis for computer programming. *Commun. ACM*, 12:576–580, October 1969.
[25] C. A. R. Hoare. Proof of correctness of data representations. *Acta Inf.*, 1:271–281, 1972.
[26] C. A. R. Hoare. The verifying compiler: A grand challenge for computing research. *J. ACM*, 50(1):63–69, 2003.
[27] C. A. R. Hoare, J. Misra, G. T. Leavens, and N. Shankar. The verified software initiative: A manifesto. *ACM Comput. Surv.*, 41:22:1–22:8, October 2009.
[28] D. Jackson. *Software Abstractions: Logic, Language, and Analysis*. The MIT Press, 2006.
[29] B. Jacobs, J. Smans, and F. Piessens. VeriFast: Imperative programs as proofs. In *Workshop Proceedings of the 3rd International Conference on Verified Software: Theories, Tools, Experiments*, August 2010.
[30] M. Kaufmann and J. S. Moore. An industrial strength theorem prover for a logic based on common lisp. *IEEE Trans. Softw. Eng.*, 23:203–213, April 1997.
[31] J. C. King. *A Program Verifier*. PhD thesis, Carnegie Mellon University, Pittsburgh, PA, USA, 1970.
[32] J. Kirschenbaum, B. Adcock, D. Bronish, H. Smith, H. Harton, M. Sitaraman, and B. W. Weide. Verifying component-based software: Deep mathematics or simple bookkeeping? In *Proceedings of the 11th International Conference on Software Reuse: Formal Foundations of Reuse and Domain Engineering*, volume 5791 of *Lecture Notes in Computer Science*, pages 31–40. Springer, 2009.
[33] J. Kirschenbaum, H. K. Harton, and M. Sitaraman. A case study in automated, modular, and full functional verification. In *Proc. AFM '08: Third Workshop on Automated Formal Methods*, pages 53–58. ACM Press, July 2008.
[34] G. Klein, J. Andronick, K. Elphinstone, G. Heiser, D. Cock, P. Derrin, D. Elkaduwe, K. Engelhardt, R. Kolanski, M. Norrish, T. Sewell, H. Tuch, and S. Winwood. seL4: formal verification of an operating-system kernel. *Commun. ACM*, 53:107–115, June 2010.
[35] J. E. Krone. *The Role of Verification in Software Reusability*. PhD thesis, Computer and Information Science, The Ohio State University, 1988.
[36] V. Kuncak, H. H. Nguyen, and M. Rinard. Deciding Boolean algebra with Presburger arithmetic. *J. Autom. Reason.*, 36:213–239, April 2006.
[37] I. Lakatos. *Proofs and Refutations: The Logic of Mathematical Discovery*. Cambridge University Press, 1976.
[38] G. T. Leavens. Tutorial on JML, the Java Modeling Language. In *Proceedings of the 22nd IEEE/ACM International Conference on Automated Software Engineering*, pages 573–573. ACM Press, 2007.
[39] D. Leinenbach, W. Paul, and E. Petrova. Towards the formal verification of a C0 compiler: Code generation and implementation correctness. In *Proceedings of the Third IEEE International Conference on Software Engineering and Formal Methods*, pages 2–12. IEEE Computer Society, 2005.
[40] K. R. M. Leino. Dafny: An automatic program verifier for functional correctness. In *16th International Conference on Logic Programming for Artificial Intelligence and Reasoning (LPAR-16)*, 2010.
[41] K. R. M. Leino, G. Nelson, and J. Saxe. ESC/Java user's manual. Technical Report 2000-002, Compaq Systems Research Center, 2000.
[42] X. Leroy. Formal verification of a realistic compiler. *Commun. ACM*, 52:107–115, July 2009.
[43] B. Liskov and J. Guttag. *Abstraction and Specification in Program Development*. MIT Press, 1986.
[44] B. Meyer. On formalism in specifications. *IEEE Software*, 2(1):6–26, Jan. 1985.
[45] G. Nelson and D. C. Oppen. Simplification by cooperating decision procedures. *ACM Trans. Program. Lang. Syst.*, 1(2):245–257, 1979.

[46] T. Nipkow, L. C. Paulson, and M. Wenzel. *Isabelle/HOL—A Proof Assistant for Higher-Order Logic*, volume 2283 of *Lecture Notes in Computer Science*. Springer, 2002.
[47] P. W. O'Hearn, H. Yang, and J. C. Reynolds. Separation and information hiding. *ACM Trans. Program. Lang. Syst.*, 31:11:1–11:50, April 2009.
[48] E. Poll, P. Chalin, D. Cok, J. Kiniry, and G. T. Leavens. Beyond assertions: Advanced specification and verification with JML and ESC/Java2. In *Formal Methods for Components and Objects (FMCO) 2005, Revised Lectures*, volume 4111 of *Lecture Notes in Computer Science*, pages 342–363. Springer, 2006.
[49] J. C. Reynolds. Separation logic: A logic for shared mutable data structures. In *Proceedings of the 17th Annual IEEE Symposium on Logic in Computer Science*, pages 55–74. IEEE Computer Society, 2002.
[50] P. Ross. The exterminators. *IEEE Spectrum*, 42:36–41, 2005.
[51] M. Sitaraman, B. Adcock, J. Avigad, D. Bronish, P. Bucci, D. Frazier, H. Friedman, H. Harton, W. Heym, J. Kirschenbaum, J. Krone, H. Smith, and B. Weide. Building a push-button RESOLVE verifier: Progress and challenges. *Formal Aspects of Computing*, 23(5):607–626, 2011.
[52] M. Sitaraman, S. Atkinson, G. Kulczycki, B. W. Weide, T. J. Long, P. Bucci, W. D. Heym, S. M. Pike, and J. E. Hollingsworth. Reasoning about software-component behavior. In *Software Reuse: Advances in Software Reusability, 6th International Conference*, volume 1844 of *Lecture Notes in Computer Science*, pages 266–283. Springer, 2000.
[53] M. Sitaraman and B. W. Weide. Component-based software using RESOLVE. *SIGSOFT Softw. Eng. Notes*, 19:21–67, October 1994.
[54] M. Sitaraman, B. W. Weide, and W. F. Ogden. On the practical need for abstraction relations to verify abstract data type representations. *IEEE Trans. Softw. Eng.*, 23:157–170, March 1997.
[55] A. Tagore, D. Zaccai, and B. Weide. To expand or not to expand: Automatically verifying software specified with complex mathematical definitions. Technical Report OSU-CISRC-5/11-TR18, Computer Science and Engineering, The Ohio State University, September 2011.
[56] B. W. Weide and W. D. Heym. Specification and verification with references. In *Proc. OOPSLA Workshop on Specification and Verification of Component-Based Systems*, 2001.
[57] B. W. Weide, W. D. Heym, and J. E. Hollingsworth. Reverse engineering of legacy code exposed. In *Proceedings of the 17th International Conference on Software Engineering*, pages 327–331. ACM, 1995.
[58] B. W. Weide, M. Sitaraman, H. K. Harton, B. Adcock, P. Bucci, D. Bronish, W. D. Heym, J. Kirschenbaum, and D. Frazier. Incremental benchmarks for software verification tools and techniques. In *Proceedings of the 2nd International Conference on Verified Software: Theories, Tools, Experiments*, volume 5295 of *Lecture Notes in Computer Science*, pages 84–98. Springer, 2008.
[59] J. Yang and C. Hawblitzel. Safe to the last instruction: automated verification of a type-safe operating system. In *Proceedings of the 2010 ACM SIGPLAN Conference on Programming Language Design and Implementation*, pages 99–110. ACM Press, 2010.
[60] K. Zee, V. Kuncak, and M. Rinard. Full functional verification of linked data structures. In *Proceedings of the 2008 ACM SIGPLAN Conference on Programming Language Design and Implementation*, pages 349–361. ACM Press, 2008.
[61] K. Zee, V. Kuncak, and M. C. Rinard. An integrated proof language for imperative programs. In *Proceedings of the 2009 ACM SIGPLAN Conference on Programming Language Design and Implementation*, pages 338–351. ACM Press, 2009.

# Absolute Infinity – A Bridge between Mathematics and Theology?

CHRISTIAN TAPP

## 1 Introduction

Mathematicians as well as theologians talk about 'absolute infinity'. In mathematics, the class of all ordinals, the class of all cardinals, and the class of all sets are examples of proper classes: assuming that one of them is a set leads to inconsistency. Even if they cannot be taken to be proper objects of mathematics, they are entities that in one way or the other *belong* to mathematics. (To what other scientific discipline could they belong? Who speaks about them if not the mathematician?) And they are infinite. Georg Cantor called their kind of infinity 'absolute infinity'. Theology and philosophy, in turn, describe God as an absolute or absolutely infinite being. Some have concluded that there is an immediate connection between mathematical and theological or philosophical conceptions of absolute infinity. Cantor's aleph series was said to be 'steps to the throne of God', absolute infinity was taken as a 'metaphysischer Grenzbegriff' (Clayton with reference to Kant).[1] The question is therefore: Does absolute infinity really bridge the gap between mathematics and theology?

I take it for granted that there is such a gap. Mathematics and theology are neither identical nor do they speak about the same objects or use the same methods.[2] The mere fact that there are people like Cantor, who have to say something of importance for both areas, hardly changes that, for biographical relations alone do not make a bridge between mathematics and theology.[3] It is an interesting question whether there are *topics* that

---

[1] See Philip Clayton: *Das Gottesproblem, Band 1: Gott und Unendlichkeit in der neuzeitlichen Metaphysik*, Paderborn: Schöningh 1996, especially Chapter 6.

[2] There are surely parallels like the one between axiomatized mathematics and some sorts of deductive theology. But these seem to be parallels of presentation, not of method; and they are parallels, not identity relations. – Ivor Grattan-Guinness offers an interesting classification of possible links between Christianity and mathematics : "Christianity and Mathematics: Kinds of Link, and the Rare Occurrences after 1750", in: *Physis* 37/2 (2000), 467–500.

[3] There are, for example, distinguished mathematicians who are very religious people,

both mathematics and theology are dealing with. One of the most promising candidates is: infinity.

At first one might observe that mathematicians use the word 'infinite' to refer to the size of sets of points or functions or functors or numbers, while theologians refer to God by it, the infinite creator of the finite world. As God is neither a set nor any other kind of mathematical object, mathematics and theology are not simply referring to the same thing or property when they use the word 'infinite'. Mathematics is not calculating within the realm of God's essence; and theology cannot make use of set theory as a means of producing new names of the unnameable.[4]

This observation does not imply that there are no links between the mathematical and the theological senses of 'infinity'. It implies only that the links, if existing, are more complicated and less explicit than a shared object, property, or method would be. I think it is promising to try to find such explicit links. But one should avoid premature claims such as 'infinity' simply means the same in the mouth of a mathematician and in the mouth of a theologian. Bluntly identifying mathematical and theological references to infinity leads into a nebular of supposed but not actual understanding.[5]

In this paper I want to focus on the case of Georg Cantor and his use of the term 'absolute infinity'. While Cantor in general clearly distinguished between the actual infinity of set theory (the 'transfinite') and the actual infinity of God, he used the expression 'actually infinite' in both cases and claimed some connections between mathematics and theology. Moreover, Cantor is considered a mathematician with deep theological inclinations and a strong sense for the frameworking and intermediative activity of philosophy. In the literature, Cantor is portrayed as someone who claims strong connections between religious, metaphysical, and mathematical realities.

One of the best known statements from Cantor in this context is that he called his series of transfinite cardinal numbers 'steps to the throne of

---

and there are theologians who are more than laymen in mathematics; there are even priests who are professors of mathematics. I do not want to focus on such biographical relations in this paper – even though they are surely worth to be studied.

For the case of the mathematician Cantor and his exchange of letters with theologians concerning the set theory and its philosophical foundations, see my *Kardinalität und Kardinäle: Wissenschaftshistorische Aufarbeitung der Korrespondenz zwischen Georg Cantor und katholischen Theologen seiner Zeit* (= Boethius, vol. 53), Stuttgart: Franz Steiner 2005.

[4]This is true at least as long as one does not 'localize' all of reality in the mind of God, as in pantheistic positions. Note, that this is *toto caelo* different from saying that God *knows* all truths.

[5]The title of the voluminous monograph by Ludwig Neidhart: *Unendlichkeit im Schnittpunkt von Mathematik und Theologie*, Göttingen: Cuvillier 2005, 'Infinity in the intersection of Mathematics and Theology', suggest that there is such an intersection; the book, however, is simply assuming that and not arguing for it.

God'. Some have taken this statement as evidence for that the late Cantor was mentally ill and held strong and strange religious views.[6] What exactly were Cantor's claims? How did he conceive of the relation of mathematical objects and God? This is what I want to deal with in this paper.

## 2 Alephs and God

As to what concerns the famous statement about Alephs as steps to the throne of God, one has first to note that this is not a quotation of Cantor's. It is somebody else's reporting about Cantor's views. The earliest evidence I was able to find dates back to 1950 when German mathematician Gerhard Kowalewski reports it in his autobiography. After having presented the definition of the Alephs as *Mächtigkeiten* of the transfinite number classes, Kowalewski writes:

> ... these powers, the Cantorian alephs, were for Cantor something holy, in a certain sense the steps which led up to the throne of the infinite, to the throne of God.[7]

This is Kowalewski's assessment of Cantor's opinion. Is that assessment right? Was it Cantor's conviction that the mathematical study of infinities leads to God? – In the following I want to analyze what Cantor really said about absolute infinity, the alephs, and God.

## 3 Absolute infinity

Cantor used the word 'absolute', i.e., its German equivalent '*absolut*', in several different ways. Some of them do not pertain to our way of inquiry as they are not directly related to the infinite. For example: 'absolute value' (often) or 'follows with absolute necessity' (GA 300),[8] Kant's 'absolute time' (*Grundlagen*, GA 192), 'absolute reality of space and time' (Cantor's Ph.D. defense thesis in Latin, GA 31), 'absolute concept of power/cardinality' ('*absoluter Mächtigkeitsbegriff*', that is, without relativization to continuity as in Steiner's works where Cantor found the notion of cardinality, see

---

[6] Compare the old caricature of Cantor receiving set theory as sort of a private revelation from God in P. Thuillier: "Dieu, Cantor, et l'infini", in: *La Recherche* 84 (1977), 1110–1116.

[7] '*Diese Mächtigkeiten, die Cantorschen Alephs, waren für Cantor etwas Heiliges, gewissermaßen die Stufen, die zum Throne der Unendlichkeit, zum Throne Gottes emporführen.*' Gerhard Kowalewski: *Bestand und Wandel. Meine Lebenserinnerungen – zugleich ein Beitrag zur neueren Geschichte der Mathematik*, München: Oldenbourg 1950, 201; translation by Michael Hallett in his *Cantorian Set Theory and Limitation of Size*, Oxford: Clarendon 1984, 44.

[8] Most of Cantor's works are accessible via the collection Georg Cantor: *Gesammelte Abhandlungen mathematischen und philosophischen Inhalts*, ed. Ernst Zermelo, Berlin: Springer 1932, reprint Hildesheim: Olms ²1962, henceforth cited as 'GA'. For a complete list of Cantor's publications see my *Kardinalität und Kardinäle* (note 3), 578–582.

*Über unendliche, lineare*, GA 151). If we sort out these usages of the word 'absolute' the remaining ones have something to do with infinity. In the following, I will try to spell out what they mean.

The main sense in which Cantor uses the predicate 'absolutely infinite' is that of what we today call 'proper classes'. He calls *On*, the class of all ordinal numbers, as well as *Card*, the class of all cardinal numbers, 'absolutely infinite'. This usage can be found all over his works, but most prominently in his letters to Dedekind from around 1900.[9] In these letters, Cantor used the expression 'absolutely infinite' synonymously with 'inconsistent' in order to denote inconsistent multiplicities that cannot be conceived of as sets, i.e., for proper classes.[10] His solution to the so-called 'antinomies of set theory' was simply to take these arguments as usual indirect proofs: the contradictions derived were disproving the assumption that *On* or *Card* or the class of all sets were sets themselves. Insted, they are absolutely infinite multitudes (*Vielheiten*). They are 'too large' to be tractable as genuine objects of set theory.[11]

This is the "technical sense" of 'absolute infinity' in Cantor's set theory. First, however, Cantor used the expression 'absolutely infinite' in a non-technical sense, for example when he called the 'aggregate' (*Inbegriff*)[12] of the integers 'absolutely infinite'. He did so already in his papers on trigonometric series in 1872, that is, at a time at which even his most charitable interpreters do not ascribe to him a clear vision of the 'antinomies'.

It is much harder to say what exactly Cantor meant by 'absolutely infinite' during the intermediate period between the early papers on trigonometric series and the late correspondence with Dedekind.

In *Über die verschiedenen Standpunkte* of 1885/6, Cantor differentiates between transfinite and absolute infinity. The difference is that transfinite infinity can still be augmented while absolute infinity can not.[13] In the *Mitteilungen* of 1887/8, Cantor gives a definition of 'actual infinity' as a

---

[9]To be more precise, Cantor used the predicate 'absolutely infinite' as a predicate for concepts (see GA 95), sequences (*Grundlagen*: GA 167, 195, 205), aggregates (*Inbegriffe*; *Grundlagen*: GA 205), and multiplicities (*Vielheiten*; Letter to Dedekind, 28.7.1899: GA 445).

[10]Cf. Cantor to Dedekind: '*Das System $\Omega$ aller Zahlen ist eine inconsistente, eine absolut unendliche Vielheit.*' Letter dated 3.8.1899 (in GA 445 erroneously edited within a letter dated 28.7.1899; quoted from Georg Cantor: *Briefe*, ed. Herbert Meschkowski and Winfried Nilson, Berlin: Springer 1991, 408).

[11]For the idea of set theory as a theory of objects with *limited* size see Michael Hallett: *Cantorian Set Theory and Limitation of Size*, Oxford: Clarendon 1984.

[12]The translation of '*Inbegriff*' as 'aggregate' is due to William B. Ewald and his translation of Cantor's *Grundlagen*, see William B. Ewald: *From Kant to Hilbert: A source book in the foundations of mathematics*, Vol. 2. Oxford: Clarendon, for example p. 916.

[13]*Über die verschiedenen Standpunkte*, GA 375.

quantity of a size exceeding all finite sizes of the same kind.[14] He sticks to the distinction of transfinite and absolute infinity as two kinds of actual infinity calling them shortly '*Transfinitum*' and '*Absolutum*'. Hence, Cantor's conception of 'the absolute' in these context is a quantitative concept. As such, it must not be confused with the absolute of, say, idealist philosophy. In Cantor's understanding, the quantitative absolute differs from the transfinite only in that it is non-augmentable.[15]

Cantor does not say what exactly he has in mind when he talks about 'augmentability'. For sure, the absolutely infinite class of transfinite cardinal numbers can in a sense be augmented by adding some ordinals to it which are not (identifiable to) cardinals. Hence, on the one hand, there is a sense in which absolutely infinite multiplicities can be augmented. On the other hand, however, it would be too restrictive to call a multiplicity 'augmentable' only if there are further objects of the same kind as the object of the multiplicity; for then the set of (finite) natural numbers – the prime example for transfinite infinity – would be absolutely infinite (trivially there are no finite integers in addition to the set of finite integers). Concludingly, it is not completely clear what Cantor meant by his definition of absolute infinity in terms of non-augmentability.

Some light may be brought to this problem by taking into account that Cantor has studied a book on natural philosophy by the 19$^{\text{th}}$ century Jesuit father Tilmann Pesch. Pesch defined infinity as '*id, quo non sit maius, nec esse possit*', 'that than which there is nothing bigger or could be'.[16] Cantor saw clearly that this definition is inadequate as it defines a maximum which does not need to be infinite. Despite this criticism, it may be the case that Pesch's definition lead Cantor to take non-augmentability as the characteristic property of absolute infinity.

As considered in itself, this usage of 'absolutely infinite' is completely unproblematic. But problems arise when it is put into an intimate relation to a philosophical concept of infinity as Kant has in his antinomies of pure reason. Cantor thought Kant's antinomies to be flawed in not distinguishing finely enough between different kinds of infinity.[17] Cantor is surely touching a huge philosophical problem when he considers some post-Kantian philosophers to be misguided in conceiving the absolute as the ideal borderline of the finite. Unfortunately, Cantor does not present his thoughts about this problem in any greater detail.

---

[14]GA 401.

[15]GA 394, 401, passim.

[16]Tilmann Pesch: *Institutiones philosophiae naturalis: Secundum principia S. Thomas Aquinatis ad usum scholasticum*, Freiburg: Herder 1883, §403.

[17]*Über die verschiedenen Standpunkte*, GA 375; cp. Zermelo's discussion in note [1], GA 377.

In *Über die verschiedenen Standpunkte*, Cantor uses a threefold distinction in order to classify the positions of other philosophers, theologians, and mathematicians with respect to the reality of the actually infinite. He distinguishes between the infinite *in Deo*, *in mundo*, and *in abstracto*.[18]

In the beginning of the *Mitteilungen*,[19] Cantor combines this distinction and the absolute/transfinite distinction to the following matrix:

| actual infinity | in Deo | in mundo | in abstracto |
|---|---|---|---|
| augmentable | | the transfinite → *Metaphysics* | transfinite numbers → *Mathematics* |
| non-augmentable | the absolute [1886], absolute infinity [1887][20] → *Theology* | | |

The blank fields in this table are really empty: neither does Cantor say that there is only augmentable infinity in nature, nor does he say there is non-augmentable infinity in nature. In the *Mitteilungen*, Cantor's point is only to make a sharp distinction between transfinite and absolute infinity and to use this distinction in order to separate the disciplines of mathematics, metaphysics, and theology.

## 4  Cantor on divine attributes

Cantor mentions several divine attributes in the *Mitteilungen* (1887/8): absolute freedom (GA 387,400), absolute omnipotence (GA 396), absolutely inscrutable power of the will (GA 404), absolute intelligence (GA 401,402), and the capability of absolutely free decisions (GA 406).[21] God and all of his attributes are '*infinitum aeternum increatum sive Absolutum*' – the eternal, uncreated infinite or absolute (GA 399). In these considerations, the term 'absolute' has no explicit quantitative connotation, nor is it explicitly related to proper classes as in the letters to Dedekind. Here, 'the absolute' is used in a philosophical or theological sense as derived from the literal meaning of the Latin '*absolutum*', stemming from '*absolvere*' – to be detached or disassociated. Using natural language predicates as predicates

---

[18] GA 372. The expressions '*in Deo*' and '*in abstracto*' are used by Cantor, the expression '*in mundo*' is added by me in order to have a handy concept available.

[19] GA 378.

[20] In *Über die verschiedenen Standpunkte* [1885/6], Cantor says 'the Absolute' (GA 372); in the *Mitteilungen* [1887/8], he says 'absolute infinity or, shortly, the absolute' (GA 378). So, at the time of the *Mitteilungen*, he surely used 'absolute infinity' and 'the absolute' synonymously.

[21] German original: '*absolute Freiheit*', '*absolute Omnipotenz*', '*absolut unermeßliche Willenskraft*', '*absolute Intelligenz*', and '*absolut freier Ratschluß*'.

for the unrestrained God requires detaching or disassociating them from limitations of their natural meanings.

## 5 A methodological parallel

There is an interesting methodological parallel between Cantor's mathematical theory and traditional theology. When Cantor analyzed traditional proofs of the impossibility of infinite numbers with the intention to disprove them, he found that the most common mistake in them was to transfer propositions holding in the domain of the finite without further qualification to the domain of the infinite. This is what he called the '*proton pseudos*', the cardinal error of those anti-infinity arguments. One may formulate this *proton pseudos* positively, as a methodological maxim: do not carelessly transfer insights from the finite domain to the domain of the infinite.

In this form, this maxim plays a major role in the traditional theological doctrine of God. In religious speech, we cannot help but use the vocabulary of our everyday language that acquired its meaning by being used in our everyday life. But in order to use this vocabulary for God, who is not an ordinary object of our everyday language, it has to run through kind of a purgatorial process traditionally called *via positiva, via negativa*, and *via eminentiae*. To put it shortly and a little laxly: part of the positive content of our concepts must be kept, the negative content of limitations, the creaturely mode of being, must be crossed out, and their meanings must be 'boosted' from the limited world of creatures to the unlimited and uncreated creator. Albeit I cannot go into the details of this maxim of theological semantics, the methodological parallel should be clear: no simple transfer of propositions or meanings from the realm of the finite to the realm of the infinite – be it the realm of infinite numbers or an infinite God.

## 6 Mathematical infinity and God

Cantor sometimes explicitly speaks about relations between the actual infinite of mathematics and God. The first place where he does so is in *Über die verschiedenen Standpunkte*. I quote from the German text and comment on it afterwards as it is hardly translatable:

> *Wenn aber aus einer berechtigten Abneigung gegen solches illegitime A. U. sich in breiten Schichten der Wissenschaft, unter dem Einflusse der modernen epikureisch-materialistischen Zeitrichtung, ein gewisser Horror Infiniti ausgebildet hat, der in dem erwähnten Schreiben von Gauß seinen klassischen Ausdruck und Rückhalt gefunden, so scheint mir die damit verbundene unkritische Ablehnung des legitimen A. U. kein geringeres Vergehen wider die Natur der Dinge zu sein, die man zu nehmen hat, wie sie sind, und es läßt sich dieses Verhalten auch als eine Art Kurzsichtigkeit auffassen, welche die Möglichkeit raubt, das A. U. zu sehen, obwohl es in seinem höchsten, absoluten Träger uns geschaffen hat und erhält und in seinen*

> *sekundären, transfiniten Formen und allüberall umgibt und sogar unserm Geiste selbst innewohnt.*[22]

The key point of this one German sentence may be paraphrased like this: In the time of Gauss there was a *Horror infiniti* in mathematics that led to rejecting not only illegitimate notions of infinity but also legitimate ones. But rejecting the legitimate forms of mathematical infinity results in a certain short-sightedness: one becomes incapable of seeing the 'highest, absolute bearer' of absolute infinity, namely God, the creator of the world. When Cantor says that depriving mathematics of actually infinite numbers means to lose a cognition of God, he indeed seems to cross a bridge between mathematics and theology.

But what is the relation between actually infinite numbers and God? According to Cantor, God is the highest, absolute bearer of actual infinity, while the secondary, transfinite forms of infinity are all around us and even in our minds. I interpret this 'forms of infinity all around us' as the *transfinitum in mundo*, and the belief in their existence as the belief that there are infinite sets of entities in our universe.[23] And I take 'being inherent to our minds' as alluding to the nature of infinite numbers as abstract objects, '*Zusammenfassungen zu einem Ganzen*', 'collections into a whole', according to Cantor's famous definition.[24] Calling both, the infinite sets of natural objects and the infinite numbers 'secondary forms' might be a hint at Spinoza, whose works Cantor has studied in some detail (especially the *Ethics*).[25]

A second example for direct relations between mathematical and theological subjects can be found in Cantor's letters to theologians. In one of these letters Cantor interprets a proposition from the dogmatic consitution '*Dei filius*' of the first Vatican council. This proposition says about God that he is

> inexpressibly loftier than anything besides himself which either exists or can be imagined[26]

---

[22] GA 374–375.

[23] For this claim (that there are infinitely many entities in our universe) to be true, it may be necessary not to restrict 'entities' to '*wirkliche*' entities in Bolzano's sense. If one admits sentences in themselves (*Sätze an sich*) or propositions or facts or whatever the like as constituents of the world, one can be sure that there are infinitely many of them 'in the world'. That does, however, not mean that one is committed to infinitely many real things in the world, where 'real' is taken in the strong, Bolzanoean sense.

[24] See *Beiträge* I (1895), GA 282; English translation of the *Beiträge* by P. E. Jourdain as *Contributions to the founding of the theory of transfinite numbers*, New York: Dover 1915.

[25] See Paolo Bussotti and Christian Tapp: "The influence of Spinoza's concept of infinity on Cantor's set theory", In: *Studies in History and Philosophy of Science* 40 (2009), 25–35.

[26] '*Super omnia, quae praeter ipsum sunt et concipi possunt, ineffabiliter excelsus*', see:

Cantor adds that God's ineffability would be the more considerable the more extended the area of things below him is. In this sense he writes in a letter to the Dominican father Thomas Esser:

> Every extension of our insight into what is possible in creation leads necessarily to an extended cognition of God.[27]

Cantor's reasoning is thus: The more cardinalities we have the more sets of things are possible, and the more sets of things are possible the more circumstances in nature can be expressed, and the more circumstances in nature can be expressed the greater for us is a God who is, in a sense, 'above' nature.

There are only very few passages in Cantor's works that suggest such an immediate link between theological and mathematical subjects. Other passages suggest more caution. So, for example, in the *Grundlagen*, Cantor says that

> the true infinite or Absolute, which is in God, permits no determination whatsoever[28]

While transfinite sets and numbers are perfectly determined or rather determinable,[29] God's infinity cannot be determined. This might mark a clear

---

[27] Denzinger, Heinrich / Hünermann, Peter: *Kompendium der Glaubensbekenntnisse und kirchlichen Lehrentscheidungen / Enchiridion symbolorum et definitionum* ..., Freiburg: Herder $^{40}$2005, no. 3001; transl. Norman Tanner (ed.): *Decrees of the Ecumenical Councils*, Vol. 2: Trent to Vatican II, London: Sheed & Ward 1990, 805.

[27] '*Jede Erweiterung unserer Einsicht in das Gebiet des Creatürlich-möglichen muß daher zu einer erweiterten Gotteserkenntnis führen.*' See my *Kardinalität und Kardinäle* (note 3), letter [CanEss96], p. 308, and the commentary on p. 86.

[28] '*Daß das wahre Unendliche oder Absolute, welches in Gott ist, keinerlei Determination gestattet*', *Grundlagen* § 5, GA 175; transl. Ewald: *From Kant to Frege* (note 12), 891.

[29] In the *Mitteilungen*, there is a similar passage about the Absolute (GA 405-406):

> *Das Transfinite [...] weist mit Notwendigkeit auf ein Absolutes hin, auf das 'wahrhaft Unendliche', an dessen Größe keinerlei Hinzufügung oder Abnahme statthaben kann und welches daher quantitativ als absolutes Maximum anzusehen ist. Letzteres übersteigt gewissermaßen die menschliche Fassungskraft und entzieht sich namentlich mathematischer Determination.*

> The transfinite [...] points with necessity to an Absolute, to the 'truly infinite', whose magnitude can neither be augmented nor diminished and which is, hence, quantitatively to be seen as an absolute maximum. In a sense, it transcends human cognitive powers and withstands mathematical determination in particular. [my transl.]

This quotation does not really help clarifying the relation between the mathematical and the metaphysical absolute. In the context of the *Mitteilungen*, 'the absolute' is quasi-defined as an inaugmentable actual infinite. It is not clear whether Cantor wants it to refer to a quantitative concept, to a general metaphysical concept, or to both. In my view, this passage from the *Mitteilungen* is compatible with either interpretation and, hence, does not help settling our question.

border line between the infinity of sets and the infinity of God if only God is undeterminable. So what about proper classes? Are they undetermined in this sense? This question is central now, for if proper classes were determined, this would show a gap between mathematical absolute infinity and God's absolute infinity; if, to the contrary, classes were also called undetermined, this would be a further parallel between absolute infinity in mathematics and in theology.

To my knowledge Cantor never says that proper classes or inconsistent multiplicities are undetermined. Rather, the examples of classes or inconsistent multiplicities Cantor discusses all fulfil one of the criteria for sets, namely that for every object it must be determined whether it belongs to the collection or not. (The difference between sets and classes lies in the fact that thinking of the elements of a proper class to be together in forming a new object, a set, leads to a contradiction, and not in that the elementhood relation would be vague in any sense.) If this kind of determination is meant in the quotation above, then one has to conclude that the difference mentioned by Cantor is not in the first place a difference between the transfinitely and the absolutely infinite mathematical objects, but a much more general difference between something determined (that only can be subject of mathematics) and something undetermined.

The most important passage in Cantor's writings is a passage from his endnotes to §4 of the *Grundlagen*. It is almost always cited if something about Cantor's views on the relation between absolutely infinite mathematical objects and the absolute infinity of God is at issue. The two best known quotations are:

> The absolute can only be acknowledged [[*anerkannt werden*]] but never known [[*erkannt werden*]][30]

and

> the absolutely infinite sequence of numbers thus seems to me to be an appropriate symbol of the absolute.

In order to grasp the exact sense of what Cantor says in these quotations we need to consider their full context. The whole passage is, however, a little dark and difficult to understand:

> I have no doubt that, as we pursue this path [of investigating transfinite numbers not only in mathematics but wherever they may occur, C. T.] ever further, we shall never reach a boundary that cannot be crossed; but that we shall also never achieve even an approximate conception of the absolute. The absolute can only be acknowledged [[*anerkannt*]] but never known [[*erkannt*]] – and not even approximately known. For just as in number-class (I) every finite number, however great, always has the same

---

[30] '*Werden*' added to the German insertions, C. T.

power of finite numbers greater than it, so every suprafinite number, however great, of any of the higher number-classes (II) or (III), etc. is followed by an aggregate of numbers and number-classes whose power is not in the slightest reduced compared to the entire absolutely infinite aggregate of numbers, starting with 1. As Albrecht von Haller says of eternity: 'I attain to the enormous number, but you, o eternity, lie always ahead of me.' The absolutely infinite sequence of numbers thus seems to me to be an appropriate symbol of the absolute; in contrast, the infinity of the first number-class (I), which has hitherto sufficed, because I consider it to be a graspable idea (not a representation [[Vorstellung]]), seems to me to dwindle into nothingness by comparison.[31]

Let me start my analysis with the less problematic point that 'the absolute can only be acknowledged [[anerkannt]] but never known [[erkannt]]'. The question is first, whether 'the absolute' refers to inconsistent multiplicities or to God. The problem is that the preceding sentence suggests that Cantor is talking about the absolute in the sense of God, while the following sentence suggests the opposite.

In the preceding sentence Cantor says that 'we shall never achieve even an approximate conception of the absolute'[32] by all our investigations in transfinite numbers and sets. If even an *approximate* conception of the absolute in question is denied, only the absolute infinity of God can be at issue. For Cantor would probably say that it is possible to gain approximative knowledge about proper classes like the class of all ordinal numbers at least in the sense in which one also has approximative knowledge about a concept by learning more and more about things falling under it. Acquisition of this kind of knowledge is also possible in case of proper classes, as Cantor would probably admit. Hence, the preceding sentence suggests to

---

[31]GA 205; translation in Ewald, *From Kant to Hilbert*, 916. The German original reads:

> Daß wir auf diesem Wege [die transfiniten Zahlen nicht nur mathematisch, sondern überall, wo sie vorkommen, zu untersuchen, C. T.] immer weiter, niemals an eine unübersteigbare Grenze, aber auch zu keinem auch nur angenäherten Erfassen des Absoluten gelangen werden, unterliegt für mich keinem Zweifel. Das Absolute kann nur anerkannt, aber nie erkannt, auch nicht annähernd erkannt werden. Denn wie man innerhalb der ersten Zahlenklasse (I) bei jeder noch so großen endlichen Zahl immer dieselbe Mächtigkeit der ihr größeren endlichen Zahlen vor sich hat, ebenso folgt auf jede noch so große überendliche Zahl irgendeiner der höheren Zahlenklassen (II) oder (III) usw. ein Inbegriff von Zahlen und Zahlenklassen, der an Mächtigkeit nicht das mindeste eingebüßt hat gegen das Ganze des von 1 anfangenden absolut unendlichen Zahleninbegriffs. Es verhält sich damit ähnlich, wie Albrecht von Haller von der Ewigkeit sagt: 'ich zieh' sie ab (die ungeheure Zahl) und Du (die Ewigkeit) liegst ganz vor mir.' Die absolut unendliche Zahlenfolge erscheint mir daher in gewissem Sinne als ein geeignetes Symbol des Absoluten; wogegen die Unendlichkeit der ersten Zahlenklasse (I), welche bisher dazu allein gedient hat, mir, eben weil ich sie für eine faßbare Idee (nicht Vorstellung) halte, wie ein ganz verschwindendes Nichts im Vergleich mit jener vorkommt.

[32]'*Auch nur angenäherten Erfassen des Absoluten*,' GA 205.

read 'the absolute' in the sense of 'God'.

The beginning of the following sentence raises some doubts, however. The sentence begins with 'for' suggesting that it will present a reason for the earlier claim. But now Cantor is explicitly talking about number classes. So either this passage is obscure, or Cantor feels entitled to this transition between the two senses of 'the absolute' because he is convinced that there are in fact relations between them allowing for that transition. We will come back to this point shortly.

The second quotation reads:

> the absolutely infinite sequence of numbers thus seems to me to be an appropriate symbol of the absolute.[33]

It is obvious but sometimes overlooked that according to this statement the absolutely infinite series of ordinal numbers is *not* said to be the absolute or to lead to the absolute (like steps lead to a throne), but to be a *symbol* of the absolute. One might express this in terms of the metaphor of steps to the throne of God, saying that numbers are steps to His throne, not to God himself. But the sentence does not stop here:

> The absolutely infinite sequence of numbers thus seems to me to be an appropriate symbol of the absolute; in contrast, the infinity of the first number-class (I), which has hitherto sufficed, because I consider it to be a graspable idea (not a representation [[*Vorstellung*]]), seems to me to dwindle into nothingness by comparison.

What Cantor has in mind here is the function or role the sequence of finite natural numbers has played in the history of theology and metaphysics. It was used as an image or a symbol for God's infinity. This role is now to be taken by the whole, absolutely infinite series of all ordinal numbers for the following reason: In the light of Cantor's theory of transfinite numbers, the series of natural numbers turns out to be intrinsically limited, namely by the ordinal number $\omega$ which is the smallest transfinite ordinal number, but greater than every natural number. It is very plausible that Cantor found such an intrinsic limitation inappropriate for a symbol of the absolute. The whole sequence of transfinite ordinal numbers, in contrast, does not suffer this limitation. The fact that the class of all ordinal numbers is not a set but a proper class, makes it more appropriate as a symbol for the unlimited God than the limited series of natural numbers. To put it in a nutshell: due to its mathematical absoluteness or absolute infinity, $On$ is a better symbol of the Absolute than the limited $\omega$.

Cantor grasps this difference between $\omega$ and $On$ also in terms of calling $\omega$ a 'graspable idea'. Tacitly that means that $On$ is not a graspable idea. This

---

[33]'*Die absolut unendliche Zahlenfolge erscheint mir daher in gewissem Sinne als ein geeignetes Symbol des Absoluten,*' GA 205.

once more confirms the thesis that already at the time of the *Grundlagen* Cantor had a perfectly clear 'solution' to the paradoxes, or better: was well aware of the arguments later used in the paradoxes. For him, these arguments were simply reductions to the absurd of the assumption that $On$ was a set or a 'graspable idea'.

This interpretation of Cantor's second famous statement about the absolute suggests a solution to a problem which remained open in interpreting Cantor's first statement about the incognizability of the absolute. One may read his statements as reasonings *via the symbolization*: the absolutely infinite series is a better symbol for the absolute infinity of God than the series of natural numbers for it is not limited in the same way (by an ordinal number like $\omega$). This symbol is convenient as one cannot climb to an epistemic position that allows one to scrutinize what it symbolizes 'from the top': there is no top above God, and there is no top above the series of ordinals. One cannot grasp the whole absolutely infinite series of transfinite ordinal numbers at once as a whole, it does not form 'a graspable idea'. Subtracting finite amounts (and even infinite amounts) does not diminish the magnitude of the class of ordinals as it does not diminish the magnitude of God. Cantor's point in the first statement is that his research in the realm of the infinite – the mathematical research about transfinite ordinal numbers as well as the metaphysical research about transfinite sets of natural objects – does *not* lead to any direct knowledge of God. Although one can 'manage' higher and higher ordinal systems (just think about the ordinal notation systems in proof theory, or the huge cardinals in contemporary set theory) with no intrinsic limit with respect to the size of the ordinals thus considered, one will not arrive at a direct cognition of the absolutely infinite multiplicity of all ordinal numbers, and *a fortiori* not at a direct cognition of God. As the class of ordinals cannot be fully embraced by mathematical thinking, so God cannot by theological thinking.

> I have no doubt that, as we pursue this path ever further, we shall never reach a boundary that cannot be crossed; but that we shall also never achieve even an approximate conception of the absolute.[34]

This does not mean that no form of knowledge about God or no cognition of God is possible. Cantor says only that his research in the realm of the transfinite does not lead to such a direct knowledge or cognition.[35]

---

[34]'*Daß wir auf diesem Wege immer weiter, niemals an eine unübersteigbare Grenze, aber auch zu keinem auch nur angenäherten Erfassen des Absoluten gelangen werden, unterliegt für mich keinem Zweifel.*'

[35]The English translation is a little bit misleading here. From the German original it is clear that 'as we pursue this path' must be read as qualification not only for 'never reach a boundary' but also for 'never achieve even an approximate conception of the absolute.'

As to what concerns absolute infinity as a possible bridge between mathematics and theology, my thesis is therefore: as Cantor has show us, there are methodological parallels between set theory and theological semantics, and there is a relation between absolute infinity of God and the absolute infinity of the series of transfinite ordinal numbers. But this relation is a symbolic one: set theory does not produce direct knowledge of God.[36]

---

[36] An earlier version of this paper was presented during the conference honoring Harvey Friedman, May 15, 2009. I am indebted to Leon Horsten and James Bradley for their helpful comments.

# Analytic Cut in Modal Logic: the System B
GRIGORI MINTS

ABSTRACT. Several modal and superintuitionistic logics do not admit complete cut-elimination in their ordinary sequent formulations. In such cases analytic cut (on all subformulas or some subclass) is often sufficient to obtain familiar consequences of cut-elimination. Generally applicable tools are illustrated here by two examples. For the propositional S5 only one application of very rudimentary "analytic cut" is needed. For the system B characterized by Kripke models with reflexive symmetric accessibility relations the analytic cut can be restricted to instances of the symmetry axiom for modal subformulas.

## 1 Introduction

Normal form theorems for derivations in modal logic are often stated as cut elimination results. In such cases the "ordinary" (Gentzen-style) rules of the system introduce logical connectives into given sequent $\Gamma \Rightarrow \alpha$ and have the subformula property. For example

$$\frac{\Gamma \Rightarrow \alpha \quad \Gamma \Rightarrow \beta}{\Gamma \Rightarrow \alpha \& \beta} \qquad \frac{\alpha, \beta, \Gamma \Rightarrow \gamma}{\alpha \& \beta, \Gamma \Rightarrow \gamma}$$

An additional rule called Cut

$$\frac{\Gamma \Rightarrow \alpha \quad \alpha, \Gamma \Rightarrow \gamma}{\Gamma \to \gamma} \; Cut$$

is used to prove equivalence of the Gentzen-style system and ordinary formulation given by axioms and inference rules. Cut does not have the subformula property: formula $\alpha$ occurs in the premises $\Gamma \Rightarrow \alpha$ and $\alpha, \Gamma \Rightarrow \gamma$ but not in the conclusion $\Gamma \Rightarrow \gamma$. Cut elimination theorem claims that Cut is admissible in the (cutfree) system consisting of the remaining rules: every proof with Cut can be transformed into a proof without Cut. Such cut-free system are known for many propositional systems, including classical and intuitionistic logic as well as superintuitionistic and modal systems, for example K, T, S4, GL (provability logic), etc. The situation with S5 is more complicated: provable modalized formulas have cut-free proofs [12] in a familiar formulation [11], but the formula $p \to \Box(q \vee \Diamond p)$ does not.

*Analytic cut* is the restriction of the Cut rule to subformulas $\alpha$ of the conclusion $\Gamma \Rightarrow \gamma$. For most of the modal propositional systems complete in Kripke semantics (including system B) completeness of the analytic cut can be easily proved as a by-product of familiar completeness and decidability proofs (cf. for example [6]) using canonical models consisting of sets of subformulas.

The restriction to analytic cuts makes possible many of the applications of complete cut elimination. For example, the familiar effective proof of the interpolation theorem by induction on the derivation (cf. [13]) still goes through with analytic cut. If $\Gamma = \Sigma, \Sigma'$, then an interpolation formula $\iota$ for the conclusion of Cut:

$$\Sigma \Rightarrow \iota; \qquad \iota, \Sigma' \Rightarrow \gamma$$

is easily obtained from the interpolation formulas for the premises of the analytic cut. Assuming $\alpha$ is in the same language as $\Sigma' \to \gamma$ and interpolants $\kappa, \lambda$ for the premises are given:

$$\Sigma \Rightarrow \kappa; \; \kappa \Sigma' \Rightarrow \alpha; \qquad \Sigma \Rightarrow \lambda; \; \lambda, \alpha, \Sigma' \to \gamma$$

define $\iota = \kappa \& \lambda$. If on the contrary $\alpha$ is in the language of $\Sigma$, then the interpolants for the premises satisfy

$$\Sigma' \to \kappa; \; \kappa, \Sigma \Rightarrow \alpha; \qquad \alpha, \Sigma \Rightarrow \lambda; \; \lambda, \Sigma' \Rightarrow \gamma.$$

Define $\iota = (\kappa \to \lambda)$.

In principle one can try to implement automated deduction with analytic cut by applying bottom-up all cut-free rules and analytic cut. This will lead to branching and may result in exponential blow-up. So it is desirable to further restrict analytic cut. This option is illustrated here in detail for the modal system B determined by the class of all reflexive symmetric Kripke frames or axiomatically by adding the schema

$$\alpha \to \Box \Diamond \alpha$$

to the modal system **T** of reflexive Kripke frames. The natural extension of familiar cut-free rules for T would be

$$\frac{\Gamma \Rightarrow \Delta, \alpha}{\Gamma \Rightarrow \Delta, \Box \Diamond \alpha} \qquad \frac{\alpha, \Gamma \Rightarrow \Delta}{\Diamond \Box \alpha, \Gamma \Rightarrow \Delta}$$

However, the formula

$$p \vee \Box(q \vee \Box(r \vee \Box \Diamond(\bar{r} \& \Diamond(\bar{q} \& \Diamond \bar{p}))))$$

is not derivable using just this rule or an obvious modification of it. It turns out that the complete set of rules for the system B is obtained by restricting applications of analytic cut to the case when one of the premises is an axiom $\beta \to \Box \Diamond \beta$:

$$\frac{\Gamma \Rightarrow \Delta, \beta \quad \Box \Diamond \beta, \Gamma \to \Delta}{\Gamma \Rightarrow \Delta} \; Sym$$

The rule $Sym$ is further restricted by the following requirement. The formula $\Box \Diamond \beta$ can be analyzed in the derivation, but the formula $\Diamond \beta$ added by this analysis cannot be analyzed further: it can be used only in an axiom $\Diamond \beta \Rightarrow \Diamond \beta$. This further cuts down possible blind search in a standard kind of proof procedure.

In fact we use a Tait-style Gentzen-type system $GB$ deriving sequents $\alpha_1, \ldots, \alpha_n$ interpreted as disjunctions. We present two completeness proofs for $GB$. The first proof is based on a familiar construction of a canonical model whose worlds are complete sets of subformulas of a given formula. However this construction would use unrestricted analytic cut to prove existence of a complete underivable extension for any underivable set of subformulas. Our proof combines such a canonical model with a trick from [4] that avoids all cuts in the case of intuitionistic logic, but leaves residual applications of $Sym$ in our case.

The second proof transforms derivations in a Kripke-style tableau formulation (known to be complete) into derivations in $GB$. This transformation can be used for example when one actually needs an interpolation formula for a given implication. One can get a derivation in a tableau system from one of existing provers for modal logic, then apply our transformation to obtain a derivation in $GB$, then construct an interpolant in a familiar way. We do not know estimates of complexity of our translation in the worst case. In small examples (section 3.5) it even shrinks derivations.

In section 2 we present the system $GB$ and prove its completeness. In section 3 we introduce a semantic tableau system $GT$ complete for B, describe a transformation of $GT$-derivations into $GB$-derivations and illustrate it by examples, including the formula mentioned above.

In section 4 we describe a simple variation of the familiar cut elimination procedure that works for propositional S5. This is a simplification of a proof for K45D given by G. Shvarts in [12] The result holds only for derivations of modalized formulas (every subformula is in the scope of a modal sign). Since $\alpha$ and $\Box \alpha$ are interderivable this results (for non-modalized $\alpha$) in one additional inference

$$\frac{\Box A}{A}$$

that can be treated as a rudimentary analytic cut with the axiom $\Box A \Rightarrow A$.

At the end of section 4 we sketch a simple completeness proof for a formulation of S5 with a restricted analytic cut.

The concluding section 5 describes a challenge: to axiomatize the purely (strict) implicational fragment of $GB$:

$$(\alpha \prec \beta) \equiv \Box(\alpha \to \beta).$$

We did not consider here predicate extensions. Similarly to the case of S5, standard axioms for quantifiers in the presence of the symmetry axiom imply the Barcan formula. However in view of the failure of interpolation for predicate modal logic [2] it is doubtful that any reasonable version of the analytic cut is complete in the predicate case.

## 2  Gentzen-style Calculus $BG$ with Analytic Cut

*Formulas* are constructed from literals $p, \bar{p}, \ldots$ by $\&, \lor, \Box, \Diamond$. *Negation* $\bar{\alpha}$ of a composite formula $\alpha$ is defined by de Morgan rules. In addition we allow now in our sequents $°$-*formulas* of the form $\Diamond(\Box \alpha)°$ and $(\Box \alpha)°$ where $\alpha$ is a formula. The superscript $°$ indicates that the formula $(\Box \alpha)°$ cannot be analyzed further. Such formulas can be used only in the axioms $\Diamond \alpha, (\Box \bar{\alpha})°, \Gamma$.

Derivable objects: *sequents*

$$\alpha_1, \ldots, \alpha_n$$

where $\alpha_i$ are formulas or $°$-formulas. This sequent is treated as a set $\{\alpha_1, \ldots, \alpha_n\}$ and read: one of $\alpha_i$ is true. In particular the order of expressions $\alpha_i$ in the sequent is irrelevant.

*Subformula* always means expression which is a formula, hence does not contain $°$.

Let
$$\Gamma^\Diamond := \{\beta : \Diamond\beta \in \Gamma\}$$

**System $GB$**
**Axioms:**   $\alpha, \bar{\alpha}, \Gamma$  and  $\Diamond\alpha, (\Box\bar{\alpha})°, \Gamma$.
**Inference Rules**

$$\frac{\alpha, \beta, \Gamma}{\alpha \lor \beta, \Gamma} \lor \qquad \frac{\alpha, \Gamma \quad \beta, \Gamma}{\alpha \& \beta, \Gamma} \&$$

$$\frac{\alpha, \Gamma}{\Diamond \alpha, \Gamma} \Diamond \qquad \frac{\beta, \Gamma \quad \Diamond(\Box\bar{\beta})°, \Gamma}{\Gamma} Sym$$

provided $\Diamond\beta$ is a formula occuring in $\Gamma$.

$$\frac{\Gamma^\Diamond, \alpha}{\Gamma, \Box\alpha} \Box$$

⊢ Γ means that Γ is derivable in $GB$. Note that the contraction rule

$$\frac{\Gamma, \alpha, \alpha}{\Gamma, \alpha}$$

is admissible in $GB$ since sequents are sets of formulas, so premise and conclusion represent the same sequent. The weakening rule is also admissible by a familiar argument.

Formula $\alpha$ in the axiom $\alpha, \bar{\alpha}, \Gamma$ can be assumed atomic. This restriction cannot be imposed on the second axiom since $\Diamond(p\&q), (\Box(\bar{p} \vee \bar{q}))^\circ$ would be underivable.

Soundness of the system $GB$ is easy: after erasing $^\circ$ marks all axioms and rules are obviously sound. Completeness needs more work.

## 2.1 Completeness of $GB$

We now prove (following standard schema, for example [6] and [7]) that any sequent underivable in GB is falsified in some world of some finite Kripke model for B. In fact we establish that there is a universal (canonical) model suitable for the falsification of all underivable sequents formed from a given finite set of formulas (each sequent falsified in its own world). It is essential that this model proves completeness of a cut-free formulation.

**Definition 1.** A formula $\alpha$ *clashes* with a sequent $\Gamma$ if both sequents $\Gamma, \alpha$ and $\Gamma, \bar{\alpha}$ are derivable:

$$\vdash \Gamma, \alpha \text{ and } \vdash \Gamma, \bar{\alpha} \qquad (1)$$

**Definition 2.** $Sub(\Gamma)$ stands for the set of all subformulas of $\Gamma$. A sequent $\Gamma$ is *complete* if it is underivable and

1. for any formula $\alpha \in Sub(\Gamma)$ either $\alpha \in \Gamma$ or $\bar{\alpha} \in \Gamma$ or ($\alpha$ *clashes* with $\Gamma$).

2. for any subformula $\Diamond \beta$ of $\Gamma$ either $\beta \in \Gamma$ or $\Diamond(\Box\bar{\beta})^\circ \in \Gamma$.

**Note.** The condition (1) is used in the proof of cut elimination which shows that (1) never holds if $\Gamma$ is underivable.

**Definition 3.** A sequent $\Gamma$ is *saturated* if it is underivable and the following conditions are satisfied for any $\phi, \psi, \beta$:

(&) $(\phi\&\psi) \in \Gamma$ implies $\phi \in \Gamma$ or $\psi \in \Gamma$.

($\vee$) $(\phi \vee \psi) \in \Gamma$ implies $\phi, \psi \in \Gamma$.

($\Diamond$) $\Diamond\phi \in \Gamma$ implies $\phi \in \Gamma$

Note that the clause $\Diamond$ includes: $\Diamond(\Box\bar{\beta})^\circ \in \Gamma$ implies $(\Box\bar{\beta})^\circ \in \Gamma$.

The following Lemma says that all rules except $\Box$ (invertible rules) were applied from the bottom up in every complete sequent.

**Lemma 1.** (saturation) If $\Gamma$ is complete, then it is saturated.

Proof. (&): Let $\phi \& \psi \in \Gamma$. We consider cases determined by three possible "positions" for each of the formulas $\phi, \psi$: the formula in $\Gamma$, or its negation in $\Gamma$, or in clash with $\Gamma$. Note that if

$$\vdash \Gamma, \phi \quad \text{and} \quad \vdash \Gamma, \psi \tag{2}$$

then $\Gamma$ is derivable by &, which contradicts completeness of $\Gamma$. Hence one of $\bar{\phi}, \bar{\psi}$ (say $\bar{\phi}$) is not in $\Gamma$ [otherwise the sequents (2) are axioms], and one of $\phi, \psi$ does not clash with $\Gamma$. If $\phi \notin \Gamma$, then $\phi$ clashes with $\Gamma$ by the completeness of $\Gamma$. Hence the sequent $\phi, \Gamma$ is derivable and $\psi$ does not clash with $\Gamma$. Also $\bar{\psi} \notin \Gamma$, since otherwise sequents in (2) are derivable; hence $\psi \in \Gamma$ as required.

($\vee$): Let $\phi \vee \psi \in \Gamma$. Then $\bar{\phi} \notin \Gamma$, since otherwise $\Gamma$ is derivable by ($\vee$). If $\phi$ clashes with $\Gamma$, then $\Gamma$ is derivable from $\Gamma, \phi$ by weakening, ($\vee$), and contraction. For $\phi$ this leaves only the possibility that $\phi \in \Gamma$. For analogous reasons, $\psi \in \Gamma$.

($\Diamond$): Let $\Diamond\phi \in \Gamma$. The sequent $\Gamma, \phi$ is underivable, since $\Gamma$ is derivable from it by the rule $\Diamond$. Hence $\phi$ does not clash with $\Gamma$, and $\bar{\phi} \notin \Gamma$, so $\phi \in \Gamma$.

**Lemma 2.** (completion). Any underivable sequent $\Gamma_0$ can be extended to a complete sequent consisting of subformulas of $\Gamma_0$ and $^\circ$-formulas $\Diamond\Box(\bar{\beta})^\circ$, $\Box(\bar{\beta})^\circ$ for $\Diamond\beta \in Sub(\Gamma^\circ)$.

Proof. Consider an enumeration $\phi_1, \phi_2, \ldots$ of all (propositional) formulas in $Sub(\Gamma_0)$. Define the sequents $\Gamma_0 \subset \Gamma_1 \subset \ldots$ such that $\Gamma_i$ is underivable and complete for all formulas $\phi_j$, $j < i$: First, either $\phi_j \in \Gamma_i$ or $\bar{\phi}_j \in \Gamma_i$ or both $\vdash \Gamma_i, \phi$ and $\vdash \Gamma_i, \bar{\phi}_j$. Second, if $\phi_j = \Diamond\beta$ then either $\beta \in \Gamma_i$ or $\Diamond(\Box\bar{\beta})^\circ \in \Gamma_i$.

If $\Gamma_i, \phi_i$ is underivable, include $\phi_i$ in $\Gamma_{i+1}$. If $\vdash \Gamma_i, \phi_i$, but $\not\vdash \Gamma_i, \bar{\phi}_i$, include $\bar{\phi}_i$ into $\Gamma_{i+1}$. Otherwise leave $\Gamma_{i+1} = \Gamma_i$. In this case $\Gamma_{i+1}$ clashes with $\phi_i$. Denote the result of these two steps by $\Gamma'_{i+1}$. It is still underivable.

If $\phi_i = \Diamond\beta$, consider the "position" of $\beta$ with respect to $\Gamma'_i$. If $\beta \in \Gamma'_i$, we are done: $\Gamma_{i+1} := \Gamma'_{i+1}$. Otherwise $\bar{\beta} \in \Gamma'_i$ or $\beta$ clashes with $\Gamma'_i$. In both cases

$$\vdash \Gamma'_i, \beta.$$

Hence $\not\vdash \Gamma'_{i+1}, \Diamond(\Box\bar{\beta})^\circ$, since otherwise $\vdash \Gamma'_{i+1}$ by $(Sym)$. Define

$$\Gamma_{i+1} := \Gamma'_{i+1}, \Diamond(\Box\bar{\beta})^\circ.$$

At the end define $\Gamma =: \bigcup \Gamma_i$.

The completeness of $\Gamma$ easily follows from the completeness of $\Gamma_i$ for $\{\phi_1, \ldots, \phi_{i-1}\}$. ⊣

**Definition 4.** Consider the following Kripke model: $\mathbf{K} = \langle W, R, \models \rangle$:

- $W$ is the set of all complete sequents.
- $R\Gamma\Sigma$ iff $\Gamma^\diamond \subseteq \Sigma$.
- $\models_\Gamma p$ iff $\bar{p} \in \Gamma$.

**Lemma 3.** $\mathbf{K}$ is a Kripke model for B : $R$ is reflexive and symmetric.

Proof. Reflexivity is obvious from the closure under ($\diamond$). To prove symmetry, assume $R\Gamma\Sigma$ and $\diamond\beta \in \Sigma$. By completeness of $\Gamma$, for every $\diamond\beta \in \Sigma$ one of $\beta$, $\diamond(\Box\bar{\beta})^\circ$ is an element of $\Gamma$. However the second case is impossible, since then $(\Box\bar{\beta})^\circ \in \Sigma$ (by $\Gamma^\diamond \subseteq \Sigma$) together with $\diamond\beta$. Hence $\Sigma^\diamond \subseteq \Gamma$ as required. ⊣

We prove that $\mathbf{K}$ falsifies every invalid subformula of $\Gamma_0$.

**Theorem 1.** For $\Gamma \in W$:

$$\theta \in \Gamma \text{ implies } \not\models_\Gamma \theta \tag{3}$$

$$\bar{\theta} \in \Gamma \text{ implies } \models_\Gamma \theta. \tag{4}$$

$$\not\models_{\Gamma_0} \Gamma_0 \tag{5}$$

Proof. Relation (5) is an immediate consequence of (3,4), which are proved by simultaneous induction on the formula $\theta$. First assume
$\theta = p$ (induction base):
In this case (4) follows from the definition of $\models$. If $\theta \in \Gamma$, then $\bar{\theta} \notin \Gamma$; otherwise $\Gamma$ is an axiom. Hence $\not\models_\Gamma p$.
The induction step ($\theta$ is a composite formula) is proved by cases.

&: $\theta = \phi \& \psi$. If $\theta \in \Gamma$, then $\phi$ or $\psi$ is in $\Gamma$ by Lemma 1, therefore $\not\models_\Gamma \phi$ or $\not\models_\Gamma \psi$ by the induction hypothesis. Hence $\not\models_\Gamma \theta$ by the truth condition for &.
If $\bar{\theta} = (\bar{\phi} \vee \bar{\psi}) \in \Gamma$, then both $\bar{\phi}$ and $\bar{\psi}$ are in $\Gamma$, by Lemma 1. Hence $\models_\Gamma \phi$, $\models_\Gamma \psi$ by the induction hypothesis; therefore $\models_\Gamma \theta$.

∨: $\theta = \phi \vee \psi$. Similarly.

$\diamond$: $\theta = \diamond\beta$. Assume $\theta \in \Gamma$. To prove $\Gamma \not\models \theta$ assume $R\Gamma\Sigma$ and prove $\Sigma \not\models \beta$. We have $\beta \in \Gamma^\diamond \subseteq \Sigma$, hence IH applies.
Now assume $\bar{\theta} = \Box\bar{\beta} \in \Gamma$. The sequent $\Gamma^\diamond, \bar{\beta}$ is underivable, since $\Gamma$ is obtained from it by rule ($\Box$). By Lemma 2 there is a complete sequent $\Sigma \supseteq \Gamma^\diamond, \bar{\beta}$. Hence $R\Gamma\Sigma$ and (by IH) $\models_\Sigma \beta$, proving $\not\models_\Gamma \theta$.

☐: $\theta = \square\beta$. Completely similar to ◊-case. If $\theta \in \Gamma$, take a complete extension $\Sigma$ of $\Gamma^{\diamond}, \beta$, proving $\not\models_\Sigma \beta$.

If $\bar\theta = \Diamond\bar\beta \in \Gamma$, then $\bar\beta \in \Sigma$ for every $\Sigma$ with $R\Gamma\Sigma$. ⊣

**Corollary 1.** *GB is complete.*

## 3 Tableau system $BT$

We follow here standard treatments using semantic tableaux like [7],[9],[3]. Derivable objects of the system $BT$ are *tableaux*, i.e. finite sequences of sequents indexed by finite sequences $\sigma$ of natural numbers $\sigma = (i_0, i_1, \ldots, i_p)$. Such finite sequnces are denoted by $\sigma, \tau, \sigma_1, \ldots$ etc. A tableau in a derivation has the form

$$\sigma_0 S_0; \sigma_1 S_1; \sigma_2 S_2; \ldots \sigma_k S_k \qquad (6)$$

where $\sigma_i$ are distinct non-empty sequences of natural numbers (*indices*). Usually $\sigma_0 = \emptyset$, the empty sequence. The *immediate successors* of $\sigma = (i_0, i_1, \ldots, i_p)$ are sequences $\sigma * \{i\} = (i_0, i_1, \ldots, i_p, i)$ or shorter $\sigma * i$. Define

$$\sigma < \tau \text{ iff } \tau = \sigma * \{i\} \text{ for some } i$$

$\sigma \leq \tau$ iff $\sigma < \tau$ or $\sigma = \tau$.

$$R\sigma\tau \text{ iff } (\sigma \leq \tau \text{ or } \tau \leq \sigma)$$

$T, T', U, V, etc.$ denote arbitrary tableaux. $lh(\sigma)$ is the length of $\sigma$:
$lh(\emptyset) = 0, \quad lh((i_0, \ldots, i_p)) = p + 1$.
The order of the indexed sequents $\sigma_i S_i$ in (6) is inessential.
**Axioms:** $T; \sigma\, A, \bar A, \Gamma; U$.
**Inference rules**

$$\frac{U; \sigma A, \Gamma}{U; \sigma(A \vee B), \Gamma} \vee \frac{U; \sigma B, \Gamma}{U; \sigma(A \vee B), \Gamma}$$

$$\frac{U; \sigma\Gamma, A \quad U; \sigma\Gamma, B}{U; \sigma\Gamma, A\&B} \&$$

$$\frac{U; \Sigma\Gamma; \sigma' A, \Pi}{U; \Sigma\Gamma, \Diamond A; \sigma'\Pi} \Diamond \qquad \text{if } R\sigma\sigma'$$

$$\frac{U; \sigma\Gamma; (\sigma * i)A}{U; \sigma\Gamma, \square A} \square$$

provided the index $\sigma * i$ is *new* (when the rule is used bottom-up), that is no sequent in the conclusion (under the line) has this index.
Note that the rule $\Diamond$ includes the case $\sigma = \sigma'$.

## 3.1 Translation of Tableaux into Formulas

The translation is essentially one introduced by Kripke [7], section 4.1, see also [9],[3].

For the tableau $T$ of the form (6) and for every $\sigma$ occurring in $T$ as an index define a formula $T^\sigma$. Let $\sigma_{i_1}, \ldots \sigma_{i_l}$ be complete list of immediate successors of $\sigma$: $\sigma < \sigma_{i_j}$ for $j = 1, \ldots l$. Then

$$T^\sigma := U_\sigma \vee \Box T^{\sigma_1} \vee \ldots \vee \Box T^{\sigma_l}.$$

The translation of the whole tableau $T$ is $T^\emptyset$. The traslation $T^s$ of a tableau $T$ into a sequent is not very different:

$$T^s := S_\emptyset, \Box T^{i_1}, \ldots, \Box T^{i_l},$$

where $i_1, \ldots, i_l$ are all one-element indices in $T$.

## 3.2 System $BT^\circ$ and one-direction transfer

The main obstacle to traslation of the tableau system $BT$ into the system $GB$ is the possibility to place the side formula $\beta$ of the rule $\Diamond$ with the main formula $\Diamond\beta$ at the "world" $\sigma$ into the world $\sigma'$ with $\sigma' < \sigma$:

$$\frac{U; \sigma'\Pi, \beta; \sigma\Gamma}{U; \sigma'\Pi; \sigma\Gamma, \Diamond\beta} \Diamond \quad \sigma' < \sigma \qquad (7)$$

Indeed in the absence of this situation (that is for a weaker modal system T determined by all reflexive Kripke frames) the transformations described in sections 3.3 and 3.4 below are sufficient for passage from a tableau proof of a sequent to a cut-free derivation of the same sequent in ordinary Gentzen-type system.

Adding the rule similar to the $Sym$-rule of the system $GB$ allows to avoid the situation in (7). This is the most important operation for systems with analytic cut.

For the new system $BT^\circ$ add formulas $(\Diamond\Box\beta)^\circ, (\Box\beta)^\circ$ to the language, add corresponging axioms and the rule $(Sym)$.

**System $BT^\circ$**
Additional axioms: $U; \Gamma, (\Box\bar\beta)^\circ, \Diamond\beta$
Additional rule:

$$\frac{U; \sigma\beta, \Gamma \quad U; \sigma\Diamond(\Box\bar\beta)^\circ, \Gamma}{U; \sigma, \Gamma} \; Sym$$

provided $\Diamond\beta$ is a formula occuring in $U, \Gamma$.

A new restriction in the rule $(\Diamond)$: the rule is applicable only if $\sigma \leq \sigma'$.

**Theorem 2.** Any $BT$-derivation can be transformed into $BT^\circ$-derivation of the same formula.

Proof. Take arbitrary $\Diamond$ inference (7) with $\sigma' < \sigma$, for example the uppermost. Apply the rule $Sym$ from the bottom up:

$$\cfrac{\cfrac{U;\sigma'\Pi,\beta;\sigma\Gamma}{U;\sigma'\Pi,\beta;\sigma\Gamma,\Diamond\beta}\ weak \quad \cfrac{\overset{\text{axiom}}{U;\sigma'\Pi;\sigma(\Box\beta)^\circ,\Gamma,\Diamond\beta}}{U;\sigma'\Pi,\Diamond(\Box\beta)^\circ;\sigma\Gamma,\Diamond\beta}\ \Diamond}{U;\sigma'\Pi;\sigma\Gamma,\Diamond\beta}\ Sym$$

⊣

## 3.3 Pruning of derivations. Disjunction property

An index $\sigma$ in a tableau $T$ is *maximal* if no proper extension of $\sigma$ occurs in $T$. $T^\sigma$ means the result of deleting from $T$ all components $\tau U_\tau$ when $\tau$ is not a prefix of $\sigma$. For example

$$T = \emptyset X; (1)Y; (2)Z; (3)U; (1,2)V; (1,3)W; (1,2,3)K; (1,2,4)L; (1,2,3,4)M$$

$$T^{(1,2,3,4)} = \emptyset X; (1)Y; (1,2)V; (1,2,3)K; (1,2,3,4)M;$$

The next statement is an extension of disjunction property to the system B. Its proof is an extension of the proof for systems like T or S4 admitting ordinary cut elimination. It goes by pruning the derivation, i.e. deletion of some formulas or entire branches.

Say that a derivation *has non-branching indices* if every two indices in every tableau are comparable (i.e., one of them is a prefix of the other one).

**Theorem 3.** Every derivation of a tableau $T$ can be pruned into a derivation of $T^\tau$ for some maximal index $\tau$ having non-branching indices.

Proof. Induction on the derivation.

Induction base is obvious. For induction step consider the last rule of a given derivation.

∨

$$\cfrac{U;\sigma A, \Gamma}{U;\sigma(A\vee B),\Gamma}\ \vee$$

By the induction hypothesis $(U;\sigma(A\vee B),\Gamma)^\tau$ is derivable for some maximal index $\tau$. If $\sigma$ is a prefix of $\tau$ then $(U;\sigma A,\Gamma)^\tau = U^\tau;\sigma A,\Gamma$ implying

$$U^\tau;\sigma(A\vee B),\Gamma\ =\ (U;\sigma(A\vee B),\Gamma)^\tau$$

If $\sigma$ is not a prefix of $\tau$ then $(U;\sigma(A\vee B),\Gamma)^\tau = U^\tau$ coinciding with $(U;\sigma A,\Gamma)^\tau = U^\tau$, so that the whole ∨-inference is pruned.

&

$$\frac{U;\sigma\Gamma, A \quad U;\sigma\Gamma, B}{U;\sigma\Gamma, A\&B} \ \&$$

Take $\tau, \tau'$ for the premises by IH. If $\sigma$ is not a prefix of $\tau$ or $\tau'$ the whole rule is pruned. Otherwise $\tau = \tau'$ since both of them are maximal and extend $\sigma$ and the & inference is preserved.

*Sym* Similarly.

$\Diamond$

$$\frac{U;\Sigma\Gamma;\sigma'A,\Pi}{U;\Sigma\Gamma,\Diamond A;\sigma'\Pi} \ \Diamond \qquad \sigma \leq \sigma'$$

If $\sigma * i$ is not a prefix of $\tau$, the rule is pruned, otherwise it is preserved.

$\Box$

$$\frac{U;\sigma\Gamma;(\sigma*i)A}{U;\sigma\Gamma,\Box A} \ \Box$$

If $\sigma'$ is not a prefix of $\tau$, the rule is pruned, otherwise it is preserved.

⊣

**Theorem 4.** Every $BT$ derivation of a formula

$$\Box\alpha_0 \vee \ldots \Box\alpha_n \tag{8}$$

can be transformed into a $BT$ derivation of a formula $\Box\alpha_i$ for some $I$.

Proof. Using inversion, get a $BT^\circ$ derivation of the tableau

$$\emptyset; (0)\alpha_0; \ldots; (n)\alpha_n$$

Then apply theorem 3 and $\Box$ rule. ⊣

### 3.4 Inversion, Focused Derivations, $BT = GB$

Next we prove standard inversion lemmas for $GT^\circ$.

**Lemma 4.** The rules $\vee, \&, \Diamond$ of the system $GT^\circ$ are invertible:

1. If $U;\sigma\Gamma, \alpha \vee \beta$ is derivable then $U;\sigma\Gamma, \alpha, \beta$ is derivable.

2. If $U;\sigma\Gamma, \alpha\&\beta$ is derivable then both $U;\sigma\Gamma, \alpha$ and $U;\sigma\Gamma, \beta$ are derivable.

3. If $U;\sigma\Gamma, \Diamond\beta;\sigma'\Pi$ is derivable then $U;\sigma\Gamma, \Diamond\beta;\sigma'\Pi, \beta$ is derivable.

Proof. The last clause is an instance of weakening. Proofs of the two first clauses are standard. ⊣

The next step is to permute the inferences so that the last component is analyzed by adjacent inferences "to the end": the analysis of the next components (generated by $\Box$ rule) does not begin until the analysis of the given component is finished. Such a restriction is sometimes called *focusing* [1].

**Theorem 5.** Every derivation in $BT^\circ$ can be transformed (preserving non-branching indices) into the focused derivation of the same tableau: all inferences with the main formula in a component $\sigma U_\sigma$ are situated below all inferences with the main formula in the components $\sigma' U'_\sigma$, $\sigma < \sigma'$.

Proof. Assume that the given derivation has non-branching indices. Use induction on the number of inferences with the main formula violating the restriction. Take the minimal index $\sigma$ for which there is a violation, and the lowermost violating inference $I$ with the main formula in the component $\sigma U_\sigma$. Let $V$ be the uppermost tableau below $I$ such that all inferences below $V$ have main formulas in $\sigma U_\sigma$.

$$\begin{array}{c} I \\ \text{non-}\sigma\text{-inferences} \\ V \\ \sigma\text{-inferences} \end{array}$$

The main formula of the inference $I$ is already present in the $\sigma$-component $V_\sigma$ of $V$, since the inferences between $I$ and $V$ do not change the $\sigma$-component. Now move $I$ down to $V$. If $I$ is & or ∨-inference, use inversion Lemma 4. In other cases the transformation is similar. The given derivation is shown on the left, the resulting one is on the right.

1. $\Diamond$:

$$\dfrac{\sigma\Gamma, \Diamond\alpha, \alpha; T}{\sigma\Gamma, \Diamond\alpha; T} I \qquad \sigma\Gamma, \Diamond\alpha, \alpha; T$$
$$\text{non-}\sigma\text{-inferences} \qquad \text{non-}\sigma\text{-inferences}$$
$$V: \sigma\Delta, \Diamond\alpha; V' \qquad \dfrac{\sigma\Delta, \Diamond\alpha, \alpha; V'}{V: \sigma\Delta, \Diamond\alpha; V'}$$

2. $\Box$:

$$\dfrac{\sigma\Gamma, \Box\alpha; \sigma\alpha; T}{\sigma\Gamma, \Box\alpha; T} I \qquad \sigma\Gamma, \Box\alpha; \sigma\alpha; T$$
$$\text{non-}\sigma\text{-inferences} \qquad \text{non-}\sigma\text{-inferences}$$
$$V: \sigma\Delta, \Box\alpha; V' \qquad \dfrac{\sigma\Delta, \Box\alpha; \sigma\alpha; V'}{V: \sigma\Delta, \Box\alpha; V'}$$

⊣

**Theorem 6.** A sequent is derivable in $BT$ iff it is derivable in $GB$.

Proof. One direction is trivial. Now assume a sequent $\emptyset\Gamma$ is derivable in $BT$. Get a focused derivation in $BT^\circ$ with non-branching indices. If needed, moving some $\Diamond$ inferences upward ensure that all $\Diamond$ inferences with the main formula in a component $\sigma U_\sigma$ transferring the side formula to the "next" component $\sigma' U'_\sigma$ with $\sigma < \sigma'$ are situated together, so that we can write all such $\Diamond$ inferences together in the form:

$$\frac{T; \sigma U_\sigma; \sigma' U'_\sigma, U_\sigma^\Diamond}{T; \sigma U_\sigma; \sigma' U_{\sigma'}} , \qquad (9)$$

and this inference is the uppermost among inferences with the main formula in $U_\sigma$. In this case $U_{\sigma'}$ consists of just one formula which was placed there when a rule $\Box$ had been applied:

$$\frac{T; \sigma U_\sigma; \sigma', \alpha}{T; \sigma U_\sigma, \Box\alpha} , \qquad (10)$$

and we assume (moving it upward if needed) that this inference immediately precedes (9). We assume also that in the axiom the component containing $\alpha, \neg\alpha$ is the maximal one, since the latter components can be ignored.

Now we use induction on given derivation $d$ to establish that for every tableau $U$ with the maximal index $\tau'$ and immediately preceding index $\tau$ one of the following sequents is derivable in $GB$:

$U_{\tau'}$ if $d$ contains an inference with the main formula in $U_{\tau'}$;

$U_\tau, \Box U_{\tau'}$ otherwise.

The induction base is now obvious. Consider induction step. In view of the focusing restriction the last inference $I$ has the main formula in $U_\tau$ or in $U'_\tau$. If the main formula is in $U'_\tau$ then by focusing the same is true for the premise, hence IH and the same inference $I$ is applicable, except the case when $I$ is a $\Box$ inference:

$$\frac{T; \tau' U_{\tau'}; \tau''\alpha}{T; \tau' U_{\tau'}, \Box\alpha}$$

In that case it is preceded (at most) by (9) from $\tau'$ to $\tau''$, then by the inferences not touching $\tau'$ further. So by IH the sequent $U_{\tau'}^\Diamond, \alpha$ is derivable and the $\Box$ rule of $GB$ (plus weakening) gives $U_{\tau'}$. Assume now the main formula of the inference $I$ is in $U_\tau$. If the previous inference is of the same kind, use IH. Otherwise $I$ is of the form (9), and we again use the $\Box$ rule of $GB$ (plus weakening). ⊣

## 3.5 Examples

Example 1. $p \to \Box\Box\Diamond\Diamond p$. First a derivation in $BT$.

$$\cfrac{\cfrac{\cfrac{\cfrac{\cfrac{\emptyset \bar{p}, p}{\emptyset \bar{p}, (0)\Diamond p} \,\Diamond}{\emptyset \bar{p}, (0,0)\Diamond^2 p} \,\Diamond}{\emptyset \bar{p}, (0)\Box\Diamond^2 p} \,\Box}{\emptyset\, \bar{p}, \Box^2\Diamond^2 p} \,\Box}{\emptyset\, \bar{p} \vee \Box^2\Diamond^2 p}$$

The translation into $BT^\circ$:

$$\cfrac{\emptyset\, \bar{p}, p \qquad \cfrac{\cfrac{(0)(\Box\bar{p})^\circ, \Diamond p \text{ axiom} \quad \cfrac{\cfrac{(0,0)(\Box^2\bar{p})^\circ, \Diamond^2 p}{(0)\Diamond(\Box\overline{\Diamond p})^\circ, (0,0)\Diamond^2 p} \text{ axiom}}{(0)\Diamond(\Box\overline{\Diamond p})^\circ, \Box\Diamond^2 p}}{(0)(\Box\bar{p})^\circ, \Box\Diamond^2 p}}{\cfrac{\emptyset\Diamond(\Box\bar{p})^\circ, (0)\Box\Diamond^2 p}{\emptyset\Diamond(\Box\bar{p})^\circ, \Box^2\Diamond^2 p}}}{\emptyset\, \bar{p}, \Box^2\Diamond^2 p}$$

Focused derivation in $BT^0$.

$$\cfrac{\emptyset\bar{p}, p \qquad \cfrac{(0)\Box\bar{p})^\circ, \Diamond p \quad \cfrac{\cfrac{(0,0)(\Box^2\bar{p})^\circ, \Box\Diamond^2 p}{(0)\Diamond(\Box\overline{\Diamond p})^\circ, \Box\Diamond^2 p}}{(0)(\Box\bar{p})^\circ, \Box\Diamond^2 p}}{\cfrac{\emptyset\Diamond(\Box\bar{p})^\circ, \Box^2\Diamond^2 p}{}}}{\emptyset\bar{p}, \Box^2\Diamond^2 p}$$

Derivation in $GB$.

$$\cfrac{\bar{p}, p \qquad \cfrac{(\Box\bar{p})^\circ, \Diamond p \quad \cfrac{\cfrac{(\Box^2\bar{p})^\circ, \Diamond^2 p}{\Diamond(\Box\overline{\Diamond p})^\circ, \Box\Diamond^2 p}}{(\Box\bar{p})^\circ, \Box\Diamond^2 p}}{\Diamond(\Box\bar{p})^\circ, \Box^2\Diamond^2 p}}{\bar{p}, \Box^2\Diamond^2 p}$$

Example 2. Proof in $BT$ of the formula

$$\gamma = p \vee \Box(q \vee \Box(r \vee \Box\Diamond\alpha))$$

where $\alpha := \bar{r}\&\Diamond\beta$, $\beta := \bar{q}\&\Diamond\bar{p}$.

$$\cfrac{\cfrac{\cfrac{(0,0)r,\bar{r}\quad\cfrac{\emptyset q,\bar{q}\quad\cfrac{\emptyset p,(0)\Diamond\bar{p}\quad\cfrac{\emptyset p,\bar{p}}{}\Diamond}{\emptyset p,(0)q,(0,0)\Diamond(\bar{q}\&\Diamond\bar{p})}\Diamond,\&}{\emptyset p,(0)q,(0,0)r,(\bar{r}\&\Diamond(\bar{q}\&\Diamond\bar{p}))}\&}{\emptyset p,(0)q,(0,0)r,(0,0,0)\Diamond\alpha}\Diamond}{\gamma}\lor,\Box\ 3\text{ times}$$

Focused proof in $BT^\circ$.

$$\cfrac{\cfrac{\emptyset p,\bar{p}\quad\cfrac{\cfrac{\cfrac{d}{(0)(\Box p)^\circ,q,\Box(r\lor\Box\Diamond(\bar{r}\&\alpha))}}{(0)(\Box p)^\circ,q\lor\Box(r\lor\Box\Diamond(\bar{r}\&\alpha))}}{\emptyset,\Diamond(\Box p)^\circ,\Box(q\lor\Box(r\lor\Box\Diamond(\bar{r}\&\alpha)))}}{\emptyset p,\Box(q\lor\Box(r\lor\Box\Diamond(\bar{r}\&\alpha)))}}{\emptyset\gamma}$$

where $d$ is the following derivation:

$$\cfrac{\cfrac{(0)q,\bar{q}\quad (0)(\Box p)^\circ,\Diamond\bar{p}}{(0)(\Box p)^\circ,q,\beta}\quad\cfrac{\cfrac{(0,0)r,\bar{r}\quad\cfrac{(0,0)r,\bar{r}\quad (0,0)(\Box\overline{\beta})^\circ,\Diamond\beta}{(0,0)(\Box\overline{\beta})^\circ,r,\alpha}}{\cfrac{(0,0)(\Box\overline{\beta})^\circ,r,\bar{r}\&\alpha}{\cfrac{(0,0)(\Box\overline{\beta})^\circ,r,\Box\Diamond(\bar{r}\&\alpha)}{\cfrac{(0,0)(\Box\overline{\beta})^\circ,r\lor\Box\Diamond(\bar{r}\&\alpha)}{(0)\Diamond(\Box\overline{\beta})^\circ,\Box(r\lor\Box\Diamond(\bar{r}\&\alpha))}}}}\text{ axiom}}{(0)(\Box p)^\circ,q,\Box(r\lor\Box\Diamond(\bar{r}\&\alpha))}$$

where the axiom is $(0,0)\Diamond(\Box\overline{\bar{r}\&\alpha})^\circ,\Box\Diamond(\bar{r}\&\alpha)$. Now it is easy to prune this into a $GB$-derivation.

Example 3. Formula $\gamma = p\lor\Box\alpha\lor\Box^2\Diamond^2(p\&\bar{q})$ where $\alpha := (q\&\bar{r})\lor\Box\Diamond r$.
$BT$-derivation.

$$\cfrac{\emptyset p,\bar{p}\quad\cfrac{\cfrac{(0)q,\bar{q}\quad (0)\bar{r},\Box\Diamond r}{(0)q\&\bar{r},\Box\Diamond r,\bar{q}}}{\emptyset\Box\alpha,\Diamond\bar{q}}}{\cfrac{\emptyset p,\Box\alpha,p\&\Diamond\bar{q}}{\cfrac{\emptyset p,\Box\alpha,(0)\Diamond(p\&\Diamond\bar{q})}{\cfrac{\emptyset p,\Box\alpha,(0,0)\Diamond^2(p\&\Diamond\bar{q})}{\cfrac{\emptyset\bar{p},\Box\alpha,\Box\Box\Diamond\Diamond(p\&\Diamond\bar{q})}{\gamma}}}}$$

## 4 A Cut-Free Version of S5

Let's reproduce a cut elimination proof for a sequent S5-system mentioned in a paper by G. Shvarts [12].

Derivable objects are sequents $\alpha_1, \ldots, \alpha_n$, where $\alpha_i$ are propositional modal formulas constructed from literals by $\&, \vee, \Box, \Diamond$ (negative normal form). A formula is *modalized* if every occurence of a propositional variable is in the scope of a modal connective.

**Inference rules.**

The standard axioms and rules for $\&, \vee, \Diamond$. The rules

$$\frac{\Gamma^m, A}{\Gamma, \Box A} \Box \qquad \frac{\alpha, \Gamma \quad \bar{\alpha}, \Delta}{\Gamma, \Delta} \text{ cut}$$

where $\Gamma^m$ consists of modalized members of $\Gamma$.

It is well known [11] that cut elimination does not hold for this formulation, for example for the following derivation:

$$\frac{\dfrac{p, \bar{p} \quad \dfrac{\Box \bar{p}, \Diamond p}{\Box \bar{p}, \Box \Diamond p} \Box}{\Diamond p, \bar{p}}}{\bar{p}, \Box \Diamond p} \tag{11}$$

**Theorem 7.** *Cut rule can be eliminated by standard reductions from proofs of modalized sequents.*

Proof. Familiar induction on cut rank (complexity of the cut formula) plus induction on the number of cuts with maximal rank. There are some twists. We use a familiar "substitution" operation: the cut is moved up the right premise (and later a dual operation of moving the cut over the left premise). We picture this as follows, with the given derivation at the left and the result of substituting $\Gamma$ for $\bar{\alpha}$ at the right:

$$\begin{array}{cc}
\overset{\cdots}{\Sigma, \bar{\alpha}} & \dfrac{\overset{\cdots}{\Gamma, \alpha} \quad \overset{\cdots}{\Sigma', \bar{\alpha}}}{\Gamma, \Sigma'} \\
\vdots & \\
\dfrac{\overset{\cdots}{\Gamma, \alpha} \quad \overset{\vdots}{\Delta, \bar{\alpha}}}{\Gamma, \Delta} & \vdots \\
& \Gamma, \Delta
\end{array}$$

where $\Sigma'$ is the result of replacing $\bar{\alpha}$ by $\Gamma$ in $\Sigma$. This transformation can fail to preserve correctness of a derivation since the restriction to modalized $\Gamma$ can be destroyed in some of the $\Box$-inferences situated over the right premise of the cut. This happens for example in the derivation (11). However this modalization restriction is preserved if $\Gamma$ in the left premise $\Gamma, \alpha$ of the cut is modalized. Similarly moving the cut up the left premise is justified when $\Delta$ is modalized. When both $\Gamma$ and $\Delta$ are modalized, the cut can be moved up so that $\alpha$ and $\bar{\alpha}$ are the main (analyzed) formulas in their premises, hence cut rank can be reduced in this case of the induction step.

Moreover, the derivation is preserved by the substitution transformation if the cut formula $\alpha$ is *not* modalized, since in that case the sequent containing (a predecessor of) a cut formula $\alpha$ or $\bar{\alpha}$ cannot be a premise of a $\Box$ inference. These remarks justify the induction base (atomic $\alpha$) and the induction step when a cut formula in one of the uppermost cuts of maximal rank is not modalized.

Consider the remaining case of the cut over a modalized formula $\alpha$. If the main connective of $\alpha$ is & or $\vee$, the familiar inversion transformations work. For the cut

$$\frac{\Gamma, \beta \& \gamma \quad \Delta, \bar{\beta} \vee \bar{\gamma}}{\Gamma, \Delta}$$

first $\bar{\beta} \vee \bar{\gamma}$ is replaced by $\bar{\beta}, \bar{\gamma}$ through the right branch, then $\beta \& \gamma$ is similarly replaced by $\beta$ and $\gamma$ in the left branch resulting in

$$\frac{\dfrac{\Gamma, \beta \quad \Gamma, \gamma}{\Gamma, \beta \& \gamma} \& \quad \dfrac{\Delta, \bar{\beta}, \bar{\gamma}}{\Delta, \bar{\beta} \vee \bar{\gamma}} \vee}{\Gamma, \Delta},$$

then in cuts of lower rank:

$$\frac{\dfrac{\Delta, \bar{\beta}, \bar{\gamma} \quad \Gamma, \alpha}{\Gamma, \Delta \bar{\beta}} \quad \Gamma, \beta}{\Gamma, \Delta}$$

The last subcase: the cut over $\alpha = \Diamond \beta$. It is in this case that the restriction to modalized last sequent is used. Before any cut-reduction for cuts of maximal rank all cuts of that rank over $\Diamond$-formulas are moved down to the endsequent, so that below such cut only other such cuts are situated. This is done by the following transformation:

$$\frac{\Gamma, \overset{\cdots}{\Diamond \beta} \quad \Delta, \overset{\cdots}{\Box \bar{\beta}}}{\Gamma, \Delta} \qquad \frac{\Gamma, \overset{\cdots}{\Delta}, \Diamond \beta \quad \Gamma, \overset{\cdots}{\Delta}, \Box \bar{\beta}}{\vdots \qquad \qquad \vdots}$$
$$\vdots \qquad\qquad\qquad \frac{\Sigma, \Diamond \beta \quad \Sigma, \Box \bar{\beta}}{\Sigma}$$
$$\Sigma$$

The left branch of the new derivation with the endsequent $\Sigma, \Diamond \beta$ is obtained from the old derivation by adding $\Diamond \beta$ to all sequents and leaving out the

$\Diamond\beta$-cuts. All other rules are preserved since $\Sigma$ is modalized, and similarly for the right branch.

Now the cuts of maximal rank on formulas not beginning with the $\Diamond$ are reduced first. When the turn of the $\Diamond$-cuts comes, their premises are modalized, hence ordinary substitution and rank-reduction go through. This concludes the proof of the theorem. ⊣

A cut elimination proof of the kind given for the system B in section 2.1 does not work for S5. The most obvious obstacle is the restriction to modal formulas: complete sequents forming worlds of a canonical model contain non-modal formulas. Another side of the same obstacle: the cut-free rules simply are not complete for non-modalized formulas.

### 4.1 A System $S5^\circ$ with $^\circ$-Cut

Let us sketch a completeness proof for a version $S5^\circ$ of S5 with a restricted analytic cut more similar to the system $BG$.

Add to the language of modal propositional logic formulas $(\Box\alpha)^\circ, (\Diamond\alpha)^\circ$ that cannot be proper part of any formulas and can be used only in additional axioms

$$\mu\alpha, (\overline{\mu\alpha})^\circ \text{ where } \mu \in \{\Box, \Diamond\} \tag{12}$$

and an additional rule:

$$\frac{\mu\alpha, \Gamma \quad (\overline{\mu\alpha})^\circ, \Gamma}{\Gamma} \; Cut^\circ$$

where $\mu\alpha$ is a subformula of $\Gamma$.

The system $S5^\circ$ is obtained by deleting the cut rule from the formulation of S5 above and adding axioms (12) and rule $Cut^\circ$.

For formulas with additional constants $\top, \bot$ consider familiar computation rules:

$$\alpha \& \top \mapsto \alpha; \; \alpha \& \bot \mapsto \bot;$$

$$\alpha \vee \top \mapsto \top; \; \alpha \vee \bot \mapsto \alpha;$$

$$\mu c \mapsto c, \text{ where } \mu \in \{\Box, \Diamond\}, \; c \in \{\top, \bot\}.$$

For arbitrary (not necessarily modalized) sequents consider also the following computation rules where notation $\alpha[\beta]$ indicates that an occurrence of a proper subformula $\beta$ of $\alpha$ is distinguished:

$$\Gamma, \alpha[\mu\beta], \overline{\mu\beta}^* \mapsto \Gamma, \alpha[\top], \overline{\mu\beta}^* \quad \Gamma, \alpha[\mu\beta], \mu\beta^* \mapsto \Gamma, \alpha[\bot], \mu\beta^* \tag{13}$$

$\beta^*$ stands for any of $\beta, \beta^\circ$. Every formula obviously reduces under repeated application of $\mapsto$ to a unique irreducible form. The same is true for sequents.

A sequent $\Gamma$ is *saturated* if for each subformula $\mu\alpha$ of $\Gamma$ one of $\mu\alpha, \overline{\mu\alpha}^\circ$ is a member of $\Gamma$.

For example, the sequent

$$\bar{p}, \Box(q \vee \Diamond p) \qquad (14)$$

is not saturated, but both extensions

$$\bar{p}, \Box(q \vee \Diamond p), \Diamond p \text{ and } \bar{p}, \Box(q \vee \Diamond p), \overline{\Diamond p}^\circ$$

are saturated. These extensions reduce under $\mapsto$ to

$$\bar{p}, \Box q, \Diamond p \text{ and } \bar{p}, \top, \overline{\Diamond p}^\circ$$

both obviously derivable. Therefore (14) is derivable by $Cut^\circ$.

A sequent has (modal) *degree 1* if it does not contain modality in the scope of other modality.

**Lemma 5.** Any saturated sequent $\Gamma$ is reducible under $\mapsto$ to a sequent $\Gamma^1$ of degree 1 which is semantically equivalent to $\Gamma$.

Proof. Let $\alpha[\mu\beta]$ be a member of $\Gamma$. By saturation $\Gamma$ contains one of $\mu\beta, \overline{\mu\beta}^\circ$. Use $\mapsto$. ⊣

**Lemma 6.** Let $\Sigma$ be a finite set of literals and

$$\Sigma, \Box\alpha_1, \Box\alpha_n, \Diamond\beta \qquad (15)$$

be a valid first degree sequent. Then one of (quantifier free) sequents

$$\Sigma, \beta; \; \alpha_1, \beta; \ldots; \alpha_n, \beta$$

is a tautology.

Proof. Suppose for contradiction that each of these sequents has a refuting truth-value assignment. Let

$$w_0, w_1, \ldots, w_n \qquad (16)$$

be these refutations. Consider the S5 Kripke model with the worlds (16) and corresponding truth-value assignments. Formula $\beta$ is false in each world, therefore $\Diamond\beta$ is false (in each world). Formula $\alpha_i$ is false in $w_i$, therefore $\Box\alpha_i$ is false. Consequently the sequent (15) is false in the world $w_0$. ⊣

**Lemma 7.** If $S5^0 \vdash \Gamma, \alpha[\bot], \mu\beta$ then $S5^0 \vdash \Gamma, \alpha[\beta], \mu\beta$;

If $S5^0 \vdash \Gamma, \alpha[\top], \overline{\mu\beta}^\circ$ then $S5^0 \vdash \Gamma, \alpha[\mu\beta], \overline{\mu\beta}^\circ$

Proof. Induction on derivation. ⊣

**Theorem 8.** $S5°$ is sound and complete

Proof. Soundness is obvious: erase °.

To prove completeness, take a valid sequent $\Gamma$. Then all its saturated extensions (in fact all extensions) are valid and $\Gamma$ is derived from them by a series of $Cut^0$. Therefore it is sufficient to assume that $\Gamma$ is saturated. Transform $\Gamma$ into a first-degree sequent $\Gamma^1$ by lemma 5. $\Gamma^1$ is still valid. By lemma 6, $\Gamma^1$ is derivable by at most two applications of modal rules plus some propositional inferences. Then a derivation of $\Gamma$ is obtained by lemma 7.  ⊣

## 5  Future Work: Axiomatization of the Strictly Implicational Fragment of B

We are interested in this section in formulas constructed from propositional variables by the connective $\Rightarrow$, where

$$\alpha \Rightarrow \beta \equiv \Box(\alpha \to \beta) \ .$$

The first step to axiomatization of this fragment is to find a suitable analog in the $\Rightarrow$-language of the basic axiom $p \to \Box\Diamond p$. The simplest choice seems to begin with the formula

$$(p \to \Box q) \to \Box(\Box p \to q)$$

The axiom $p \to \Box\Diamond p$ is obtained from it by substituting $q, p$ by $\bot, \bar{p}$ respectively. Howeever simply replacing $\to$ by $\Rightarrow$ leads to a formula $\Box(p \to \Box q) \to \Box(\Box p \to q)$ which is derivable in T (from $(p \to \Box q) \to (\Box p \to q)$).

One can try adding one more $\Box$ to the conclusion,

$$\Box(p \to \Box q) \to \Box\Box(\Box p \to q)$$

but is this axiom sufficient for the $\Rightarrow$-fragment? The next attempt is to express $\Diamond$ in $p \to \Box\Diamond p$, in terms of $\Box$, i.e., consider $p \to \Box(\Box(p \to \bot) \to \bot)$, then replace $\bot$ by the new variable $q$. The result $p \to \Box(\Box(p \to q) \to q)$ is underivable in B , since it is not derivable even in S5: translation into predicate logic yields $Pa \to (\forall x(Px \to Qx) \to \forall x Qx)$.

Howewer adding a $\Box$ to the first $q$ produces a derivable formula

$$p \to \Box(\Box(p \to \Box q) \to q) \ .$$

Here is a derivation in $BT$:

$$\frac{\emptyset\bar{p},p \quad \dfrac{(0)\bar{q},q}{\emptyset\Diamond\bar{q};(0)q}}{\dfrac{\emptyset\bar{p},p\&\Diamond\bar{q};(0)q}{\dfrac{\emptyset\bar{p};(0)\Diamond(p\&\Diamond\bar{q}),q}{\emptyset \ \bar{p}\vee\Box(\Box(\bar{p}\vee\Box q)\vee q)}}}$$

It remains to see whether this formula provides an axiomatization of the fragment in question after being added to an axiomatization of the strictly implicational fragment of T [9],[5].

## BIBLIOGRAPHY

[1] Andreoli J.-M. Logic programming with focusing proofs in linear logic. Journal of Logic and Computation, 2(3):197-347, 1992.
[2] Fine, K., Failures of the Interpolation Lemma in Quantified Modal Logic, Journal of Symbolic Logic, 1979,44, no. 2, 201-206
[3] Fitting M., Proof Methods for Modal and Intuitionistic Logics, Dordrecht, 1983
[4] Fitting, M.: Intuitionistic Logic, Model Theory, and Forcing, North-Holland, Amsterdam, 1969
[5] Hacking J., What is strict implication. JSL, 28, 1963, 51-71
[6] G. E. Hughes and M. J. Cresswell, A new Introduction to Modal Logic, Routledge, 1996
[7] Kripke S., Semantical Analysis of Modal Logic I, Zeitschr.f. math. Logik und Grundl. d. Mathematik, 9, 1963, 67-96
[8] G. Mints, On Some Calculi of Modal Logic. Proc. Steklov Inst. of Mathematics, 98, 1971, 97-122 (Translated by AMS)
[9] Mints G.,Lewis Systems and the System T. In: G. Mints, Select Papers in Proof Theory. North-Holland-Bibliopolis, 1993, 221-294 (Russian Original 1974)
[10] Mints, G. Indexed systems of sequents and cut-elimination, Journal of Philosophical Logic, v. 26, no 6, 1997, p. 671-696
[11] Ohnishi, M. and Matsumoto, K., Gentzen Method in Modal Calculi, Osaka Mathematical Journal, 9 (1957),113-130
[12] Shvarts, Grigori F. Gentzen style systems for K45 and K45D, Logic at Botik, Proc. Symposium on logical foundations of computer science, Pereslavl-Zalessky/USSR 1989, Lect. Notes Comput. Sci. 363, 245-256 (1989).
[13] Takeuti, G. Proof Theory, North Holland, 1987

# Disguising Induction:
# Proofs of the Pigeonhole Principle for Trees

Jeffry L. Hirst

ABSTRACT. We examine the relationship between a pigeonhole principle for trees and induction on $\Sigma_2^0$ formulas. This analysis is carried out in the framework of reverse mathematics utilizing a hierarchy of axiom systems formulated by Harvey Friedman.

Let $2^{<\mathbb{N}}$ denote the set of all finite sequences of zeros and ones. We often use $\sigma$ to denote both a finite sequence and the associated finite function, so $\sigma(0)$ is the first element of the sequence, and $\sigma(\mathrm{lh}(\sigma)-1)$ is the last. If $\tau$ consists of $\sigma$ with appended elements we write $\sigma \subseteq \tau$, and write $\sigma \subset \tau$ when $\tau$ is a proper extension of $\sigma$. Viewing $2^{<\mathbb{N}}$ as a partial order ordered by the $\subseteq$ relation, we can think of any subset of $2^{<\mathbb{N}}$ as a subtree. A bijection between a subset $S \subseteq 2^{<\mathbb{N}}$ and $2^{<\mathbb{N}}$ that preserves extension is an order isomorphism. Using this terminology, we can formulate the following pigeonhole principle on binary trees.

TT(1): Suppose $f : 2^{<\mathbb{N}} \to n$ for some $n \in \mathbb{N}$. Then there is a subtree $S \subseteq 2^{<\mathbb{N}}$ order isomorphic to $2^{<\mathbb{N}}$ and a $c < n$ such that $f(\sigma) = c$ for every $\sigma \in S$.

This pigeonhole principle follows immediately from a version of Hindman's theorem. If we let FIN denote the collection of all nonempty finite subsets of $\mathbb{N}$, then the familiar finite sum form of Hindman's theorem [9] is equivalent to the following statement. (See [1].)

HT: Suppose $f : \mathsf{FIN} \to n$ for some $n \in \mathbb{N}$. Then there is a sequence $\langle X_i \rangle_{i \in \mathbb{N}}$ of elements of FIN and a $c < n$ such that

- if $i < j$ then $\max(X_i) < \min(X_j)$, and
- for every finite nonempty set $J \subset \mathbb{N}$, $f(\cup_{j \in J} X_j) = c$.

Here is a proof that HT implies TT(1). Suppose $f : 2^{<\mathbb{N}} \to n$. For each $X \in \mathsf{FIN}$, let $\sigma_X$ be a sequence in $2^{<\mathbb{N}}$ of length $\max(X) + 1$ such that for each $i$, $\sigma_X(i) = 1$ if and only if $i \in X$. Define $g : \mathsf{FIN} \to n$ by $g(X) = f(\sigma_X)$. Apply HT to $g$, and let $\langle X_i \rangle_{i \in \mathbb{N}}$ be the resulting sequence of finite subsets and let $c$ be the associated color. Define $Y_{\langle\rangle} = \emptyset$, and for each nonempty $\tau \in 2^{<\mathbb{N}}$, let $Y_\tau$ be the union of $X_0$ or $X_1$, $X_2$ or $X_3$, $X_4$ or $X_5$ and so on, where the set $X_{2i}$ is included if $\tau(i) = 0$ and $X_{2i+1}$ is included if $\tau(i) = 1$. More formally, for nonempty $\tau \in 2^{<\mathbb{N}}$, let $Y_\tau = \bigcup_{i < \mathrm{lh}(\sigma)} X_{2i+\tau(i)}$. Then the set $S = \{\sigma_{Y_\tau} \mid \tau \in 2^{<\mathbb{N}}\}$ is a subtree of $2^{<\mathbb{N}}$ and the map taking $\tau$ to $\sigma_{Y_\tau}$ is an order isomorphism between $2^{<\mathbb{N}}$ and $S$. Furthermore, for each $\tau$, $f(\sigma_{Y_\tau}) = g(Y_\tau) = c$, so $S$ is the desired monochromatic subtree.

Timothy McNicholl [11] asked if the use of Hindman's theorem is actually necessary to prove TT(1). Reverse mathematics, based on the axiom systems formulated by Harvey Friedman [6, 7], provides an excellent tool set for addressing this type of question. Indeed, HT is not needed to prove TT(1), as was shown in [3]. We will present a proof of this result below.

We will formalize our proof in the subsystem $\mathsf{RCA}_0$. This theory has variable types for natural numbers and sets of natural numbers, basic arithmetic axioms including induction restricted to $\Sigma^0_1$ formulas, and the recursive comprehension axiom, which (naïvely) asserts the existence of computable sets. A very detailed discussion of $\mathsf{RCA}_0$ can be found in Simpson's book [12]. Since TT(1) implies the infinite pigeonhole principle, it cannot be proved in $\mathsf{RCA}_0$ [10]. However, if we append the induction scheme for $\Sigma^0_2$ formulas, denoted by $\Sigma^0_2 - \mathsf{IND}$, then TT(1) can be proved, as stated in the following result.

THEOREM 1. ($\mathsf{RCA}_0$) $\Sigma^0_2 - \mathsf{IND}$ *implies* TT(1).

**Proof.** We carry out the proof in $\mathsf{RCA}_0$. Suppose $f : 2^{<\mathbb{N}} \to n$. Let $\{X_j \mid j < 2^n\}$ be the collection of all subsets of $\{0, 1, \ldots n-1\}$, enumerated so that $X_i \subseteq X_j$ implies $i \leq j$. Since the entire range of $f$ is included in some $X_j$, there is a $j$ such that

$$\exists \sigma \; \forall \tau \supseteq \sigma \; (f(\tau) \in X_j),$$

which is clearly a $\Sigma^0_2$ formula when the finite sequences are identified with their natural number codes. $\Sigma^0_2 - \mathsf{IND}$ implies that there is a least such $j$. (For equivalent formulations of induction schema, see Theorem 2.4 of [8].) Call this least element $j_0$, and choose $\sigma_0$ such that $\forall \tau \supseteq \sigma_0 \; (f(\tau) \in X_{j_0})$. If $c = f(\sigma_0)$, then every $\tau \supseteq \sigma_0$ can be extended to an element $\tau'$ with $f(\tau') = c$. We can construct a monochromatic subtree order isomorphic to $2^{<\mathbb{N}}$ by the following process. Let $\sigma_0$ be the root node. Fix a level-by-level enumeration of $2^{<\mathbb{N}}$. If $\sigma$ has been added to the tree, for each $i \in \{0, 1\}$

let $\tau_i$ be the first enumerated extension of $\sigma^\frown i$ that has color $c$. Note that recursive comprehension suffices to prove the existence of this subtree. □

Imitating [3], we will call the monochromatic tree of the preceding proof the standard tree of color $c$ based at $\sigma_0$. The fact that HT is not required for the proof of TT(1) is now an easy corollary.

COROLLARY 2. $\mathsf{RCA}_0 + \mathsf{TT}(1)$ *does not prove* HT.

**Proof.** The natural numbers together with the computable sets form a model of $\mathsf{RCA}_0 + \Sigma^0_2 - \mathsf{IND}$, but not a model of arithmetical comprehension (since not all arithmetically definable sets are computable.) This model is not a model of HT, since $\mathsf{RCA}_0 + \mathsf{HT}$ implies arithmetical comprehension (Theorem 2.6 of [1]). □

Although this shows that HT is not necessary to prove TT(1), it leads us to ask if $\Sigma^0_2 - \mathsf{IND}$ is necessary to prove TT(1). We can address this question by looking at various combinatorial principles that imply TT(1) and exploring their relationships to $\Sigma^0_2 - \mathsf{IND}$.

# 1 Eventually constant tails

The core of the proof of Theorem 1 is locating a node $\sigma$ such that for every extension $\alpha$ of $\sigma$ the spectrum of colors appearing above $\alpha$ is exactly the same as the spectrum of colors appearing above $\sigma$. We can formalize the existence of such a $\sigma$ as follows:

$\mathsf{ECT}(2^{<\mathbb{N}})$: If $f : 2^{<\mathbb{N}} \to n$ then $\exists \sigma \, \forall \tau \supseteq \sigma \, \forall \alpha \supseteq \sigma \, \exists \beta \supset \alpha \, (f(\tau) = f(\beta))$.

This same principle can be expressed for colorings of $\mathbb{N}$.

$\mathsf{ECT}(\mathbb{N})$: If $f : \mathbb{N} \to n$ then $\exists b \, \forall x \geq b \, \exists y > x \, (f(x) = f(y))$.

Informally, $\mathsf{ECT}(\mathbb{N})$ asserts that there is a $b$ such that whenever $[x, \infty) \subseteq [b, \infty)$ then the range of $f$ restricted to $[x, \infty)$ is identical to the range of $f$ restricted to $[b, \infty)$. That is, we are saying that the range of $f$ on tails is eventually constant. The label ECT is a mnemonic for eventually constant tails.

$\mathsf{ECT}(2^{<\mathbb{N}})$ can be substituted for the use of $\Sigma^0_2 - \mathsf{IND}$ in the proof of Theorem 1. The next three lemmas explore the relationships between the two forms of ECT and $\Sigma^0_2 - \mathsf{IND}$, and the consequences are collected in Theorem 6.

LEMMA 3. $(\mathsf{RCA}_0) \, \mathsf{ECT}(2^{<\mathbb{N}})$ *implies* $\mathsf{ECT}(\mathbb{N})$.

**Proof.** Working in $\mathsf{RCA}_0$, fix $f : \mathbb{N} \to n$. Define $g : 2^{<\mathbb{N}} \to n$ by $g(\sigma) = f(\mathrm{lh}(\sigma))$. Apply $\mathsf{ECT}(2^{<\mathbb{N}})$ to $g$ and obtain $\sigma$. Let $b = \mathrm{lh}(\sigma)$ and choose $x \geq b$. Choose any $\tau \supseteq \sigma$ with $\mathrm{lh}(\tau) = x$. By $\mathsf{ECT}(2^{<\mathbb{N}})$, there is a $\beta \supset \tau$ such that $g(\beta) = g(\tau)$. Thus $\mathrm{lh}(\beta) > x$ and $f(\mathrm{lh}(\beta)) = g(\beta) = g(\tau) = f(x)$. □

LEMMA 4. $(\mathsf{RCA}_0)$ $\mathsf{ECT}(\mathbb{N})$ *implies* $\Sigma_2^0 - \mathsf{IND}$.

**Proof.** By Exercise 11.3.13 of [12], over $\mathsf{RCA}_0$ the scheme $\Sigma_2^0 - \mathsf{IND}$ is equivalent to the bounded $\Sigma_2^0$ comprehension scheme, which is the assertion that

$$\forall n \exists X \forall i (i \in X \leftrightarrow (i < n \land \varphi(i)))$$

where $\varphi(i)$ is a $\Sigma_2^0$ formula not containing $X$ free. We will use $\mathsf{ECT}(\mathbb{N})$ to derive bounded $\Sigma_2^0$ comprehension.

Suppose $\varphi(i)$ is $\exists x \forall y \theta(i, x, y)$, where $\theta$ is quantifier free. Fix $n$. Define $f : \mathbb{N} \to n + 1$ by writing each natural number as $mn + i$ with $i < n$ and setting

$$f(mn + i) = \begin{cases} i & \text{if } \mu x < m \ \forall y < m \ \theta(i, x, y) \text{ is not equal to} \\ & \quad \mu x < m + 1 \ \forall y < m + 1 \ \theta(i, x, y) \\ n & \text{otherwise.} \end{cases}$$

In the preceding, when $\forall x < m \ \exists y < m \ \neg \theta(i, x, y)$, we define the expression $\mu x < m \ \forall y < m \ \theta(i, x, y)$ to be equal to $m$.

Note that if there is an $x$ such that $\forall y \theta(i, x, y)$ then $\Sigma_1^0 - \mathsf{IND}$ implies there is a least such element; call it $x_0$. Applying $\mathsf{B}\Sigma_0^0$, which is also a consequence of $\Sigma_1^0 - \mathsf{IND}$ (see [8], Chapter I, section 2), we can find an $m_0$ so large that $\forall t < x_0 \ \neg \forall y < m_0 \ \theta(i, t, y)$. Then for any $m > m_0$,

$$\mu x < m \ \forall y < m \ \theta(i, x, y) = x_0 = \mu x < m + 1 \ \forall y < m + 1 \ \theta(i, x, y),$$

so $f(mn + i) = n$. Consequently, on any final segment of $[m_0 n + i, \infty)$, $i$ is not in the range of $f$. Summarizing, if $\exists x \forall y \theta(i, x, y)$, then eventually $i$ is omitted from all tails of $f$.

On the other hand, suppose $\neg \exists x \forall y \theta(i, x, y)$ and fix an element $b > 0$. If for every $m > b$ we have $\forall x < m \ \exists y < m \ \neg \theta(i, x, y)$, then for all $m > b$ we have $f(mn+i) = i$ and so eventually $i$ is in every tail. Suppose that for some $m > b$ we have $\exists x < m \ \forall y < m \ \theta(i, x, y)$. Let $x_0$ be the least element less than $m$ satisfying $\forall y < m \ \theta(i, x_0, y)$. Since $\neg \exists x \forall y \ \theta(i, x, y)$, let $y_0$ be the least element such that $\neg \theta(i, x_0, y_0)$. Then $\mu x < y_0 \forall y < y_0 \ \theta(i, x, y) = x_0$, but $\mu x < y_0 + 1 \ \forall y < y_0 + 1 \ \theta(i, x, y) > x_0$. Hence $f(y_0 n + i) = i$ for arbitrarily large values of $n$. Summarizing, if $\neg \exists x \forall y \theta(i, x, y)$ then eventually $i$ is in every tail.

Apply ECT($\mathbb{N}$) to $f$ to find a $b$ such that the range of $f$ is constant on final segments of $[b, \infty)$. By bounded $\Sigma_1^0$ comprehension (which is a consequence of $\Sigma_1^0 -$ IND [12]) the set

$$Y = \{i < n \mid \exists t > b \ f(t) = i\}$$

exists. By recursive comprehension, the set $X = \{i < n \mid i \notin Y\}$ also exists, and by the preceding paragraphs, $i \in X$ if and only if $i < n$ and $\exists x \forall y \theta(i, x, y)$, as desired. □

LEMMA 5. (RCA$_0$) $\Sigma_2^0 -$ IND *implies* ECT($2^{<\mathbb{N}}$).

**Proof.** Suppose $f : 2^{<\mathbb{N}} \to n$. Let $\langle X_i \mid i < 2^n \rangle$ enumerate the subsets of $n$ in an order that preserves containment. By the $\Sigma_2^0$ least element principle (which is equivalent over RCA$_0$ to $\Sigma_2^0 -$ IND [8]), there is a least $j$ such that we can find a node $\sigma$ so that $\forall \tau \supseteq \sigma (f(\tau) \in X_j)$. Since $j$ is the least such integer, for every $\tau \supseteq \sigma$, the spectrum of colors appearing above $\tau$ must match that above $\sigma$. □

THEOREM 6. (RCA$_0$) *The following are equivalent:*

1. ECT($2^{<\mathbb{N}}$).

2. ECT($\mathbb{N}$).

3. $\Sigma_2^0 -$ IND.

**Proof.** Immediate from Lemmas 3, 4, and 5. □

Thus, the ECT principles are essentially disguised forms of $\Sigma_2^0$ induction. This is a common attribute among many principles that imply TT(1), as shown in the next section.

## 2 A result of Corduan, Groszek, and Mileti

Doctors Corduan, Groszek, and Mileti reveal a strong connection between TT(1) and $\Sigma_2^0 -$ IND in the following conservation result, which appears in [4].

THEOREM 7. *If $\mathfrak{T}$ is any extension of RCA$_0$ by $\Pi_1^1$ axioms, then $\mathfrak{T}$ proves TT(1) if and only if $\mathfrak{T}$ proves $\Sigma_2^0 -$ IND.*

Since the usual infinite pigeonhole principle, RT(1), can be expressed as a $\Pi_1^1$ formula and is known to be equivalent to the scheme B$\Pi_1^0$ and therefore strictly weaker than $\Sigma_2^0 -$ IND, an immediate corollary of Theorem 7 is that RT(1) does not imply TT(1) over RCA$_0$. This result appears as Corollary

3.8 in [4] and provides the current best strict lower bound on the strength of TT(1).

Using Theorem 7, we can show that many informal proofs of TT(1) make use of disguised forms of $\Sigma_2^0 - \mathsf{IND}$. For example, in the next paragraph we present an alternative proof of a portion of Theorem 6. This new argument sidesteps the technical details of Lemma 4, but also does not yield information about ECT(ℕ).

**Proof.** [Alternative proof that $\mathsf{ECT}(2^{<\mathbb{N}})$ implies $\Sigma_2^0 - \mathsf{IND}$ over $\mathsf{RCA}_0$.] By Theorem 7, since $\mathsf{ECT}(2^{<\mathbb{N}})$ is a $\Pi_1^1$ sentence it suffices to prove that $\mathsf{ECT}(2^{<\mathbb{N}})$ implies TT(1) over $\mathsf{RCA}_0$. Working in $\mathsf{RCA}_0$, suppose $f : 2^{<\mathbb{N}} \to n$ and apply $\mathsf{ECT}(2^{<\mathbb{N}})$ to find a $\sigma$ such that every extension of $\sigma$ can be further extended to a node $\beta$ with $f(\beta) = f(\sigma)$. As in the proof of Theorem 1, $\mathsf{RCA}_0$ proves the existence of the standard tree of color $f(\sigma)$ based at $\sigma$. □

The following limit principle arises as a natural intermediate step in a proof of TT(1) from stable Ramsey's Theorem for pairs (denoted $\mathsf{SRT}^2$).

**L:** Given $f : \mathbb{N}^2 \to a$ such that $\lim_n f(x,n)$ exists for every $x \in \mathbb{N}$, there is a least $b$ such that for some $x$, $\lim_n f(x,n) = b$.

Let $\mathsf{L}^+$ denote the stronger version of L resulting from replacing "every" in the hypothesis by the word "some." Rather than deducing L from $\mathsf{SRT}^2$, we will prove $\mathsf{L}^+$ from $\Sigma_2^0 - \mathsf{IND}$. This is sharper, since $\mathsf{SRT}^2$ is strictly stronger than $\Sigma_2^0 - \mathsf{IND}$ [2].

LEMMA 8. ($\mathsf{RCA}_0$) $\Sigma_2^0 - \mathsf{IND}$ *implies* $\mathsf{L}^+$. *Consequently,* $\Sigma_2^0 - \mathsf{IND}$ *also proves* L.

**Proof.** Suppose $f : \mathbb{N}^2 \to a$ and $\lim_n f(x,n)$ exists for some $x \in \mathbb{N}$. Thus there is a $b < a$ such that $\exists x \exists t \forall n (n > t \to f(x,n) = b)$. By the $\Sigma_2^0$ least element principle, which is equivalent to $\Sigma_2^0 - \mathsf{IND}$, there is a least such $b$. Thus $\mathsf{L}^+$ holds. Since predicate calculus proves that $\mathsf{L}^+$ implies L, the last sentence of the lemma follows immediately. □

By proving that L implies TT(1), we create an opportunity for applying Theorem 7.

LEMMA 9. ($\mathsf{RCA}_0$) L *implies* TT(1).

**Proof.** Let $f : 2^{<\mathbb{N}} \to a$ and let $\langle \sigma_i \rangle_{i \in \mathbb{N}}$ be an enumeration of $2^{<\mathbb{N}}$. Let $\langle Y_i \rangle_{i < 2^a}$ be an enumeration of the power set of $\{0, 1, \ldots a-1\}$ such that $Y_i \subseteq Y_j$ implies $i \leq j$. Define $g : \mathbb{N}^2 \to 2^a$ by

$$g(m,n) = \mu t(\{f(\tau) \mid \tau \supseteq \sigma_m \wedge \mathrm{lh}(\tau) \leq n\} = Y_t).$$

Note that recursive comprehension suffices to prove the existence of $g$ and that for fixed $m$, the function $g(m,n)$ is increasing. By the $\Pi_1^0$ least element principle (which is provable in $\mathsf{RCA}_0$), for each $m$ there is a least upper bound on the range of $g(m,n)$. Consequently, for each $m$, $\lim_n g(m,n)$ exists. By L, there is a least $b$ and a $\sigma_m \in 2^{<\mathbb{N}}$ such that $\lim_n g(m,n) = b$. Since $b$ is least, for every $\sigma_j$ extending $\sigma_m$, we have $\lim_n g(j,n) = b$ also. In particular, every node extending $\sigma_m$ can be extended to a node of color $f(\sigma_m)$. Thus, the standard tree of color $f(\sigma_m)$ extending $\sigma_m$ (as constructed in the proof of Theorem 1) is isomorphic to $2^{<\mathbb{N}}$. □

Combining Theorem 7 with the preceding lemmas, we see that L and L$^+$ are disguised forms of $\Sigma_2^0 - \mathsf{IND}$.

COROLLARY 10. ($\mathsf{RCA}_0$) *The following are equivalent:*

1. $\Sigma_2^0 - \mathsf{IND}$.

2. L$^+$.

3. L.

**Proof.** Lemma 8 shows that over $\mathsf{RCA}_0$, item 1 implies item 2, and item 2 implies item 3. Since L is $\Pi_1^1$, Theorem 7 applied to Lemma 9 shows that $\mathsf{RCA}_0$ proves that item 3 implies item 1. □

We close the section by noting that any proof of $\mathsf{TT}(1)$ relying on the standard tree construction from the proof of Theorem 1 inherently uses $\Sigma_2^0 - \mathsf{IND}$.

THEOREM 11. ($\mathsf{RCA}_0$) *The following are equivalent:*

1. $\Sigma_2^0 - \mathsf{IND}$.

2. If $f : 2^{<\mathbb{N}} \to a$ then there is a node $\sigma$ such that the standard tree of color $f(\sigma)$ based at $\sigma$ is isomorphic to $2^{<\mathbb{N}}$.

**Proof.** The proof of Theorem 1 shows that item 1 implies item 2. To prove the reversal, note that over $\mathsf{RCA}_0$, item 2 is equivalent to the statement "if $f : 2^{<\mathbb{N}} \to a$ then there is a node $\sigma$ such that for every $n$ there is a stage $t$ such that the algorithm for constructing the standard tree of color $f(\sigma)$ based at $\sigma$ halts and produces a tree containing an initial segment order isomorphic to the full binary tree of height $n$." Since this statement is $\Pi_1^1$ and implies $\mathsf{TT}(1)$ over $\mathsf{RCA}_0$, by Theorem 7 it implies $\Sigma_2^0 - \mathsf{IND}$. Thus, item 2 implies item 1 over $\mathsf{RCA}_0$. □

## 3  Null stable colorings

In this section, we present one more disguised form of induction, a related though weaker combinatorial principle that is actually equivalent to TT(1), and a proof of TT(1) from a stable version of Ramsey's theorem for pairs in trees. All these results depend on the following notion. Suppose we have $f : 2^{<\mathbb{N}} \to 2$. We say that $f$ is a *null stable* coloring if for every $\sigma \in 2^{<\mathbb{N}}$ there is a $\sigma' \supseteq \sigma$ such that for every $\tau \supseteq \sigma'$ we have that $f(\tau) = 0$. Note that by definition, every null stable coloring is a 2-coloring. Intuitively, $f$ is null stable if above each node we can find a node above which $f$ is constantly 0. Given a finite sequence of null stable colorings, we could sequentially apply the definition of null stable for each coloring and eventually arrive at a node which is colored 0 for all of the colorings. The next theorem shows that the existence of such a node is a disguised form of $\Sigma^0_2 - \mathsf{IND}$.

THEOREM 12. (RCA$_0$) *The following are equivalent:*

1. $\Sigma^0_2 - \mathsf{IND}$.

2. *Suppose $\langle f_i \rangle_{i<n}$ is a sequence of null stable colorings of $2^{<\mathbb{N}}$. Then there is a $\sigma \in 2^{<\mathbb{N}}$ such that $f_i(\sigma) = 0$ for every $i < n$.*

**Proof.** We work in RCA$_0$. First, assume $\Sigma^0_2 - \mathsf{IND}$ and suppose $\langle f_i \rangle_{i<n}$ is a sequence of null stable 2-colorings. Define $f : 2^{<\mathbb{N}} \to 2^n$ by $f(\sigma) = \sum_{i<n} f_i(\sigma) \cdot 2^i$. By Lemma 5 and $\Sigma^0_2 - \mathsf{IND}$, we may apply $\mathsf{ECT}(2^{<\mathbb{N}})$ to $f$ and find a $\sigma_0$ such that $\forall \alpha \supseteq \sigma_0 \, \exists \beta \supseteq \alpha \, f(\sigma_0) = f(\beta)$. Suppose, by way of contradiction, that $f(\sigma_0) \neq 0$. Fix $i < n$ such that $f_i(\sigma_0) = 1$. Since $f_i$ is null stable, we can find an $\alpha_0 \supseteq \sigma_0$ such that for all $\beta \supseteq \alpha_0$, $f_i(\beta) = 0$. Since $\sigma_0$ was chosen using $\mathsf{ECT}(2^{<\mathbb{N}})$, for some $\beta_0 \supset \alpha_0$, $f(\sigma_0) = f(\beta_0)$. However, $f_i(\sigma_0) = 1$ and $f_i(\beta_0) = 0$, so $f(\sigma_0) \neq f(\beta_0)$, yielding the desired contradiction. Thus we must have $f(\sigma_0) = 0$. Since $f(\sigma_0) = 0$, $f_i(\sigma_0) = 0$ for every $i < n$.

To prove that item 2 implies item 1, we will use item 2 to deduce TT(1) and apply Theorem 7. Assume item 2 and let $f : 2^{<\mathbb{N}} \to n$. Define the sequence $\langle f_i \rangle_{i<n}$ by setting $f_i(\sigma) = 1$ if $f(\sigma) = i$ and $f_i(\sigma) = 0$ if $f(\sigma) \neq i$. If each $f_i$ was null stable, then by item 2 we could locate a $\sigma$ such that $f_i(\sigma) = 0$ for all $i < n$, contradicting the fact that $f(\sigma) = i$ for some $i < n$. Thus there is an $i_0$ such that $f_{i_0}$ is not null stable. For this $i_0$, we can locate a $\sigma_0$ such that for every $\sigma' \supseteq \sigma_0$ there is a $\tau \supseteq \sigma'$ such that $f_{i_0}(\tau) = 1$. Choose any $\tau \supseteq \sigma_0$ with $f_{i_0}(\tau) = 1$. Then $f(\tau) = i_0$ and every node extending $\tau$ has an extension of color $i_0$. Consequently the standard tree for $f$ of color $i_0$ based at $\tau$ witnesses TT(1) for $f$. Since item 2 is $\Pi^1_1$, $\Sigma^0_2 - \mathsf{IND}$ follows by Theorem 7. (We could substitute a use of Theorem 11 for Theorem 7 here, if we liked.) □

The construction in the preceding proof of a single coloring from many 2-colorings suggests a way to exchange one application of TT(1) for many colors for many simultaneous applications of TT(1) restricted to 2-colorings. In fact, we will see that the corresponding formulations are provably equivalent. This is of some interest, since TT(1) restricted to any standard number of colors is a theorem of $\mathsf{RCA}_0$. The next theorem capitalizes on this notion, and even restricts the simultaneous applications to TT(1) for null stable 2-colorings, which for single applications is very clearly a theorem of $\mathsf{RCA}_0$.

THEOREM 13. ($\mathsf{RCA}_0$) *The following are equivalent:*

1. TT(1).

2. *Suppose that for each $i < n$, $f_i : 2^{<\mathbb{N}} \to 2$. Then there is a subtree $S \subseteq 2^{<\mathbb{N}}$ order isomorphic to $2^{<\mathbb{N}}$ such that for each $i < n$, $f_i$ is constant on $S$.*

3. *Item 2 holds in the case where each $f_i$ is null stable.*

**Proof.** Assuming $\mathsf{RCA}_0$, item 2 can be proved by applying TT(1) to the function $f$ constructed as in the first paragraph of the proof of Theorem 12. Since item 3 is a restricted form of item 2, it remains only to prove TT(1) from item 3.

Suppose $f : 2^{<\mathbb{N}} \to n$. Construct $f_i$ for $i < n$ as in the reversal for Theorem 12. If one of $f_i$ functions is not null stable, then the standard tree construction as in the proof of the reversal of Theorem 12 witnesses TT(1) for $f$, completing the proof. If all of the $f_i$ functions are null stable, then we may apply item 3 to find a subtree $S$ isomorphic to $2^{<\mathbb{N}}$ and monochromatic for all the $f_i$ functions. Let $\sigma$ be the root node of $S$. Now $f(\sigma) = i_0$ for exactly one $i_0 < n$, and for every $\tau$ in $S$, $f_i(\tau) = 1$ if and only if $i = i_0$. Consequently, $f(\tau) = i_0$ for every $\tau \in S$, so $S$ witnesses TT(1) for $f$. □

Using the ideas from this section, we close by proving TT(1) from a stable version of Ramsey's theorem in trees for pairs and two colors. This is a tree analog of the proof of RT(1) from $\mathsf{SRT}^2$ in [2].

We adopt the notation from [5] for the following. A function on pairs of comparable tree nodes $f : [2^{<\mathbb{N}}]^2 \to k$ is said to be 3-*stable* if for each $\sigma \in 2^{<\mathbb{N}}$ there is a $c < k$ such that for every $\sigma' \supseteq \sigma$ there exists $\tau \supseteq \sigma'$ with $f(\sigma, \rho) = c$ for all $\rho \supseteq \tau$. The principle $\mathsf{S}^3\mathsf{TT}^2_2$ asserts that every 3-stable two coloring of comparable pairs from $2^{<\mathbb{N}}$ has a monochromatic subtree isomorphic to $2^{<\mathbb{N}}$.

THEOREM 14. ($\mathsf{RCA}_0$) $\mathsf{S}^3\mathsf{TT}^2_2$ *implies* TT(1).

**Proof.** Assume $\mathsf{RCA}_0$ and suppose $f : 2^{<\mathbb{N}} \to n$. Define $g : [2^{<\mathbb{N}}]^2 \to 2$ for $\sigma \subset \tau$ by setting $g(\sigma, \tau) = 1$ if and only if $f(\sigma) = f(\tau)$. If $\sigma$ witnesses that $g$ is not 3-stable, then the standard tree for $f$ based at $\sigma$ with color $f(\sigma)$ witnesses $\mathsf{TT}(1)$ for $f$. If $g$ is 3-stable, we may apply $\mathsf{S}^3\mathsf{TT}_2^2$ to find a monochromatic subtree $S$ for $g$. Select a sequence of $n+1$ comparable nodes in $T$. Some pair in this sequence must be colored identically by $f$. Thus $g([S]^2) \equiv 1$, and $S$ is a monochromatic subtree for $f$. □

It is not known whether or not $\mathsf{SRT}_2^2$ implies $\Sigma_2^0 - \mathsf{IND}$ [2]. Similarly, it is not known whether or not $\mathsf{S}^3\mathsf{TT}_2^2$ implies $\Sigma_2^0 - \mathsf{IND}$. Since $\mathsf{S}^3\mathsf{TT}_2^2$ is a $\Pi_2^1$ sentence, Theorem 7 is not applicable. Thus, Theorem 14 is a candidate for a proof of $\mathsf{TT}(1)$ that does not rely on a disguised use of $\Sigma_2^0 - \mathsf{IND}$.

## Acknowledgements

This publication was made possible in part through the support of a grant (ID# 20800) from the John Templeton Foundation. The opinions expressed in this publication are those of the author and do not necessarily reflect the views of the John Templeton Foundation.

## Bibliography

[1] Andreas R. Blass, Jeffry L. Hirst, and Stephen G. Simpson, *Logical analysis of some theorems of combinatorics and topological dynamics*, Logic and combinatorics (Arcata, Calif., 1985), Contemp. Math., vol. 65, Amer. Math. Soc., Providence, RI, 1987, pp. 125–156.

[2] Peter A. Cholak, Carl G. Jockusch, and Theodore A. Slaman, *On the strength of Ramsey's theorem for pairs*, J. Symbolic Logic **66** (2001), no. 1, 1–55, DOI 10.2307/2694910.

[3] Jennifer Chubb, Jeffry L. Hirst, and Timothy H. McNicholl, *Reverse mathematics, computability, and partitions of trees*, J. Symbolic Logic **74** (2009), no. 1, 201–215, DOI 10.2178/jsl/1231082309.

[4] Jared R. Corduan, Marcia J. Groszek, and Joseph R. Mileti, *A note on reverse mathematics and partitions of trees*, J. Symbolic Logic. To appear.

[5] Damir D. Dzhafarov, Jeffry L. Hirst, and Tamara J. Lakins, *Ramsey's theorem for trees: the polarized tree theorem and notions of stability*, Arch. Math. Logic **49** (2010), no. 3, 399–415, DOI 10.1007/s00153-010-0179-6.

[6] Harvey Friedman, *Some systems of second order arithmetic and their use*, Proceedings of the International Congress of Mathematicians (Vancouver, B. C., 1974), Vol. 1, Canad. Math. Congress, Montreal, Que., 1975, pp. 235–242.

[7] Harvey Friedman, *Abstracts: Systems of second order arithmetic with restricted induction, I and II*, J. Symbolic Logic **41** (1976), no. 2, 557–559.

[8] Petr Hájek and Pavel Pudlák, *Metamathematics of first-order arithmetic*, Perspectives in Mathematical Logic, Springer-Verlag, Berlin, 1998. Second printing.

[9] Neil Hindman, *Finite sums from sequences within cells of a partition of N*, J. Combinatorial Theory Ser. A **17** (1974), 1–11, DOI 10.1016/0097-3165(74)90023-5.

[10] Jeffry L. Hirst, *Combinatorics in subsystems of second order arithmetic*, Ph.D. Thesis, The Pennsylvania State University, 1987.

[11] Timothy H. McNicholl. Private communication.

[12] Stephen G. Simpson, *Subsystems of second order arithmetic*, 2nd ed., Perspectives in Logic, Cambridge University Press, Cambridge, 2009.

# Friedman and the Axiomatization of Kripke's Theory of Truth

JOHN P. BURGESS

ABSTRACT. What is the simplest and most natural axiomatic replacement for the set-theoretic definition of the minimal fixed point on the Kleene scheme in Kripke's theory of truth? What is the simplest and most natural set of axioms and rules for truth whose adoption by a subject who had never heard the word "true" before would give that subject an understanding of truth for which the minimal fixed point on the Kleene scheme would be a good model? Several axiomatic systems, old and new, are examined and evaluated as candidate answers to these questions, with results of Harvey Friedman playing a significant role in the examination.

## 1

Small though it is, the area of logic concerned with axiomatic theories of truth is large enough to have two distinguishable sides. These go back to contrasting early reactions of two eminent logicians to Saul Kripke's "Outline of a Theory of Truth" [1975]. One side originates with Harvey Friedman, who first wrote Kripke in the year of the publication of the "Outline", but whose published contributions are contained in a joint paper with Michael Sheard from over a decade later, Friedman and Sheard [1987]. (There was also a sequel, Friedman and Sheard [1988], but I will not be discussing it.) The questions raised in that paper are these: First, which combinations of naive assumptions about the truth predicate are consistent? Second, what are the proof-theoretic strengths of the consistent combinations?

In the Friedman-Sheard paper, combinations of items from a menu of a dozen principles are added to a fixed base theory that includes first-order Peano arithmetic PA. A variety of model constructions are presented to show various combinations consistent, and a number of deductions to show various other combinations inconsistent, and complete charts of the status of all combinations worked out. There turn out to be nine maximal consistent sets.

In a portion attributed in the paper to Friedman alone (§7), two sample results on proof-theoretic strength are presented, showing one combination very weak and another very strong. Later additional results on proof-theoretic strength were obtained by a number of workers, and most recently Graham Leigh and Michael Rathjen [forthcoming] have finished the job, so that we now have a complete determination of the proof-theoretic strengths of all nine maximal consistent sets.

Though the questions addressed in Friedman's work are purely mathematical, and the paper with Sheard explicitly declares its philosophical neutrality, the notion of truth is so philosophically fraught that one naturally expects some of the formal results will turn out to have some bearing on questions of interest to philosophers. This expectation is not disappointed, and I will be making use of Friedman's proofs of both his sample results in the course of this paper.

## 2

I follow the example of Friedman and Sheard by describing in advance the base language and theory to be considered, and in listing and naming the various candidate principles of truth. (See the table of "Principles of Truth".) The base language will be that of arithmetic with a truth predicate T. Formulas not involving the new predicate are called *arithmetical*. Sometimes it will be convenient to have also a falsehood predicate F, where *falsehood* is truth of the negation (as *denial* is assertion of the negation, and *refutation* is proof of the negation), while the negation of truth is *untruth*. F need not be thought of as a primitive but may be thought of as defined. (Some truth principles that are nontrivial when it is taken as primitive become trivial when it is taken as defined.) $T(x)$ literally means "$x$ is the code number for a true sentence". The coding of sentences and formulas may as usual be carried out so that simple syntactic operations on sentences and formulas correspond to primitive recursive functions on their code numbers. I write $T[A]$ to mean $T(a)$, where $a$ is the numeral for the code number of $A$. Otherwise I follow the relaxed attitude towards notation in Sheard's "Guide to Truth Theories" [1994], and in consequence can and will only give sketches of proofs of some results (since fully rigorous and detailed proofs would require more pedantic correctness about notational matters).

The base theory will be first-order Peano arithmetic PA, with the understanding that when new predicates are added to the language, the instances of the scheme of mathematical induction for formulas involving them are added as well. The underlying logic will be classical, and where it makes a difference it may be assumed that the deduction system for classical logic is one in which proofs do not involve open formulas, and the only rule is modus ponens. Even in weak subtheories of PA, notions of *correctness* and *erroneousness* can be defined for atomic arithmetic sentences, which are equations between closed terms, and proved to have the properties one would expect for truth and falsehood restricted to such sentences. And even in such weak subtheories, construction of self-referential examples is possible by the usual diagonal procedure. These include *truth-tellers*, asserting their own truth, and two kinds of *liars*, namely, *falsehood-tellers* asserting their own falsehood, and *untruth-tellers*, asserting their own untruth. (More precisely, a truth-teller is a sentence saying that any number satisfying a certain arithmetical condition is the code number of a true sentence, where provably the one and only number satisfying the condition is the code number of that very sentence; and analogously for liars. Thus what a truth-teller literally *is*, is a sentence $\forall z(\sim C(z) \vee \mathsf{T}(z))$ with code $b$, where $C$ is arithmetical and where $C(b)$ and $\forall n(n = b \vee \sim C(n))$ are provable in PA.) Unlike Friedman and Sheard I will not count any truth principles — they count truth distribution and truth classicism — as part of the base theory. Comments on some individual principles will be in order.

As to the four rules, these are, like the rule of necessitation in modal logic, to be applied only in categorical demonstrations, not hypothetical deductions. For instance, with truth introduction, if we have *proved* that $A$, we may infer "$A$ is true". If we have merely deduced $A$ from some hypothesis, we may not infer "$A$ is true" under that hypothesis. Allowing introduction or elimination to be used hypothetically would amount to adopting truth appearance and disappearance, and hence truth transparency, as axiom schemes applicable to all sentences, and that would be inconsistent. Indeed, the usual reasoning in the liar paradox shows that allowing either one of introduction or elimination to be used hypothetically, while allowing the other to be used at least categorically, leads to contradiction.

As to the axioms and schemes, the composition and decomposition axioms, even without those for atomic truth and falsehood, imply truth transparency for *arithmetical* sentences and formulas, arguing by induction on logical complexity of the sentence or formula in question. With composition and decomposition for atomic truth and falsehood as well, truth transparency extends to *truth-positive* sentences and formulas, those built up from arithmetical formulas and atomic formulas involving the new predicates by conjunction, disjunction, and quantification. With the further addition of truth consistency, one would get truth distribution and truth disappearance for all formulas.

Table 1. Principles of Truth

| RULES | |
|---|---|
| Truth Introduction | from a sentence to infer the truth of the sentence |
| Truth Elimination | from the truth of a sentence to infer the sentence |
| Untruth Introduction | from the negation of a sentence to infer the untruth of the sentence |
| Untruth Elimination | from the untruth of a sentence to infer the negation of the sentence |

| AXIOMS | |
|---|---|
| Truth Consistency | No sentence is both truth and false. |
| Truth Completeness | Every sentence is either true or false. |
| Positive Composition | |
|   Equational | If an equation is correct, it is true. |
|   Negational | If a sentence is false, its negation is true. |
|   Conjunctive | If two sentences are true, their conjunction is true. |
|   Disjunctive | *analogous* |
|   Universalizing | *analogous* |
|   Existentializing | *analogous* |
|   Atomic Truth | If a sentence is true, it is true that it is true |
|   Atomic Falsehood | If a sentence if false, it is true that it is false |
| Negative Composition | *correlates for falsehood of Positive Composition* |
|   e.g. Conjunctive | If either of two sentences is false, their conjunction is false. |
| Positive Decomposition | *converses of Positive Composition* |
|   e.g. Conjunctive | If a conjunction of two sentences is true, both of them are true. |
| Negative Decomposition | *converses of Negative Composition* |
|   e.g. Conjunctive | If a conjunction of two sentences if false, at least one of them is false. |
| Truth Distribution | If a conditional and its antecedent are true, so is the consequent. |
| Truth Classicism | Any instance of excluded middle (or other classical law) is true. |

| SCHEMES | |
|---|---|
| Truth Transparency | |
|   Sentential | $A$ is true iff $A$ |
|   Formulaic | For all $n$, $A(n)$ is true iff $A(n)$ |
| Truth Appearance | |
|   Sentential | If $A$, then $A$ is true. |
|   Formulaic | For all $n$, if $A(n)$, then $A(n)$ is true |
| Truth Disappearance | |
|   Sentential | If $A$ is true, then $A$. |
|   Formulaic | For all $n$, if $A(n)$ is true, then $A(n)$ |

| THEORIES | | STRENGTH |
|---|---|---|
| PA* | truth rules | PA |
| FS | truth rules, consistency & completeness, composition & decomposition except atomic truth & falsehood | $RA(<\omega)$ |
| KF | all composition and decomposition | $RA(<\varepsilon_0)$ |
| KF⁺ | KF + consistency | $RA(<\varepsilon_0)$ |
| KHH = PKF | partial logic variant of KF | $RA(<\omega^\omega)$ |
| KF$\mu$ | all composition + minimality | $ID_1$ |

# 3

The other side of axiomatic truth theory originates with Solomon Feferman. The background here is his well-known work on predicative analysis (Feferman [1964]). The idea of predicative analysis is that one starts with the natural numbers, and then considers a first round of sets of natural numbers defined by formulas involving quantification only over natural numbers, and then considers a second round of sets of natural numbers defined by formulas involving quantification only over natural numbers and sets of the first round, and so on. The process can be iterated into the transfinite, up to what has come to be called the Feferman-Schütte ordinal $\Gamma_0$.

Instead of considering round after round of sets, those of each round defined in terms of those of earlier rounds, one could consider instead round after round of satisfaction predicates, each applying only to formulas involving only earlier ones. Instead of speaking of definable sets and elementhood one would speak of defining formulas and satisfaction. But in arithmetic formulas can be coded by numbers, and the notion of the satisfaction of a formula by a number reduced to that of the truth of sentence obtained by substituting the numeral for the number for the variable in the formula. So in the end all that is really needed is round after round of truth predicates, each applicable only to sentences containing only earlier ones. Feferman [1991] finds that the process iterates only up to the ordinal $\varepsilon_0$, though by introducing what he calls "schematic" theories it can be extended up to $\Gamma_0$.

Kripke gives a set-theoretic construction of a model for a language with a self-applicable truth predicate, and this raises the question whether the hierarchy of truth predicates could be replaced by a single self-applicable one. To pursue this possibility it would be necessary to replace the set-theoretic construction of a model by an axiomatic theory. Thus arose the question of *axiomatizing Kripke's theory of truth*.

Feferman proposed a candidate axiomatization (which became known from citations of his work in the literature well before its publication in Feferman [1991]) with all the composition and decomposition axioms. In the literature the label KF (for Kripke-Feferman) is sometimes used for this theory, as it will be here, but is sometimes used for this theory plus truth consistency, which here will be called $KF^+$. Later Volker Halbach and Leon Horsten [2006] produced a variant of KF based on partial logic, which they called PKF but which I will call KHH. They give a sequent-calculus formulation, but a natural deduction formulation will be given in a book by Horsten [2011].

## 4

This past semester an undergraduate philosophy major at my school, Dylan Byron, asked me to direct him in a reading course on the literature on axiomatic theories of truth. Over the semester he expressed increasing disappointment at the scarcity in the literature of articulations of just what the philosophical aims and claims of axiomatic truth theories are supposed to be, and hearing his complaints I became convinced that there was a need for more philosophical discussion of just what is meant by "an axiomatization of Kripke's theory of truth".

There are at least three potential sources of ambiguity, two generally recognized and the other perhaps not. To begin with, Kripke has not just one construction, but several, differing in two dimensions. On the one hand, one can choose among different underlying logical schemes: the Kleene trivalent scheme, the van Fraassen supervaluation scheme, and others. On the other hand, for any given scheme, one can choose among different fixed points: the minimal one, the intersection of all maximal ones, and others. The multiplicity of fixed-points is what allows Kripke to distinguish the outright paradoxical examples like liar sentences from merely ungrounded examples like truth-teller sentences, the former being true in no fixed points, the latter in some but not others. These two sources of ambiguity in the notion of "Kripke's theory of truth" are generally recognized. It is the minimal fixed point on the Kleene scheme that has received the most attention, from Kripke's original paper to the present day — I set aside work of Andrea Cantini [1990] on the van Fraassen scheme — and I will concentrate on it.

Beyond this, though it would be difficult to overstate how guarded are Kripke's philosophical formulations in his "Outline", one passage does suggest that there may be two levels or stages of understanding the concept of truth, earlier and later:

> If we think of the minimal fixed point, say under the Kleene valuation, as giving a model of natural language, then the sense in which we can say, in natural language, that a Liar sentence is not true must be thought of as associated with some later stage in the development of natural language, one in which speakers reflect on the generation process leading to the minimal fixed point. It is not itself a part of that process. (Kripke [1975], 714)

Thus there is a further ambiguity in the notion of "axiomatizing Kripke's theory of truth", and a need to distinguish the problem of codifying in axioms a prereflective understanding of truth from the problem of doing the same for a post-reflective understanding.

## 5

Early in Kripke's exposition of his proposal (§III of Kripke [1975]), he invites us to join him in imagining trying to explain the meaning of "true" to someone who does not yet understand it. Herein lies what for me is a crucial question for the problem of axiomatizing the earlier, pre-reflective understanding, which I would state as follows:

> *Internal Axiomatization.* What is the simplest and most natural set of axioms and rules whose adoption by a subject who had never heard the word "true" before would give that subject an understanding of truth for which the minimal fixed point on the Kleene scheme would be a good model?

If we had an answer to this question, the question whether the minimal fixed point on the Kleene scheme really provides a good "model of natural language" would largely reduce to the question whether it is plausible to suggest that speakers of natural language first acquire an understanding of truth by adopting something like the indicated system of axioms and rules. Needless to say, the notion of "good model" here is an intuitive, not a rigorously defined one.

The internal axiomatization question is essentially the question of what we would have to tell a subject who had never heard the word "true" before to help him acquire a pre-reflective understanding of Kripkean truth. One might be inclined to think, "We could just tell *him* what Kripke tells *us*." But Kripke, as he repeatedly emphasizes, is speaking to *us* in a metalanguage, describing his fixed points from the outside, saying things that cannot be said in the object language, or recognized as true from the inside. Kripke says, for instance, that neither untruth-teller sentences nor truth-teller sentences are true, thus asserting what an untruth-teller sentence asserts and denying what a truth-teller sentence asserts. If we told the subject what Kripke tells *us*, we'd be skipping right over the pre-reflective to the post-reflective stage.

The problem of axiomatizing the later, *post*-reflective understanding, is a separate problem, which I would state as follows:

> *External Axiomatization.* What is the simplest and most natural axiomatic replacement for Kripke's set-theoretic definition of the minimal fixed point on the Kleene scheme?

The notion of "simplest and most natural axiomatic replacement" is no more rigorously defined than that of "good model", but this does not mean that we cannot recognize examples when we see them. A paradigm would be PA itself, arguably the simplest and most natural set axiomatic replacement for the set-theoretic definition of the natural numbers as the elements of the smallest set containing zero and closed under successor.

# 6

Beginning with the internal question, let us return to Kripke's discussion of the subject being taught the meaning of "true" (Kripke [1975], 701). Kripke supposes the subject has knowledge of various empirical facts: for instance, meteorological facts, such the fact that snow is white, and historical facts about what is said in what texts, perhaps the fact that "Snow is white" appeared in the *New York Times* on such-and-such a date. But the subject has initially no knowledge about truth. Kripke then imagines us telling the subject "that we are entitled to assert (or deny) of any sentence that it is true precisely under the circumstances when we can assert (or deny) the sentence itself", which I take to amount to giving him the four categorical rules of inference in the table.

Kripke then explains how his subject, having already been in a position to assert "Snow is white", is now in a position to assert "'Snow is white' is true", and how, having already been also in a position to assert "'Snow is white' appears in the *New York Times* of such-and-such a date", he is now in a position to infer and assert "Some true sentence appears in the *New York Times* of such-and-such a date". Kripke concludes "In this manner, the subject will eventually be able to attribute truth to more and more statements involving the notion of truth itself."

Kripke's discussion can be adapted to the situation where the base theory to which the truth predicate is being added is PA. We suppose the subject initially knows and speaks of nothing but numbers and their arithmetical properties, and of sentences and their syntactic properties insofar as statements about the latter can be coded as statements about the former. Now suppose we introduce a truth predicate and give the subject the four categorical rules in the table. Let us call the resulting theory PA*.

Then what Kripke said about "Snow is white" and "...appears in the *New York Times* of such-and-such a date" applies to, say, "Seventeen is prime" and "...is provable in Robinson arithmetic Q". The subject will be able to assert — the theory PA* will be able to prove — that Robinson arithmetic proves some true sentence, and beyond that "more and more statements involving the notion of truth itself".

# 7

The paper of Friedman and Sheard contains information about the scope and limits of what PA* can prove. In the first place, it can't prove contradictions: it is consistent, as is the theory, now called FS (for Friedman-Sheard), which adds truth consistency and completeness, and the composition and decomposition axioms except those for atomic truth and falsehood. Consistency is proved by a model-theoretic construction that represents an independent discovery of the principle of "revision" theories of truth.

To recall how revision works in a fairly general form (as in various works of Anil Gupta and Hans Herzberger), we can construct a sequence of models, indexed by ordinals, each of which consists of the standard model of arithmetic plus an assignment of an extension to the truth predicate. At stage zero the truth predicate may be assigned any extension we please. At stage one its extension consists of the (code numbers for) sentences that are true in Tarski's sense at stage zero. Stage two is obtained from stage one as stage one was from stage zero, and so on. At stage $\omega$, anything that has always been true from some point on is put in the extension, and anything that has always been false from some point on is left out of the extension. Other sentences are put in or left out according as they were put in or left out originally, at stage zero. And so on.

The consistency proof in the paper of Friedman and Sheard involves only the finite stages. One considers the set that contains a sentence $A$ just in case $A$ has always been true from some point on. This set is closed under logical consequence and under the four categorical truth rules, and contains all the axioms and theorems of PA* and indeed of FS, but does not contain $0 = 1$. In the section of the paper on proof-theoretic strength, a refinement of the method of the consistency proof is used to show that PA* is a *conservative extension* of PA. It proves no new arithmetical sentences. (Indeed, this is proved for PA* plus the axioms of truth consistency and truth completeness.)

The revision method can be adapted to show that PA* by itself does not imply various additional axioms of FS. For instance, let $L$ be a liar sentence. Start at stage zero with $L$ in the extension of the truth predicate and its conjunction with itself, $L$ & $L$, out of it. Continue the construction through all the ordinals less than $\omega^2$. Consider the set of sentences that have always been true from some point on. These will not contain the sentence "if $L$ is true then $L$ & $L$ is true", because this will fail at stage zero and at every subsequent limit stage, and positive conjunctive composition therefore fails. Something similar can be done for truth consistency and completeness and the composition and decomposition axioms for the connectives and quantifiers.

## 8

PA*, though suggested by a literalistic reading of some of Kripke's initial heuristic remarks, does not correspond very directly to Kripke's eventual set-theoretic construction of the minimal fixed point on the Kleene scheme, and this for a double reason.

First, PA*, like all the systems considered by Friedman and Sheard, is based on classical logic, and has every instance of truth classicism as a theorem, including for example $\mathsf{T}[L \vee \sim L]$ where $L$ is a liar sentence. Kripke represents himself as adhering throughout to classical *propositional* logic, but allowing that departures from classical *sentential* logic may be needed if our language contains sentences that do not express propositions (as departures from classical sentential logic would also be needed if our language contained ambiguous sentences expressing multiple propositions). Even where there are sentences that do not express propositions, use of classical logic will be appropriate if one follows the van Fraassen scheme; but on the Kleene scheme, $\mathsf{T}[L \vee \sim L]$, where $L$ is a liar sentence, does *not* hold in any fixed point, and the internal axiomatization question as I formulated it was a question about the Kleene scheme.

Presumably this first problem could be resolved by replacing PA* with a version pPA* based on partial logic. And in any case it is of interest to consider the internal axiomatization question for the van Fraassen scheme.

Second, however, even considering the question for the van Fraassen scheme, the minimal fixed point does not provide a good model for PA*. Though as acknowledged earlier, the notion of "good model" is not a rigorously defined one, that does not prevent us from recognizing that the minimal fixed isn't one, simply because it is far more complicated than is needed. We get a model of PA* already even if we only carry out the finite stages of Kripke's inductive construction. Even the simplest and most natural examples of sentences that don't get evaluated as true or false until some transfinite stages turn out to be neither provable nor refutable PA*. For example, if we let $\tau_0, \tau_1, \tau_2, \ldots$, be the sentences $0 = 0, \mathsf{T}[0 = 0], \mathsf{T}[\mathsf{T}[0 = 0]], \ldots, \tau_\omega$ be the sentence saying that all the $\tau_n$ are true, then PA* cannot prove $\tau_\omega$. (It fails in the model used to prove consistency.)

Thus we have not found in the considerations advanced so far an answer to the internal axiomatization question, the question what to tell the subject who does not know the meaning of "true". Telling him everything Kripke tells us is too much, and telling him just the four categorical rules is not enough.

# 9

At this point, a fantasy suggests itself. Suppose that instead of starting with a human being and giving *him* the four categorical truth rules, we start with a Superhuman Being, and give *Her* those rules. We suppose She has enormous cognitive abilities about all matters *not* involving the notion of truth, which in the test case of arithmetic might be represented by the ability to draw inferences using the omega rule.

For any arithmetic sentence $A$ that is true in Tarski's sense, She can prove it using the $\omega$-rule, and She can then infer "$A$ is true". If $A$ is unprovable in PA, then the arithmetical sentence saying so will be true in Tarski's sense, so She can prove that, too, using the $\omega$-rule. So She can prove "Some true arithmetical statement is not provable in PA", and so on.

Formally we might represent Her by a theory PA$\omega$* consisting of Peano arithmetic plus the omega rule and the four categorical truth rules. It is not too hard to see (using the basic result about $\omega$-logic that a sentence follows by $\omega$-logic from a set of first-order sentences if and only if it is true in all $\omega$-models of that set) that what is provable is precisely what holds in the minimal fixed point on the van Fraassen scheme.

Presumably by replacing our theory with a variant pPA$\omega$* based on partial logic we can get an equivalent characterization of the minimal fixed point on the Kleene scheme. But needless to say, none of this gives us an answer to the internal axiomatization question as I formulated it, as a question about natural language as spoken by human beings, not Superhuman Beings.

## 10

Another thought may now suggest itself. Perhaps we could tell our human subject about the foregoing fantasy, and then in addition specify that he is entitled to assert a sentence himself if and only if he is entitled to assert that She of the fantasy would be entitled to assert it. Formally, we could add a predicate ]S for "The Superhuman Subject could assert", with appropriate axioms and rules.

The question which principles are appropriate for S must be approached with caution, however. We cannot, for instance, assume "If She can assert that $A$, then $A$" as an unrestricted axiom scheme, since contradiction results upon applying that principle to a self-referential sentence of the kind "This very sentence is something She cannot assert". The most cautious approach would assume that $S[A]$ or $T[A]$ hold only for sentences $A$ not containing S (as with a Tarski-style truth-predicate).

Some principles that seem appropriate are the following: (a) the rules permitting inference in categorical demonstrations from "She can assert that $A$" to $A$ and from $A$ to "She can assert that $A$", as per our imagined specifications to the human subject; (b) the axiom that She can assert any axiom of logic or arithmetic; (c) the axiom that She can make inferences from assertion to assertion using modus ponens; (d) ditto for the $\omega$-rule; (e) ditto for the four truth rules. Let us call the system given by these principles SPA$\omega$*.

SPA$\omega$* is consistent. This can be established by showing that a fixed point on the van Fraassen scheme provides a model (à la van Fraassen). The arithmetical part of the model is standard. The predicates T and S have the same extension, the set of (codes for) sentences valued true in the fixed point, but T is treated as a partial predicate, with anti-extension the set of (codes for) sentences valued false, whereas S is treated as a total predicate, whose anti-extension is simply the complement of its extension.

SPA$\omega$* provides more than enough in the way of axioms and rules to prove the test sentence $\tau_\omega$ mentioned earlier as unprovable in PA*. We may reason as follows. We can assert $0 = 0$ or $\tau_0$, hence so can She. But then since She can reason using the truth rules, for any $n$, if She can insert $\tau_n$, then She can assert $T[\tau_n]$ or $\tau_{n+1}$. Hence, by induction, for every $n$, She can assert $\tau_n$, and that $\tau_n$ is true. Hence, since She can reason by the $\omega$-rule, She can assert that for every $n$, $\tau_n$ is true. But that is to assert $\tau_\omega$, and since we have just deduced that She can assert it, we can assert it, too.

SPA$\omega$* provides enough in the way of axioms and rules to prove the consistency of PA as well. The argument is that She can assert each axiom, and She can reason by modus ponens, so She can assert each theorem, and so through Her ability to use the truth rules, She can assert *the truth of* each theorem of PA, and since — skipping some details here — She can also assert for each nontheorem that it is *not* a theorem, She then can, for each sentence, assert that if it is a theorem it is true, and then through Her ability to use the $\omega$-rule, She can assert that every theorem of PA is true. But $0 \neq 1$, and since we have just asserted it, She can assert that $0 \neq 1$ as well, and then through Her ability to use the truth rules, She can infer that $0 = 1$ is untrue. Hence She can assert that $0 = 1$ is a nontheorem, and since we have just deduced that She can assert it, we can assert it, too.

I will not pursue the development of the theory further here. In particular, I leave the determination of the exact proof-theoretic strength of SPA$\omega$*, and that of the variant pSPA$\omega$* based on partial logic, to the experts. Presumably pSPA$\omega$* would represent one candidate answer to what I have called the internal question, the question of what to tell the human subject who has never heard the word "true" before. But having brought this kind of answer, involving a new predicate over and above the truth predicate, to your attention, let me now set it aside.

## 11

Returning to theories involving no new predicates but T, there lie near at hand two further conceivable answers to the question what to tell our subject: "We can tell him the axioms of KF" or "We can tell him the axioms of KHH". (I do not mean to imply that the originators of either theory advocated it as an answer to the internal question as I have posed it, but only that it is natural to take up the issue whether one or the other of them might be a good answer.)

The difference between the two is that KF is based on classical, and KHH on partial logic. This difference results in a difference in proof theoretic strength. For Halbach and Horsten [2006] show that their system, though stronger than FS, which Halbach [1994] had shown to have the same strength as RA$(< \omega)$, is weaker than KF, which Feferman [1992] had shown to have the same strength as RA$(< \varepsilon_0)$. Its strength is that of RA$(< \omega^\omega)$.

But though it is the issue of partial *versus* classical logic that distinguishes PA* or PA$\omega$* from pPA* or pPA$\omega$*, and the same issue that distinguishes KF from KHH, the issue apparently distinguishes in different ways in the two cases. The difference between PA$\omega$* and pPA$\omega$* looks like the difference between an object language based on a van Fraassen fixed point and an object language based on a Kleene fixed point. The difference between KF and KHH looks like the difference between an object language based on a Kleene logic and a metalanguage based on classical logic.

The mere fact that it is based on classical rather than partial logic means that KF has theorems, such as instances of excluded middle for liar sentences, that are not valued true in the minimal or indeed any fixed point on the Kleene scheme. For that matter, many of the composition and decomposition axioms of KF are not valued true in any fixed point on *either* the Kleene *or* the van Fraassen scheme. Moreover KF proves sentences such as "a liar sentence is not true" that Kripke explicitly classifies as post-reflective rather than pre-reflective. All this seems to disqualify KF as an answer to the internal axiomatization question, leaving KHH as the only surviving candidate.

If we accept KHH as tentative answer to the internal question, we now face the question whether it is plausible to suggest that the way we actually learn the meaning of "true" is by coming to internalize something like that system of axioms and rules. Perhaps the sheer number of axioms and rules involved is enough to make the defensibility of the claim seem doubtful, but in addition there is fact that, in Feferman's quotable phrase, "nothing like sustained ordinary reasoning can be carried out" in the kind of partial logic on which KHH is based. It was this fact that led Feferman seek a system like KF based on classical logic in the first place. Perhaps the availability of the natural deduction formulation in Horsten [2011] may soften this judgment, but this is not an issue I can pursue further here.

# 12

I turn instead from the internal to the external question, the question of characterizing by axioms and rules, without explicit set-theoretic apparatus, Kripke's model construction, as viewed from the outside rather than the inside. I should first consider whether KF, which was disqualified as an answer to the internal question, looks any better as an answer to external question, since after all the reason KF was disqualified as an answer to the internal question was precisely that it seemed to be describing a fixed point from without rather than from within.

But KF does not look good as answer to the external axiomatization question, either, at least not as I formulated the question. KF provides a simple and natural axiomatization, from an external perspective, of the properties of an arbitrary fixed point; but as I formulated the axiomatization question, it was not about *an arbitrary* fixed point but rather about *the minimal* fixed point. KF shares with KHH the feature that, by design, we get a model for it from *any* Kleene fixed point. This feature seems less appropriate for a representation of a *post*-reflective understanding, than for a representation of a *pre*-reflective understanding, when we are just going forward with whatever specifications we have been given, without thinking about the scope and limits of how far they will take us.

This feature is what is responsible for the limited proof-theoretic strength of KF. Earlier I mentioned that Feferman showed its strength to be the same as that of the theory known as RA($< \varepsilon_0$). But that is known (using results in Feferman [1982] in one direction and a combination of Aczel [1980] with Friedman [1970] in the other) to be the same as the proof-theoretic strength of the theory asserting for each positive arithmetic inductive operator the existence of *some* fixed point; and the more illuminating comparison is between KF and this last theory, in the literature called $\widehat{\mathrm{ID}}_1$.

KF is interpretable in it because the inductive operator involved in Kripke's construction is positive arithmetic, and because *any* fixed point for that operator provides a model of KF. Interpretability holds in the opposite direction also (not literally in KF itself, but in the conservative extension related to KF as the theory known as ACA$_0$ is related to PA), because, as noted in Cantini [1989], for any positive arithmetic inductive operator, KF proves that a certain associated self-referential formula defines a fixed point for it. This may be shown as follows. Let $\Phi(X,x)$ be a formula with no quantification over set-variables in which the all appearances of the free set-variable $X$ are positive. We may take it to be of the form

(1) $\quad \forall y_1 \ldots \exists y_k \bigvee\limits_{i=1 \text{ to } r} \bigwedge\limits_{j=1 \text{ to } s} \varphi_{ij}(x, y_1, \ldots, y_k)$

where each $\phi_{ij}$ is either arithmetical or of the form $X(u_{ij})$ where $u_{ij}$ is one of the variables $y_1, \ldots, y_k$. In the usual way — that is, by a "parametrized" version of the Gödel diagonal construction used to produce truth-tellers, liars, and so on — introduce $B(x)$, with free variable $x$, that "says"

(2) $\quad \forall y_1, \ldots, \exists y_k \bigvee\limits_{i} \bigwedge\limits_{j} \varphi^*_{ij}(x, y_1, \ldots, y_k)$

where $\varphi^*_{ij}$ is $\varphi_{ij}$ if $\varphi_{ij}$ is arithmetical, and $\mathsf{T}[B(u_{ij})]$ if $\varphi_{ij}$ is $X(u_{ij})$. In other words, $B(x)$ "says" $\Phi(\mathsf{T}[B()], x)$. Then KF proves

(3) $\quad \forall x(b(x) \leftrightarrow \forall y_1, \ldots, \exists y_k \bigvee\limits_{i} \bigwedge\limits_{j} \varphi^*_{ij}(x, y_1, \ldots, y_k))$

The formula involved is truth-positive, and so as remarked at the end of §2, transparency applies, and KF proves

(4) $\quad \forall x(\mathsf{T}[B(x)] \leftrightarrow \forall y_1, \ldots, \exists y_k \bigvee\limits_{i} \bigwedge\limits_{j} \varphi^*_{ij}(x, y_1, \ldots, y_k))$

In other words, if we give the formula $\mathsf{T}[B(x)]$ the name $\Phi(x)$, then KF proves for pertinent $\Phi$ that the associated $\Phi^*$ gives a fixed point for the operator given by $\Phi$.

## 13

The only attempt in the literature known to me to frame an axiomatic theory of truth that would incorporate minimality, as KF and KHH do not, occurs in work of Cantini [1989]. He introduces a system KF$^+$ + GID, where GID is a certain general scheme of inductive definition that he shows to hold for the minimal fixed point. The incorporation of minimality through such a scheme is somewhat indirect, and to that extent complicated and artificial, so I would like to propose another system, which I will call KF$\mu$ (with $\mu$ for minimal), that incorporates minimality more straightforwardly.

Let us think of our theories as formulated with only the truth predicate T as primitive (the falsehood predicate F being treated as a defined). KF takes the composition and decomposition laws in the table as axioms. By contrast, KF$\mu$ takes only the composition laws as axioms, but adds an axiom scheme of minimality. For each formula $\tau(x)$ it is an axiom that if the set of truths satisfying $\tau$ is closed under the composition laws, then every truth satisfies $\tau$. What the closure assumption here amounts to is a conjunction of conditions, one for each composition axiom. Thus corresponding to positive equational composition, for instance, we have the condition that any correct equation satisfies $\tau$. KF$\mu$ provides one *obvious* candidate for an answer to the external axiomatization question, and I will devote the remainder of this talk to exploring its consequences.

The first thing to be noted is that every theorem of KF and indeed of KF$^+$ is a theorem of KF$\mu$. First, to get KF, we note that the decomposition axioms can be derived using the minimality scheme. Taking the positive equational decomposition axiom for example, let $\tau(x)$ say: if $x$ is (the code number for) an equation, it is correct. It is easily seen that the set of truths satisfying $\tau$ is closed under the composition laws. The only such law that could conceivably be a problem is the one whose consequent mentions the specific kind of sentence mentioned in $\tau$ (namely, equations), since all other sentences *vacuously* satisfy $\tau$. So we need only check the positive equational composition axiom; and the condition corresponding to this axiom is (as noted above) simply that every correct equation should satisfy $\tau$, which it trivially does. The minimality principle now applies to tell us that every truth satisfies $\tau$, which is to say, that every true sentence fulfills the condition that if it is an equation it is correct, or more simply, that every true equation is correct. That is the positive equational decomposition axiom we wanted. Exactly the same method can be used with all the other decomposition axioms.

Further, to get KF$^+$, the same method can be used to prove the truth consistency axiom, that no sentence is both true and false (where falsehood is truth of the negation). Let $\tau(x)$ say: $x$ is not (the code number for) of a false sentence. It is then not hard to prove in KF$\mu$ closure under the composition laws. For instance the composition law for addition, for instance, what we must check is that if $A$ and $B$ are true and not false, then their conjunction is true and not false. But this is easy using the positive composition and negative decomposition axioms for conjunction (which respectively tell us that if both conjuncts are true, their conjunction is true, and that if a conjunction is false, one of its conjuncts is false). Once we have all the composition axioms, minimality applies to tell us that every true sentence satisfies the condition $\tau$ of not being false, which is the truth consistency axiom. That axiom is known to be unprovable in KF, though it has also been shown by Cantini that KF$^+$ is not of greater proof-theoretic strength than plain KF.

## 14

There are also theorems of KF$\mu$ that are not theorems of KF$^+$. Consider a truth-teller sentence, a sentence $B$ constructed by the usual diagonal method so that $B$ "says" $\mathsf{T}[B]$. It cannot be proved in KF$^+$ that $B$ is not true, because the axioms of KF$^+$ hold for any fixed point, and though no truth-teller is true in the *minimal* fixed point, each is true in some fixed point or other. But it can be proved in KF$\mu$ that $B$ is not true.

What $B$ literally *is*, is a sentence $\forall z(\sim C(z) \vee \mathsf{T}(z))$ with code $b$, where $C$ is arithmetical and where $C(b)$ and $\forall n(n = b \vee \sim C(n))$ are provable in PA, so that PA proves $\mathsf{T}(b) \to B$ and $B \to \mathsf{T}(b)$. Note that these are provable in KF$\mu$ not only because KF$\mu$ extends PA, but also simply because the formula involved is truth-positive, and so as remarked at the end of §2, transparency applies. Now let $\tau(x)$ say: $x$ is not (the code number for) any of the following formulas:

> $\mathsf{T}[B]$, which is to say, $\mathsf{T}(b)$
> $\sim C(b) \vee \mathsf{T}(b)$
> $B$, which is to say, $\forall p(\sim C(p) \vee \mathsf{T}(p))$

Let $\mathsf{T}^*(x)$ abbreviate the conjunction of $\mathsf{T}(x)$ and $\tau(x)$. Checking the closure of the set of sentences satisfying $\tau$ under the composition laws amounts to checking that the composition axioms hold with truth* (or $\mathsf{T}^*$) in place of truth (or $\mathsf{T}$). The only cases that could possibly cause trouble are those where the consequent of the axiom is the truth* of one of the three exceptional sentences mentioned in $\tau$:

1. if $B$ is true*, then $\mathsf{T}[B]$ is true*

2. if $\sim C(b)$ is true* or $\mathsf{T}(b)$ is true*, then $\sim C(b) \vee \mathsf{T}(b)$ is true*

3. if for all $p$, $\sim C(p) \vee \mathsf{T}(p)$ is true*, then $\forall p(\sim C(p) \vee \mathsf{T}(p))$ is true*

In all these cases the consequent of the axiom fails by definition of T*, so we must show, working in KF$\mu$, that the antecedent fails as well. And indeed, the antecedent of (1) fails by definition of truth*. The first disjunct of the antecedent of (2) fails since $C(b)$ holds and we have truth transparency for $\sim C(b)$ because it is arithmetical, while the second disjunct fails by definition of truth*. The antecedent of (3) fails in the instance $p = b$ by definition of T*. Once we have the all the composition axioms, minimality applies and tells us that every true sentence is true*, which is to say, is not one of the sentences mentioned in T*, or in other words, that those sentences, including $B$, are not true. In a similar way it can be proved that $B$ is not false, either.

## 15

KF$\mu$ is essentially a subtheory of the theory known in the literature as ID$_1$, which for each positive arithmetic inductive operator asserts the existence of a minimal fixed point. In the joint paper with Sheard, Friedman proves that a certain axiomatic truth theory H is proof-theoretically as strong as ID$_1$.

(It may be mentioned that though KF$^+$ has all the truth principles from the list of Friedman and Sheard for the system they call H, as well as truth distribution, which is part of the Friedman-Sheard base theory, and though Friedman proves that H is of impredicative proof-theoretic strength, and so of much greater strength than KF, still Friedman's proof for H does not apply to KF$^+$, because it makes use of truth classicism, which is part of the Friedman-Sheard base theory, and so not mentioned by them explicitly in their definition of H, but is not available in KF$^+$.)

Friedman's proof uses truth principles only to get a certain lemma, after which they are not referred to again. So it will be enough for us to prove for $T = \text{KF}\mu$ the lemma that Friedman proves for $T = \text{H}$. It reads as follows:

> *Friedman's Lemma.* For any arithmetical formula $R(x, y)$ and any formula $A(x)$, the theory $T$ proves the following:
>
> If transfinite induction along $R$ holds for $\mathsf{T}[B(x)]$ for every formula $B(x)$ then transfinite induction along $R$ holds for $A(x)$.

The claim that transfinite induction holds for all formulas of form $\mathsf{T}[B(x)]$, unlike the claim that transfinite induction holds for *all* formulas $A(x)$, can be stated in a single sentence, and that is what makes the lemma key to Friedman's proof.

There is in fact a single formula $B$, depending only on $R$, such that for any formula $A$, KF$\mu$ proves that if transfinite induction along $R$ holds for $\mathsf{T}[B(x)]$, then in holds for $A(x)$. The formula is suggested by Kripke's proof that the set of truths in the minimal fixed point is a complete $\prod_1^1$ set.

Take $B$ constructed by the usual diagonal method so that $B(n)$ "says" $\forall m(R(m,n) \to \mathsf{T}[B(m)])$. What $B(x)$ literally *is*, is a formula

$$\forall z \forall y \forall w(\sim C(z) \vee \sim R(y,x) \vee \sim I(z,y,w) \vee \mathsf{T}(w))$$

with code $b$, where $C$ is arithmetical, where $C(b)$ and $\forall n(n = b \vee \sim C(n))$ are provable in PA, and where $I$ is the usual arithmetical formula expressing "$w$ is the code for the sentence resulting from substituting the numeral for $y$ for the free variable in the formula with code $z$", with its usual properties. Then PA proves

$$\forall n(\forall m(R(m,n) \to \mathsf{T}[B(m)]) \to B(n))$$

and its converse. Note that KF$\mu$ then proves

$$\forall n(\forall m(R(m,n) \to \mathsf{T}[B(m)]) \to \mathsf{T}[B(n)])$$

by truth transparency, since $B$ is T-positive. This is the antecedent of transfinite induction along $R$ for $\mathsf{T}[B(x)]$, so by the assumption of Friedman's lemma the consequent $\forall n \mathsf{T}[B(n)]$ holds.

Suppose now for contradiction that the conclusion of Friedman's lemma fails, that the antecedent of transfinite induction along $R$ for $A(x)$, namely,

$$\forall n(\forall m(R(m,n) \to A(m)) \to A(n))$$

holds, but the consequent $\forall n A(n)$ fails. To deduce a contradiction and complete the proof that Friedman's lemma holds, we prove, working in KF$\mu$, that for any $n$, $\sim A(n)$ implies $\sim \mathsf{T}[B(n)]$ or equivalently $\sim B(n)$.

To this end, writing $b_n$ for the code of $B(n)$, let $\mathsf{T}^*(x)$ be the conjunction of $\mathsf{T}(x)$ and the formula $\tau(x)$ saying that $x$ is not (the code of) any of the following sentences:

$\mathsf{T}[B(n)] = \mathsf{T}(b_n)$, where $\sim A(n)$
$\sim C(b) \vee \sim R(m,n) \vee \sim I(b,m,b_m) \vee \mathsf{T}(b_m)$, where $R(m,n)$ and $\sim A(m)$
$B(n) = \forall z \forall y \forall w(\sim C(z) \vee \sim R(y,x) \vee \sim I(z,y,w) \vee \mathsf{T}(w))$, where $\sim A(n)$

In verifying the composition axioms for $\mathsf{T}^*$, the only cases that could conceivably cause trouble are those where the consequent of the axiom is the truth of one of the three sentences mentioned in $\mathsf{T}^*$:

1. if $B(n)$ is true*, then $\mathsf{T}[B(n)]$ is true*

2. if $\sim C(b) \vee \sim I(b,m,b_m) \vee \sim R(m,n)$ is true* or $\mathsf{T}(b_m)$ is true* then $\sim C(b) \vee \sim R(m,n) \vee \sim I(b,m,b_m) \vee \mathsf{T}(b_m)$ is true*

3. if for all $p,m,q, \sim C(p) \vee \sim R(m,n) \vee \sim I(p,m,q) \vee \mathsf{T}(q)$ is true* then $B(n) = \forall p \forall m \forall q (\sim C(p) \vee \sim R(m,n) \vee \sim I(p,m,q) \vee \mathsf{T}(q))$ is true*

Here (1) will be troublesome when $\sim A(n)$, (2) will be troublesome when $R(m,n)$ and $\sim A(m)$, and (3) will be troublesome when $\sim A(n)$ again, since these are the cases in which the consequent of the axiom fails by the definition of $\mathsf{T}^*$ and we must prove in these cases that the antecedent fails also. For (3), if $\sim A(n)$, then by the antecedent of transfinite induction along $R$ for $A(x)$, there is some $m$ with $R(m,n)$ and $\sim A(m)$, and the thing to prove is that the antecedent of (3) fails for $p = b$ and this $m$ and $q = b_m$. The argument is much as in the case of the truth-teller example.

Once we have the all the composition axioms, minimality applies and tells us that every true sentence is true*, which is to say, is not one of the sentences mentioned in $\mathsf{T}^*$, or in other words, that those sentences, including $B(n)$ whenever $\sim A(n)$, are not true. So Friedman's lemma is established, and applies to complete the proof.

## 16

The method of proof here suggests a strategy proving in KF$\mu$ for *any* positive inductive operator that a certain associated self-referential formula defines a minimal fixed point for it. The result is that ID$_1$ is interpretable in KF$\mu$ (or more precisely, in the conservative extension related to KF$\mu$ as ACA$_0$ is related to PA).

The execution of the strategy goes as follows. Let $\Phi(X,x)$ be an $X$-positive formula with free variable $x$. We may take it to be of the form

1. $\forall y_1 \ldots \exists y_k \bigvee_{i=1}^{\text{to } r} \bigwedge_{j=1}^{\text{to } s} \varphi_{ij}(x, y_1, \ldots y_k)$

where each $\varphi_{ij}$ is either arithmetical or of the form $X(u)$ where $u$ is one of the variables.

In the usual way — again, by a "parametrized" version of the Gödel diagonal construction used to produce truth-tellers, liars, and so on — introduce $B(x)$, with free variable $x$, that "says"

2. $\forall y_1 \ldots \exists y_k \bigvee_i \bigwedge_j \varphi_{ij}{}^*(x, y_1, \ldots, y_k)$

where $\varphi_{ij}{}^*$ is $\varphi_{ij}$ if $\varphi_{ij}$ is arithmetical, and $\mathsf{T}[B(u)]$ if $\varphi_{ij}$ is $X(u)$. In other words, $B(x)$ "says" $\Phi(\mathsf{T}[B()], x)$. Then KF proves

3. $\forall x(B(x) \leftrightarrow \forall y_1,\ldots,\exists y_k \bigvee_i \bigwedge_j \varphi_{ij}{}^*(x,y_1,\ldots,y_k))$

The formula involved is truth-positive, so as remarked at the end of §2, transparency applies, and KF proves

4. $\forall x(\mathsf{T}[B(x)] \leftrightarrow \forall y_1,\ldots,\exists y_k \bigvee_i \bigwedge_j \varphi_{ij}{}^*(x,y_1,\ldots,y_k))$

In other words, KF proves that $\mathsf{T}[B(x)]$ gives a fixed point for the operator given by $\Phi$.

We now wish to show that $\mathrm{KF}\mu$ further proves that this fixed point is minimal. In other words, for each $A(x)$ we are to prove, working in $\mathrm{KF}\mu$, that if $A(x)$ gives a fixed point for the operator given by $\Phi$, which is to say, if we have

5. $\forall x(A(x) \leftrightarrow \forall y_1 \ldots \exists y_k \bigvee_i \bigwedge_j \varphi_{ij}\dagger(x,y_1,\ldots,y_k))$

where $\varphi_{ij}\dagger$ is $\varphi_{ij}$ if $\varphi_{ij}$ is arithmetical, and $A(u)$ if $\varphi_{ij}$ is $X(u)$, then we have $\forall x(B(x) \to A(x))$. That is to say, we must show that we have $\sim B(n)$ or equivalently $\sim \mathsf{T}[B(n)]$ whenever $\sim A(n)$.

It is enough to obtain a formula $\mathsf{T}^*(x)$ of the form "$\mathsf{T}(x)$ and $x$ is not one of the formulas ... " where the excluded formulas include $B(n)$ for those $n$ such that $\sim A(n)$, for which we can prove the composition axioms. For then minimality tells us that truth ($\mathsf{T}$) implies truth* ($\mathsf{T}^*$), so that no true formula is one of the excluded formulas, or equivalently, neither $B(n)$ for $n$ such that $\sim A(n)$ nor any other of the excluded formulas is true.

To this end we note that what $B(x)$ really *is*, is a T-positive formula

6. $\forall z(\sim C(z) \vee \forall y_1,\ldots,\exists y_k \bigvee_i \bigwedge_j \varphi_{ij}\ddagger(z,x,y_1,\ldots,y_k))$

with code $b$ where $C(z)$ is arithmetical and $C(b)$ and $\forall z(\sim C(z) \vee z = b)$, where $\varphi_{ij}\ddagger$ is $\varphi_{ij}$ if $\varphi_{ij}$ is arithmetical, and is $\forall w(\sim I(z,u,w) \vee \mathsf{T}(w))$ if $\phi_{ij}$ is $X(u)$, where $I$ is the usual formula saying "$w$ is the code for the sentence that results when the numeral for $u$ is substituted for the free variable in the formula with code $z$", with the usual properties.

Writing $b_m$ for the code of $B(m)$, the formulas to be excluded by $\mathsf{T}^*(x)$ are the following:

7. (a) $\mathsf{T}[B(m)]$, which is to say, $\mathsf{T}(b_m)$ if $\sim A(m)$
   (b) $\sim I(b,m,b_m) \vee \mathsf{T}(b_m)$
       if $\mathsf{T}(b_m)$ is excluded by (i)
   (c) $\forall w(\sim I(b,m,w) \vee \mathsf{T}(w))$
       if $\sim I(b,m,b_m) \vee \mathsf{T}(b_m)$ is excluded by (ii)
   (d) $\bigwedge_j \varphi_{ij}\ddagger(b,n,m_1,\ldots,m_k)$
       if any conjunct is excluded by (iii)

(e) $\bigvee_i \bigwedge_j \varphi_{ij} \ddagger (b, n, m_1, \ldots, m_k)$
if any disjunct is excluded by (iv)

(f) $\exists y_k \bigvee_i \bigwedge_j \varphi_{ij} \ddagger (b, n, m_1, \ldots, y_k)$
if every instance $y_k = m_k$ is excluded by (v)

(g) $\forall y_1, \ldots, \exists y_k \bigvee_i \bigwedge_j \varphi_{ij} \ddagger (b, n, y_1, \ldots, y_k)$
if any instance $y_1 = m_1$ is excluded by (vi)

(h) $\sim C(b) \vee \forall y_1, \ldots, \exists y_k \bigvee_i \bigwedge_j \varphi_{ij} \ddagger (b, n, y_1, \ldots, y_k)$
if the second disjunct is excluded by (vii)

(i) $B(n)$, which is to say, $\forall z (\sim C(z) \vee \forall y_1, \ldots, \exists y_k \bigvee_i \bigwedge_j \varphi_{ij} \ddagger (z, n, y_1, \ldots, y_k))$
if the instance $z = b$ is excluded by (viii)

In proving composition for T*, the exclusion in (7i), for example, does not result in of a violation of the composition law that if all instances are true* a universal generalization is true*, since the only universal generalization it excludes is one for which an instance is also excluded. The same applies to any of (7b)-(7h). With (7a) we need to show that we do not get a violation of the composition law that if $D$ is true*, then T$[D]$ is true*.

So we need to show that if $\sim A(n)$, so that (7a) excludes T$[B(n)]$ from being true*, then $B(n)$ is also excluded from being true*. Applying (7b)-(7i) in reverse order, a sufficient condition for exclusion, which is to say for $\sim$ T*$[B(n)]$, is that we should have

8. $\exists y_1, \ldots, \forall y_k \bigwedge_i \bigvee_j \sim \varphi_{ij}\ddagger^*(b, n, y_1, \ldots, y_k))$

where $\varphi_{ij}\ddagger^* = \varphi_{ij}\ddagger = \varphi_{ij}$ if $\varphi_{ij}$ is arithmetical, and $\varphi_{ij}\ddagger^* = $ T*$[\varphi_{ij}\ddagger] = $ T*$[$T$[B(u)]]$ if $\varphi_{ij}$ is of form $X(u)$ and hence $\varphi_{ij}\ddagger$ of form T$[B(u)]$.

For this case we apply $\sim A(n)$ and the fact (5) that $A(x)$ is a fixed point for $\Phi$. Together these give us

9. $\exists y_1, \ldots, \forall y_k \bigwedge_i \bigvee_j \sim \varphi_{ij} \dagger (n, y_1, \ldots, y_k))$

where $\phi_{ij}\dagger = \varphi_{ij}$ if $\varphi_{ij}$ is arithmetical and $\varphi_{ij}\dagger = \sim A(u)$ if $\varphi_{ij}$ is of form $X(u)$. Comparing (8) and (9), it is enough to show that for all $m$, if $\sim A(m)$ then $\sim$ T*$[$T$[B(m)]]$. But that is immediate from (7a), to complete the proof.

To compare this result with the work of Cantini, one could say that the fact that the minimal fixed point provides a model for KF$\mu$ is true almost by definition, while the interpretability of ID$_1$ in KF$\mu$ requires proof, whereas with Cantini's KF+ + GID it is the other way around. What the strength would be of a variant pKF$\mu$ = KHH$\mu$ of KF$\mu$ based on partial logic is a question I leave to the experts, along with the prior question how one would even formulate such a theory.

## 17

Having taken so much from Friedman, it may be bad form for me to close by asking for more, but I am tempted to do so. It would be nice to have the Friedman-Sheard project redone in a way that does not make truth classicism or any truth principles as part of the base theory. It would be nice to have the Friedman-Sheard project done again in a version based on partial rather than classical logic. I would be nice to have the Friedman-Sheard project redone for theories with *two* predicates T and S. The task is enormous, and so one may think, "If only Friedman could be lured back to the subject to do some of this work for us!"

But perhaps philosophers should first do a bit more in the way of distinguishing, within the enormous range of combinations, those potentially of the most philosophical interest. Something like that is what I have been attempting to do here.

## Acknowledgements

The author is grateful to Jeremy Avigad, Andrea Cantini, Graham Leigh, Volker Halbach, Leon Horsten, and Michael Rathjen for useful comments on earlier drafts of this paper, and background information on proof-theoretic matters.

## BIBLIOGRAPHY

[1] Aczel, Peter, 1980, "Frege structures and the notion of proposition, truth and set", The Kleene Symposium, Jon Barwise et al. (editors), Amsterdam: North-Holland, 31-59.
[2] Cantini, Andrea, 1989,"Notes on Formal Theories of Truth", *Zeitschrift für Mathematische Logik und Grundlagen der Mathematik* 35:97-130.
[3] Cantini, Andrea, 1990, "A Theory of Formal Truth Arithmetically Equivalent to $ID_1$", *Journal of Symbolic Logic* 55:244-59.
[4] Feferman, Solomon, 1964, "Systems of Predicative Analysis", *Journal of Symbolic Logic* 29:1-30.
[5] Feferman, Solomon, 1982, "Iterated Inductive Fixed-Point Theories: Application to Hancock's Conjecture", in G. Metakides (ed.), *Patras Logic Symposium*, Amsterdam: North Holland, 171-196.
[6] Feferman, Soloman, 1991,"Reflecting on Incompleteness", *Journal of Symbolic Logic* 56:1-49.
[7] Friedman, Harvey, 1970, "Iterated Inductive Definitions and $\sum_1^1$-AC", in J. Myhill (ed.), *Intuitionism and Proof Theory*, Amsterdam: North Holland, 435-442.
[8] Friedman, Harvey and Michael Sheard, 1987, "An Axiomatic Approach to Self-Referential Truth", *Annals of Pure and Applied Logic* 33:1-21.
[9] Friedman, Harvey and Michael Sheard, 1988, "The Disjunction and Existence Properties for Axiomatic Systems of Truth", *Annals of Pure and Applied Logic* 40:1-10.
[10] Halbach, Volker, 1994, "A System of Complete and Consistent Truth", *Notre Dame Journal of Formal Logic* 35:311-27.
[11] Halbach, Volker and Leon Horsten, 2006 "Axiomatizing Kripke's Theory of Truth", *Journal of Symbolic Logic* 71: 677-712.
[12] Horsten, Leon, 2011, *Axiomatic Truth Theories and Deflationism*, MIT Press.

[13] Leigh, Graham, and Michael Rathjen, forthcoming, "An Ordinal Analysis of Self-Referential Truth".
[14] Sheard, Michael, 1994, "A Guide to truth Predicates in the Modern Era", *Journal of Symbolic Logic* 59:1032-54.

# Xeno Semantics for Ascending and Descending Truth

Kevin Scharp

## Introduction

As part of an approach to the liar paradox and the other paradoxes affecting truth, I have proposed replacing our concept of truth with two concepts: ascending truth and descending truth.[1] I am not going to discuss why I think this is the best approach or how it solves the paradoxes; instead, I concentrate on the theory of ascending and descending truth. I formulate an axiomatic theory of ascending truth and descending truth (ADT) and provide a possible-worlds semantics for it (which I dub *xeno semantics*). Xeno semantics is a generalization of the familiar neighborhood semantics, which itself is a generalization of the standard relational semantics. Once the details of ADT have been presented, it is easy to show that neither relational semantics nor neighborhood semantics will work for it; thus, the move to a more general framework is required. The main result is a fixed point theorem that guarantees the existence of an acceptable first-order constant-domain xeno model. From this result it follows that ADT is sound with respect to the class of such models. The upshot is that ADT is consistent relative to the background set theory.

## 1 A Theory of Ascending Truth and Descending Truth: ADT

Let $\mathcal{L}^-$ be a first-order language with the usual syntax (individual constants, variables, predicate letters, function letters, sentential operators, and quantifiers), the usual logical connectives ($\neg, \wedge, \vee, \rightarrow, \leftrightarrow$), existential and universal quantifers ($\forall, \exists$), and the vocabulary of Peano Arithmetic (0, successor function, addition and multiplication functions). Let $L$ extend $\mathcal{L}^-$ by adding three one-place predicates, '$A(x)$', '$D(x)$', and '$S(x)$' to be interpreted as an ascending truth predicate, a descending truth predicate, and a safety predicate, respectively. In what follows, I use lowercase Greek letters

---

[1] [4] and [5].

for variables ranging over well-formed formulas of $\mathcal{L}$ and angle brackets for names in $\mathcal{L}$ of well-formed formulas of $\mathcal{L}$; e.g., '$\phi$ is a sentence of $\mathcal{L}$' and '$\langle\phi\rangle$ refers to $\phi$'.

Consider the following list of axiom schemata.

D1  $D(\langle\phi\rangle) \to \phi$

D2  $D(\langle\sim\phi\rangle) \to \sim D(\langle\phi\rangle)$

D3  $D(\langle\phi\wedge\psi\rangle) \to D(\langle\phi\rangle) \wedge D(\langle\psi\rangle)$

D4  $D(\langle\phi\rangle) \vee D(\langle\psi\rangle) \to D(\langle\phi\vee\psi\rangle)$

D5  $D(\langle\phi\rangle)$ if $\phi$ is a tautology (logical truth) of first order predicate calculus.

D6  $D(\langle\phi\rangle)$ if $\phi$ is a theorem of PA.

D7  $D(\langle\phi\rangle)$ if $\phi$ is an axiom of ADT (i.e., if $\phi$ is an instance of D1–D6, A1–A6, M1–M4)

A1  $\phi \to A(\langle\phi\rangle)$

A2  $\sim A(\langle\phi\rangle) \to A(\langle\sim\phi\rangle)$

A3  $A(\langle\phi\rangle) \vee A(\langle\psi\rangle) \to A(\langle\phi\vee\psi\rangle)$

A4  $A(\langle\phi\wedge\psi\rangle) \to A(\langle\phi\rangle) \wedge A(\langle\psi\rangle)$

A5  $\sim A(\langle\phi\rangle)$ if $\phi$ is a contradiction (logical falsehood) of first order predicate calculus.

A6  $\sim A(\langle\phi\rangle)$ if $\phi$ is the negation of a theorem of PA.

M1  $D(\langle\phi\rangle) \leftrightarrow \sim A(\langle\sim\phi\rangle)$

M2  $S(\langle\phi\rangle) \leftrightarrow (D(\langle\phi\rangle) \vee \sim A(\langle\phi\rangle))$

M3  $\phi \wedge S(\langle\phi\rangle) \to D(\langle\phi\rangle)$

M4  $A(\langle\phi\rangle) \wedge S(\langle\phi\rangle) \to \phi$

E1  If $\sigma = \tau$ and $\psi$ results from replacing some occurrences of $\sigma$ with $\tau$ in $\phi$, then $D(\langle\phi\rangle) \leftrightarrow D(\langle\psi\rangle)$.

E2  If $\sigma = \tau$ and $\psi$ results from replacing some occurrences of $\sigma$ with $\tau$ in $\phi$, then $A(\langle\phi\rangle) \leftrightarrow A(\langle\psi\rangle)$.

E3    If $\sigma = \tau$ and $\psi$ results from replacing some occurrences of $\sigma$ with $\tau$ in $\phi$, then $S(\langle\phi\rangle) \leftrightarrow S(\langle\psi\rangle)$.

Let ADT be the theory that consists of all the classical consequences of the instances of the above schemata. Note that there are redundancies in this list. There are reasons for including each of these principles, but elaborating them would take us too far from our current topic.

The ascending truth predicate obeys one direction (A1) of the familiar Tarskian schema-T, which is often taken to be constitutive of truth; the descending truth predicate obeys the other direction (D1). In addition, because of (M1), $D(x)$ and $A(x)$ are *dual* predicates—they have the same relationship that obtains between possibility and necessity, between obligation and permission, between provability and consistency, etc.[2]

## 2  Xeno Semantics

There are two major differences between the semantics I give for ADT and more familiar possible-worlds semantics. The first is that the most common form of possible worlds semantics, relational semantics, validates certain formulas and rules that are inconsistent with ADT; indeed, even the more general neighborhood semantics validates a rule that is inconsistent when paired with ADT. Therefore, neither of these semantics will work for ADT. The second difference is that ADT is a theory of three predicates, while modal logics are almost always theories of sentential operators, like the familiar necessity operator ($\Box$). I discuss these two differences in order.

In relational semantics, the extension of $\Box$ at each world is determined by a binary accessibility relation on the set of worlds; we can think of this as a function that assigns each world a set of worlds (i.e., those accessible from it). Moreover, each sentence is assigned a proposition, which is a set of worlds (i.e., those in which it is true). In neighborhood semantics, the extension of at each world is determined by a function that assigns each world a set of sets of worlds; as in relational semantics each sentence is assigned a proposition, which is a set of worlds. However, the problem with using relational semantics is that every relational model validates the so-called K axiom, $\Box(\phi \rightarrow \psi) \rightarrow (\Box\phi \rightarrow \Box\psi)$. Richard Montague proved that the predicate version of the K axiom (i.e., $\forall\phi\forall\psi(\Box(\langle\phi \rightarrow \psi\rangle) \rightarrow (\Box(\langle\phi\rangle) \rightarrow \Box(\langle\psi\rangle)))$ is inconsistent with the combination of D1, D5, D6, and D7 from ADT. Thus, no relational semantics can serve as a semantics for ADT.

Moreover, the problem with using neighborhood semantics is that every neighborhood model validates the so-called E-Rule:

(E)    If $\vdash \phi \leftrightarrow \psi$, then $\vdash \Box\phi \leftrightarrow \Box\psi$.

---
[2] Dana Scott emphasized the importance of duality for ADT in private communication.

However, the result of adding the predicate version of (E) to ADT is inconsistent. We know that when we make the move to first order logic and predicates, Godel's Diagonalization lemma guarantees that if our language can express Peano Arithmetic or its own theory of syntax (these are pretty minimal expressive constraints), then it will have sentences $\phi$ s.t. $\sim D(\langle \phi \rangle)$ is provably equivalent to $\phi$. Let $H$ be a sentence of $\mathcal{L}$ s.t. $H \leftrightarrow \sim D(\langle H \rangle)$. It is easy to show that ADT $\vdash \sim D(\langle H \rangle)$ (assume $D(\langle H \rangle)$; if $D(\langle H \rangle)$ then $H$; if $H$, then $\sim D(\langle H \rangle)$; so if $D(\langle H \rangle)$ then $\sim D(\langle H \rangle)$; thus, $\sim D(\langle H \rangle)$). Let $J$ be any classical first-order tautology.[3] Since ADT $\vdash \sim D(\langle H \rangle)$, which is just $H$, ADT $\vdash H \leftrightarrow J$. By the predicate version of (E), ADT $\vdash D(\langle H \rangle) \leftrightarrow D(\langle J \rangle)$. However, ADT $\vdash D(\langle J \rangle)$. Thus, ADT $\vdash D(\langle H \rangle)$. $\bot$. This argument shows that the predicate version of rule E is incompatible with D1 and D5; similar arguments show that it is also incompatible with D1 and D6, and that it is incompatible with D1 and D7. These results show that no neighborhood semantics can serve as a semantics for ADT.[4]

In the new semantics, which I call *xeno semantics*[5], the extension of $\square$ at each world is determined by both an accessibility relation *and* a neighborhood function. As before, each sentence is assigned a set of worlds as its proposition. However, although the neighborhood function is unchanged, the key to xeno semantics is that the accessibility relation is relative to each type of sentence in the language. Indeed, we can think of xeno semantics as involving as many binary accessibility relations as there are syntactic types of sentences.

In *relational* semantics $\square \phi$ is true at a world $w$ if and only if $\phi$ is true at all worlds accessible from $w$. In *neighborhood* semantics, $\square \phi$ is true at a world $w$ if and only if the set of worlds in which $\phi$ is true is a neighborhood of $w$. In *xeno* semantics, $\square \phi$ is true at a world $w$ if and only if the set of worlds in which $\phi$ is true is a neighborhood of all worlds accessible$_\phi$ from $w$, where 'accessible$_\phi$' is the accessibility relation assigned to $\phi$'s syntactic type. So one can think of xeno semantics as a blend of relational semantics and neighborhood semantics with a relativization to syntactic types. In xeno semantics, each sentence is assigned a proposition (a set of worlds) and a relation on the set of worlds. We can think of this as a sentence granting accessibility from one world to others, or we can say that the accessibility relation is relative to each sentence. Moreover, the accessibility relation alone does not determine the extension of $\square$ at each world; rather, together the accessibility relation and the neighborhood relation determine

---

[3] Just to be clear, '$H$' and '$J$' are terms of the metalanguage, not $L$.

[4] Dana Scott first noticed this problem.

[5] Xeno semantics is named after our dog— thanks to Alison Duncan Kerr for the suggestion.

the extension of □ *for that particular sentence* at each world. Alternatively, we can think of a proposition as a pair of a subset of $W$ and a relation on $W$. But neighborhoods of a world are still just subsets of $W$. □'s extension at a world is then is an operation on propositions, and it is determined by the whole neighborhood function, not just the neighborhoods of that world. The details are given in the next section. The following diagram might illuminate the three kinds of semantics:

### Relational Semantics

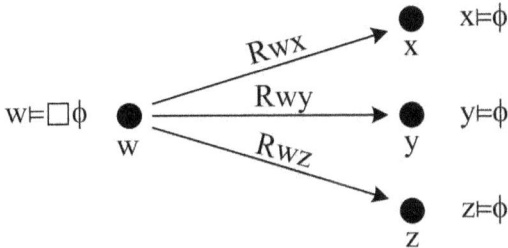

### Neighborhood Semantics

$w \models \Box \phi$  ●  $N(w) = \{N_{w1}, N_{w2}, \ldots\}$  $P(\phi) \in N(w)$
       $w$

### Xeno Semantics

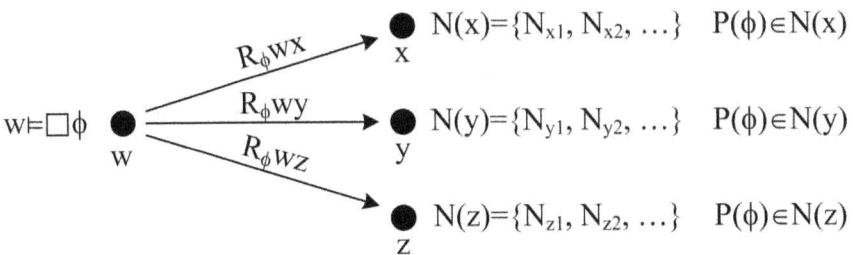

Figure 1.

The second major difference between ADT and other theories for which possible-worlds semantics have been given (e.g., normal modal logics) is that ADT is a theory of three predicates, while possible-worlds semantics are almost always given for theories of sentential operators (e.g., a necessity operator). For example, let $L_N$ be a sentential language with the usual syntax, the usual logical operators, and a sentential operator, $\Box$. Let $W$ be a set of worlds and let $R$ be a relation on $W$ (called the *accessibility relation*). Together, $W$ and $R$ are called a *relational frame*, $\mathfrak{F} = \langle W, R \rangle$. A valuation function, $V$, assigns to each sentential variable of $L_N$ a truth value at each world in $W$. Together, $\mathfrak{F}$ and $V$ are called a *relational model*, $\mathfrak{M} = \langle \mathfrak{F}, V \rangle$ fpr $:_N$. Each world in $W$ is classical in that a classical scheme determines the value of truth-functionally compound sentences. That gives us the following clauses for defining truth at a world in a model (i.e., $\langle \mathfrak{M}, w \rangle \vDash \phi$):

($\phi$)  $\langle \mathfrak{M}, w \rangle \vDash \phi$ if and only if $w \in V(\phi)$ for $\phi$ atomic.

($\sim$)  $\langle \mathfrak{M}, w \rangle \vDash \sim \phi$ if and only if it is not the case that $\langle \mathfrak{M}, w \rangle \vDash \phi$

($\wedge$)  $\langle \mathfrak{M}, w \rangle \vDash \phi \wedge \psi$ if and only if $\langle \mathfrak{M}, w \rangle \vDash \phi$ or $\langle \mathfrak{M}, w \rangle \vDash \psi$

($\vee$)  $\langle \mathfrak{M}, w \rangle \vDash \phi \vee \psi$ if and only if $\langle \mathfrak{M}, w \rangle \vDash \phi$ or $\langle \mathfrak{M}, w \rangle \vDash \psi$

($\rightarrow$)  $\langle \mathfrak{M}, w \rangle \vDash \phi \rightarrow \psi$ if and only if if $\langle \mathfrak{M}, w \rangle \vDash \phi$, then $\langle \mathfrak{M}, w \rangle \vDash \psi$

($\leftrightarrow$)  $\langle \mathfrak{M}, w \rangle \vDash \phi \leftrightarrow \psi$ if and only if $\langle \mathfrak{M}, w \rangle \vDash \phi$ iff $\langle \mathfrak{M}, w \rangle \vDash \psi$

The clause for sentences of the form $\phi$ is:

($\Box$)  $\langle \mathfrak{M}, w \rangle \vDash \phi$ if and only if $\forall u \in W$ if $Rwu$, then $\langle \mathfrak{M}, u \rangle \vDash \phi$

(i.e., $\Box \phi$ is true at $w$ if and only if $\phi$ is true at all worlds accessible from $w$). One can use these clauses to provide an inductive definition of $\langle \mathfrak{M}, w \rangle \vDash \phi$ based on the complexity of $\phi$. A sentence $\phi$ is *valid in a model* $\mathfrak{M}$ (i.e., $\mathfrak{M} \vDash \phi$) if and only if $\forall w \in W \langle \mathfrak{M}, w \rangle \vDash \phi$. A sentence $\phi$ is *valid on a frame* $\mathfrak{F}$ (i.e., $\mathfrak{F} \vDash \phi$) if and only if for all $\mathfrak{M}$ based on $\mathfrak{F}$, $\forall w \in W, \langle \mathfrak{M}, w \rangle \vDash \phi$.

However, in the case of ADT, we cannot adopt this kind of semantics since the inductive definition of $\langle \mathfrak{M}, w \rangle \vDash \phi$ would not go through. The problem, of course, is that while $\Box \phi$ is syntactically more complex than $\phi$ since '$\Box$' is a sentential operator, $D(\langle \phi \rangle)$ need not be more complex than $\phi$ because '$D(x)$' is a predicate.

There has been some work done on using modal logic for predicates instead of operators, and one way to do it involves revision sequences. Revision sequences were originally designed to handle circular definitions, in which the definiens occurs as part of the definiendum. They can be adapted to modal logics for predicates by thinking of the definition of truth at a world in a model as a circular definition by virtue of the modal clauses. For example, '$D(\langle\phi\rangle)$' can occur in the definiens for $\langle\mathfrak{M}, w\rangle \models D(\langle\phi\rangle)$, which makes the overall definition circular. We can then use a revision sequence to arrive at particular frames and models.[6]

A revision sequence begins with an interpretation of the circularly defined term in question, and then one generates a sequence of interpretations through a revision rule, which is based on the circularly defined term. In our case, we start with a first order language that contains a predicate $D(x)$, which will serve as our descending truth predicate (we will worry just about $D(x)$ first, and then see if we can define $A(x)$ and $S(x)$ in terms of it). The revision sequence begins with a model of the language that is similar to the first-order xeno models discussed above, except this model will not satisfy the (D) clause. Instead, we use the (D) clause to generate a new model of the language, but it won't satisfy the (D) clause either; by repeating this process over and over, we generate a sequence of models of the language. The goal is to reach a fixed point—i.e., a point in the sequence where it stops changing. If we can reach such a point, we would then have a legitimate definition of truth at a world in a xeno model for our descending truth *predicate*—a model of the language that satisfies the (D) clause.

## 3 A Fixed Point Theorem

Here we construct a revision sequence of xeno models and prove that it reaches a fixed point. Actually, our construction will be a bit more complicated—we first construct one revision sequence, $\Omega_0$, using *neighborhood* semantics; we can think of this as our characterization sequence. It does not reach a fixed point, but it does classify our sentences in an illuminating way. We then use the results of this characterization sequence to construct the initial *xeno* model for a second revision sequence, $\Omega_1$. The second revision sequence will eventually reach a fixed point. So we use a sequence of neighborhood models to construct a sequence of xeno models, and we prove that the sequence of xeno models reaches a fixed point. The fixed point for the sequence of xeno models will be a xeno model and it is our intended model for ADT.

### 3.1 The Characterization Sequence $\Omega_0$

Again, $\mathcal{L}^-$ is a first order language with the usual connectives, quantifiers, individual constants, individual variables, and $n$-place predicates. We want $\mathcal{L}^-$ to have the resources to express its own syntax. The usual way of ensuring this is to stipulate that PA (Peano Arithmetic) is expressible in $\mathcal{L}^-$; however, there are some complications with this method that we explore

---

[6] Stewart Shapiro suggested this strategy to me. It is used in [1] and [2].

below. We stipulate that PA is expressible in $\mathcal{L}^-$, but we also make sure that it's individual constants can directly refer to its own closed formulas by including all its closed formulas in the domain of any model for it.[7] Let $\mathcal{L}$ be the result of adding the predicate $D(x)$ to $\mathcal{L}^-$. Let $L^-$ be the set of well-formed formulas of $\mathcal{L}^-$ and let $L$ be the set of well-formed formulas of $\mathcal{L}$.

We consider a neighborhood frame $MF = \langle W, N, \mathfrak{D} \rangle$, where $W$ is a set of worlds, $N$ is a neighborhood function from $W$ to $2^{2^W}$, and $\mathfrak{D}$ is the domain—a non-empty set. Let $\mathfrak{F}$ be a *suitable frame* if and only if:

(i) every neighborhood of every world in $W$ is non-empty,

(ii) every world in $W$ has a neighborhood,

We consider a neighborhood model $\mathfrak{M} = \langle \mathfrak{F}, I \rangle$, where $F$ is a suitable neighborhood frame, and $I$ is an interpretation function.

Let $M_0 = \langle W_0, N_0, D_0, I_0 \rangle$ be a neighborhood model based on a suitable frame, where:

(i) $\mathbb{N} \subset \mathfrak{D}_0$ (i.e., the domain contains the natural numbers),

(ii) $L \subset \mathfrak{D}_0$ (i.e., the domain contains the sentences of $L$),

(iii) $\forall w \in W_0, I_0$ assigns the arithmetic vocabulary in $L$ to its standard interpretation in $D_0$,

(iv) $\forall \phi \in L \exists \sigma$ $\sigma$ is an individual constant of $\mathcal{L}$ and $I_0(\sigma) = \phi$.

(v) $I_0(D(x), w) = \emptyset$ for all $w \in W_0$.

Let $\nu$ be a valuation (i.e., an assignment of elements from the domain to each individual variable of $L$).

(F) $\langle \mathfrak{M}, w \rangle \vDash_\nu F(a_1, \ldots, a_n)$ (where $a_i$ is either an individual constant or an individual variable) if and only if $\langle f(a_1), \ldots, f(a_n) \rangle \in I(F, w)$, where if $a_i$ is a variable $x_i$, then $f(a_i) = \nu(x_i)$, and if $a_i$ is an individual constant $c_i$, then $f(a_i) = I(c_i)$ (for each $n$-place predicate $F$).

($\sim$) $\langle \mathfrak{M}, w \rangle \vDash_\nu \sim \phi$ if and only if it is not the case that $\langle \mathfrak{M}, w \rangle \vDash_\nu \phi$

---

[7]The reason is that Gödel's diagonalization lemma guarantees the existence of sentences like H above that are provably equivalent to $\sim D(\langle H \rangle)$. Note that '$H$' need not refer to '$\sim D(\langle H \rangle)$' for the two to be provably equivalent; nevertheless, provable equivalence is enough for most purposes. However, since the predicate version of rule $E$ fails in ADT and xeno semantics, we need to require that '$H$' actually refers to the formula '$\sim D(\langle H \rangle)$' of $\mathcal{L}$. See [3] for discussion.

(∧)  $\langle \mathfrak{M}, w \rangle \vDash_\nu \phi \wedge \psi$ if and only if $\langle \mathfrak{M}, w \rangle \vDash_\nu \phi$ and $\langle \mathfrak{M}, w \rangle \vDash_\nu \psi$

(∨)  $\langle \mathfrak{M}, w \rangle \vDash_\nu \phi \vee \psi$ if and only if $\langle \mathfrak{M}, w \rangle \vDash_\nu \phi$ or $\langle \mathfrak{M}, w \rangle \vDash_\nu \psi$

(→)  $\langle \mathfrak{M}, w \rangle \vDash_\nu \phi \to \psi$ if and only if if $\langle \mathfrak{M}, w \rangle \vDash_\nu \phi$, then $\langle \mathfrak{M}, w \rangle \vDash_\nu \psi$

(↔)  $\langle \mathfrak{M}, w \rangle \vDash_\nu \phi \leftrightarrow \psi$ if and only if $\langle \mathfrak{M}, w \rangle \vDash_\nu \phi$ iff $\langle \mathfrak{M}, w \rangle \vDash_\nu \psi$

(∀)  $\langle \mathfrak{M}, w \rangle \vDash_\nu \forall x \phi(x)$ if and only if for each $x$-variant $\nu' \langle \mathfrak{M}, w \rangle \vDash_{\nu'} \phi(x)$

(∃)  $\langle \mathfrak{M}, w \rangle \vDash_\nu \exists x \phi(x)$ if and only if there is an $x$-variant $\nu'$ s.t. $\langle \mathfrak{M}, w \rangle \vDash_{\nu'} \phi(x)$

We can say $\langle \mathfrak{M}, w \rangle \vDash \phi$ if and only if $\phi$ is a closed formula and for all valuations $\nu, \langle \mathfrak{M}, w \rangle \vDash_\nu \phi$. Notice that the extension of the descending truth predicate, $D(x)$, is stipulated to be empty in every world in $M_0$. Accordingly, $\mathfrak{M}_0$ has no clause for

$\mathfrak{M}_0$ will serve as the initial model for our first revision sequence. Before presenting the revision sequence, a few definitions are in order.

> A *revision rule* $\rho$ is an operation on the set of functions from $ffD(x)g \times Dg$ to $ft, fg$. The members of this set of functions are **hypotheses**. Each hypothesis interprets $D(x)$. We focus on revision sequences $\Omega$ whose length, $\mathrm{lh}(\Omega)$, is a limit ordinal or On, the class of all ordinals. Let $\Omega @.\alpha$ be the $\alpha$th member of $\Omega$. Let $\Omega \mid \alpha$ be the restriction of $\Omega$ to ordinal $\alpha$.
> 
> If $x \in ft, fg$ and $d \in \mathfrak{D}$, then $d$ is **stably** $x$ in $\Omega$ if and only if $\exists \beta$ s.t. $\beta < \mathrm{lh}(\Omega)$ and for all ordinals $\gamma$, if $\beta \leq \gamma < \ln(\Omega)$ then $[\Omega @ \gamma](d) = x$; the least such $\beta$ is the **stabilization point** of $d$ in $\Omega$. Say $d$ is **stable in** $\Omega$ if and only if for some $x \in ft, fg$, $d$ is stably $x$ in $\Omega$.
> 
> A hypothesis $h$ **coheres with a sequence** $\Omega$ if and only if for all $d \in \mathfrak{D}$ and all $x \in ft, fg$, if $d$ is stably $x$ in $\Omega$ then $h(d) = x$.
> 
> $\Omega$ is a **revision sequence** for $\rho$ if and only if for all $\alpha < \mathrm{lh}(\Omega)$:
> (i)  if $\alpha = \beta + 1$, then $\Omega @ \alpha = \rho(\Omega @ \beta)$, and
> (ii) if $\alpha$ is a limit ordinal then $\Omega @ \alpha$ coheres with $\Omega \mid \alpha$ (i.e., for all $d \in \mathfrak{D}$ and all $x \in ft, fg$ if $d$ is stably $x$ in $\Omega \mid \alpha$, then $\Omega @ \alpha(d) = x$).[8]

These definitions are based on those in Gupta and Belnap (1993), which is the standard reference for revision sequences.

---

[8][1] chapter 5.

Our revision rule, which will generate the revision sequence, is based on the clause (D) that we would have wanted in our first order neighborhood semantics. Let $\Omega_0$ be the revision sequence of length On with initial model $\mathfrak{M}_0$ generated by the following revision rule $\rho_0$:

($\rho_0 - 1$) If $\alpha$ is not a limit ordinal, then $\forall w \in W$, if $\exists X \in N(w)$, s.t. $\forall x \in X, \langle \Omega_0 @ \alpha, x \rangle \vDash \phi$, then $\phi \in I(D, w)$ for $\Omega_0 @ \alpha + 1$; otherwise, $\phi \notin I(D, w)$ for $\Omega_0 @ \alpha + 1$.

($\rho_0 - 2$) If $\alpha$ is a limit ordinal and $D(\langle \phi \rangle)$ is stably true in $\Omega_0 \mid \alpha$, then $\forall w \in W, \phi \in I(D, w)$ for $\Omega_0 @ \alpha$.

($\rho_0 - 3$) If $\alpha$ is a limit ordinal and $D(\langle \phi \rangle)$ is stably false in $\Omega_0 \mid \alpha$, then $\forall w \in W, \phi \notin I(D, w)$ for $\Omega_0 @ \alpha$.

($\rho_0 - 4$) If $\alpha$ is a limit ordinal and $D(\langle \phi \rangle)$ is unstable in $\Omega_0 \mid \alpha$, then $\forall w \in W, \phi \notin I(D, w)$ for $\Omega_0 @ \alpha$

The revision sequence based on this rule will have a fixed set of worlds and a fixed neighborhood function on that set. Obviously, the interpretation, $I$, changes from step to step, but the only difference between steps will be the interpretation of $D(x)$. The interpretation of all other expressions in $L$ does not change. One can think of this as a set of revision sequences, one for the extension of $D(x)$ at each world. Of course, at $\mathfrak{M}_0$, and indeed at each step throughout $\Omega_0$, every world satisfies the same formulas.

## 3.2 The Primary Sequence $\Omega_1$

Remember, $\Omega_0$ is not the sequence we ultimately care about—its role is to help us assign accessibility relations to the sentences of L in a xeno semantics. That is, we use the results of $\Omega_0$ to construct a new revision sequence of xeno models that *does* eventually reach a fixed point.

Not just any xeno frame and xeno model will do for these purposes. We need to define acceptable xeno frame and acceptable xeno model. There is one additional complication in the construction—we distinguish between traditional worlds (the set $C \subseteq W$) and non-traditional worlds (the set $C'$); the clause for $D(x)$ is defined only on traditional worlds and validity is defined as truth at all traditional worlds. The extension of $D(x)$ at non-traditional worlds is stipulated below.[9]

We will consider a constant domain xeno frame $\mathfrak{F} = \langle W, C, N, R, \mathfrak{D} \rangle$, where $W$ is a set of worlds, $C \subseteq W, N$ is a neighborhood function from

---

[9]There is a sense in which the logic determined by the particular xeno semantics I provide could be called a *non-traditional* modal logic in the spirit of non-normal modal logics and non-classical modal logics. I do not know if it is possible to avoid this aspect of the construction.

$W$ to $2^{2^W}$, $R$ is a denumerable set of binary relations on $W$, and $\mathfrak{D}$ is a non-empty set. Let $\mathfrak{F}$ be an *acceptable constant domain xeno frame* if and only if:

1. $C \subset W$ [non-traditional worlds]
2. $\forall w \in W, N(w) \neq \emptyset$ [all worlds have neighborhoods]
3. $\forall w \in W \forall X \in N(w), X \neq \emptyset$ [non-empty neighborhoods]
4. $\forall w \in W \forall X \in N(w), w \in X$ [inclusive neighborhoods]
5. $\forall v \in C' \forall X \in N(v) \forall x \in X, x \in C'$ [non-traditional neighborhoods]
6. $\forall u \in C, C \in N(u)$ [$C$ is a traditional neighborhood]
7. $\forall w \in C, X \in N(w) \vee Y \in N(w) \rightarrow X \cup Y \in N(w)$ [supplemented neighborhoods]
8. If $X \subset C$ then $\forall u \in C, X \notin N(u)$ [no proper subset of $C$ is a traditional neighborhood]

It should be obvious that acceptable constant domain xeno frames exist.

We consider a xeno model $\mathfrak{M} = \langle \mathfrak{F}, \mathfrak{R}, I \rangle$ where $\mathfrak{F}$ is a xeno frame, $\mathfrak{R}$ is an accessibility function ($\mathfrak{R}$ is a function from $L$ to $R$, so it assigns each sentence $\phi$ of $L$ a binary relation on $W$, designated $R_\phi$), and $I$ is an interpretation function ($I$ assigns each individual constant a member of the domain at each world and each $n$-place predicate a set of ordered n-tuples from the domain at each world). We use the following definitions for accessibility relations:

$R_\phi$ is *reflexive* if and only if $\forall w \in W R_\phi ww$
$R_\phi$ is *coreflexive* if and only if $\forall u \in W \forall w \in W, R_\phi wu \rightarrow w = u$
$R_\phi$ is *closed* if and only if $\forall u \in C \forall w \in W, R_\phi uw \rightarrow w \in C$
$R_\phi$ is *open* if and only if $\forall u \in C \exists v \in C', R_\phi uv$

Note that if an accessibility relation is coreflexive, then it is closed (but the converse fails). All accessibility relations in the xeno models we consider are reflexive. Intuitively, the accessibility relations assigned to the instances of axioms of ADT are coreflexive, as are those assigned to sentences of $\mathcal{L}^-$.

Let $\mathfrak{M}$ be an *acceptable xeno model* if and only if:

1. $\mathfrak{F}$ is an acceptable constant domain xeno frame
2. $\forall \phi \in L R_\phi$ is reflexive
3. If $I(\sigma) = I(\tau) \in L$, and $\psi$ results from replacing occurrences of $\sigma$ with $\tau$ in $\phi$, then $R_\phi = R_\psi$.

4. $R_{\phi \wedge \psi} = R_{\psi \wedge \phi}$

5. $R_{\phi \vee \psi} = R_{\psi \vee \phi}$

6. $R_{\sim\sim\phi} = R_\phi$

7. $R_\phi$ is coreflexive for $\phi \in L^-$.

8. If $R_\phi$ is coreflexive then $R_{D\langle\phi\rangle}$ is coreflexive

9. $R_{D\langle\phi\rangle \to \phi}$ is coreflexive

10. $R_{D\langle\sim\phi\rangle \to \sim D\langle\phi\rangle}$ is coreflexive

11. $R_{D\langle\phi\wedge\psi\rangle \to D\langle\phi\rangle \wedge D\langle\psi\rangle}$ is coreflexive

12. $R_{D\langle\phi\rangle \vee D\langle\psi\rangle \to D\langle\phi\vee\psi\rangle}$ is coreflexive

13. $R_\phi$ is coreflexive for $\phi$ a first order classical logical truth.

14. $R_\phi$ is coreflexive for $PA \vdash \phi$

15. If $R_\phi$ is coreflexive and $R_\psi$ is coreflexive then $R_{\phi \to \psi}, R_{\phi \wedge \psi}, R_{\phi \vee \psi}$ are coreflexive.

16. $R_\phi$ is coreflexive if and only if $R_{\sim\phi}$ is coreflexive.

17. If $R_{\phi \wedge \psi}$ is closed and $C \subseteq P_\nu(\phi \wedge \psi)$, then $R_\phi$ is closed and $R_\psi$ is closed.[10]

18. If $R_\phi$ is closed and $R_\psi$ is closed then $R_{\phi \vee \psi}$ is closed.

19. $L \subset \mathfrak{D}$ (i.e., the domain contains all closed sentences of $L$)

20. $\mathbb{N} \subset \mathfrak{D}$ (i.e., the domain contains the natural numbers)

21. $\forall w \in W$, I assigns the arithmetic vocabulary in $\mathcal{L}$ to their standard interpretation in $\mathfrak{D}$

22. All individual constants have the same denotation in every world.

23. All predicates (except possibly $D$) have the same extension in every world.

24. All predicates have the same extension in all non-traditional worlds.

25. $\forall \phi \in L, \exists \sigma \sigma$ is an individual constant of $\mathcal{L}$ and $I(\sigma) = \phi$

---

[10] For each $\phi$ in $L$, $P_\nu(\phi) = \{w \in W : \langle \mathfrak{M}, w \rangle \vDash_\nu \phi\}$

26. If $R_\phi$ is coreflexive then $C \subseteq P_\nu(\phi)$ or $C \subseteq P_\nu(\sim \phi)$.

Let $\Delta_M$ be the set of sentences of $\mathcal{L}$ closed under the following rules:

$(\Delta_\mathfrak{M} - 1)$ $\forall \phi \in L$ if $\phi = D\langle\psi\rangle \to \psi$ then $\phi \in \Delta_\mathfrak{M}$

$(\Delta_\mathfrak{M} - 2)$ $\forall \phi \in L$ if $\phi \in \Delta_\mathfrak{M}$ and $\sigma$ is an individual constant and $I(\sigma) = \phi$, then $D\sigma \in \Delta_\mathfrak{M}$

That is, $\Delta_\mathfrak{M}$ is the set of instances of axiom schema $D1$ closed under applications of $D$.

Let $\mathfrak{M}_1 = \langle W_1, C_1, N_1, R_1, \mathfrak{D}_1, \mathfrak{R}_1, I_1 \rangle$ be an acceptable constant domain xeno model, where:

(i) $I_1(D(x), w) = \emptyset$ for all $w \in W_1$.

(ii) if $\phi$ is stably true in $\Omega_0$, then $R_{1\phi}$ is closed in $\mathfrak{M}_1$.

(iii) if $\phi$ is stably false in $\Omega_0$, then $R_{1\phi}$ is closed in $\mathfrak{M}_1$

(iv) if $\phi$ is unstable in $\Omega_0$ and $\phi \notin \Delta_{M1}$ then $R_{1\phi}$ is open in $\mathfrak{M}_1$.

There are several issues to be settled before we can be sure that $\mathfrak{M}_1$ exists.

First, we need to show that there are acceptable constant domain xeno models. That is easy—let $\mathfrak{M}$ be a constant domain xeno model based on an acceptable xeno frame such that the natural numbers and the closed formulas of $\mathcal{L}$ are members of its domain, the arithmetic vocabulary of $|CL$ receives its standard interpretation in every world, there is a name in $\mathcal{L}$ for every member of the domain, the interpretation is the same at every world, and every relation in $R$ is the identity relation. Then $\mathfrak{M}$ is an acceptable constant domain xeno model.

Second, we need to show that $\mathfrak{R}_1$ is well-defined. Given the interpretation function $I_1$, we define $\Delta_{\mathfrak{M}1}$ as above. There are eighteen conditions under the definition of an acceptable xeno model that pertain to $\mathfrak{R}_1$. None of them conflict with the above specification that defines $\mathfrak{M}_1$. For example, condition 4 is: $R_{1\phi\wedge\psi} = R_{1\psi\wedge\phi}$. It is obvious that $\phi \wedge \psi$ is stable in $\Omega_0$ if and only if $\psi \wedge \phi$ is stable in $\Omega_0$; thus the specification of $R_1$ does not conflict with this condition. The same holds for all the others. One might worry about condition 9, but all instances of $D1$ are in $\Delta_{\mathfrak{M}1}$, so that does not pose a problem. Thus, the above specification of the accessibility relations in $_M 1$ does not conflict with the definition of an acceptable xeno model. Therefore, $\mathfrak{R}_1$ is well-defined.

Let $\nu$ be a valuation (i.e., n assignment of elements from the domain to each individual variable of $\mathcal{L}$).

(F) $\langle \mathfrak{M}, w \rangle \vDash_\nu F(a_1, \ldots, a_n)$ (where $a_i$ is either an individual constant or an individual variable) if and only if $\langle f(a_1), \ldots, f(a_n) \rangle \in I(F, w)$, where if $a_i$ is a variable $x_i$, then $f(a_i) = \nu(x_i)$, and if $a_i$ is an individual constant $c_i$, then $f(a_i) = I(c_i)$ (for each $n$-place predicate $F$).

($\sim$) $\langle \mathfrak{M}, w \rangle \vDash_\nu \sim \phi$ if and only if it is not the case that $\langle \mathfrak{M}, w \rangle \vDash_\nu \phi$

($\wedge$) $\langle \mathfrak{M}, w \rangle \vDash_\nu \phi \wedge \psi$ if and only if $\langle \mathfrak{M}, w \rangle \vDash_\nu \phi$ and $\langle \mathfrak{M}, w \rangle \vDash_\nu \psi$

($\vee$) $\langle \mathfrak{M}, w \rangle \vDash_\nu \phi \vee \psi$ if and only if $\langle \mathfrak{M}, w \rangle \vDash_\nu \phi$ or $\langle \mathfrak{M}, w \rangle \vDash_\nu \psi$

($\rightarrow$) $\langle \mathfrak{M}, w \rangle \vDash_\nu \phi \rightarrow \psi$ if and only if if $\langle \mathfrak{M}, w \rangle \vDash_\nu \phi$, then $\langle \mathfrak{M}, w \rangle \vDash_\nu \psi$

($\leftrightarrow$) $\langle \mathfrak{M}, w \rangle \vDash_\nu \phi \leftrightarrow \psi$ if and only if $\langle \mathfrak{M}, w \rangle \vDash_\nu \phi$ iff $\langle \mathfrak{M}, w \rangle \vDash_\nu \psi$

($\forall$) $\langle \mathfrak{M}, w \rangle \vDash_\nu \forall x \phi(x)$ if and only if for each $x$-variant $\nu'$ $\langle \mathfrak{M}, w \rangle \vDash_{\nu'} \phi(x)$

($\exists$) $\langle \mathfrak{M}, w \rangle \vDash_\nu \exists x \phi(x)$ if and only if there is an $x$-variant $\nu'$ s.t. $\langle \mathfrak{M}, w \rangle \vDash_{\nu'} \phi(x)$

We can say $\langle \mathfrak{M}, w \rangle \vDash \phi$ if and only if $\phi$ is a closed formula and for all valuations $\nu$, $\langle \mathfrak{M}, w \rangle \vDash_\nu \phi$.

Notice that the extension of the descending truth predicate, $D(x)$, is empty in every world in $\mathfrak{M}_1$. Accordingly, $\mathfrak{M}_1$ has no clause for $D(x)$. Let $\Omega_1$ be the revision sequence of length On with initial model $\mathfrak{M}_1$ generated by the following revision rule $\rho_1$:

($\rho_1 - 1$) $\alpha$ is not a limit ordinal: if $\forall u \in C, \forall w \in W, R_\phi uw \rightarrow P_{\alpha-1}(\phi) \in N(w)$, then $\forall w \in W, \phi \in I(D, w)$ for $\Omega_1 @ \alpha$; otherwise, $\forall w \in W, \phi \notin I(D, w)$ for $\Omega_1 @ \alpha$.

($\rho_1 - 2$) $\alpha$ is a limit ordinal: if $D\langle \phi \rangle$ is stably true in $\Omega_1 \mid \alpha$, then $\forall w \in C, \phi \in I(D, w)$ for $\Omega_1 @ \alpha + 1$.

($\rho_1 - 3$) $\alpha$ is a limit ordinal: if $D\langle \phi \rangle$ is stably false in $\Omega_1 \mid \alpha$, then $\forall w \in C, \phi \notin I(D, w)$ for $\Omega_1 @ \alpha + 1$.

($\rho_1 - 4$) $\alpha$ is a limit ordinal: if $D\langle \phi \rangle$ is unstable in $\Omega_1 \mid \alpha$, then $\forall w \in C, \phi \notin I(D, w)$ for $\Omega_1 @ \alpha + 1$.

The revision sequence based on this rule will have a fixed set of worlds and a fixed neighborhood function on that set. As before, the interpretation, $I$, changes from step to step, but the only difference between steps will be the interpretation of $D(x)$. The interpretation of all other expressions in $\mathcal{L}$ does not change. The assignment of accessibility relations to sentences of $\mathcal{L}$ does not change.

## 3.3 A Fixed Point for $\Omega_1$

Now we prove that $\Omega_1$ reaches a fixed point. Before we do that, we need several more definitions and results pertaining to revision sequences (I omit the proofs).

A hypothesis $h$ is **cofinal in a sequence** $\Omega$ if and only if for all ordinals $\alpha < ln(\Omega)$ there is a $\beta$ s.t. $\alpha \leq \beta < ln(\Omega)$ and $\Omega@\beta =$h.

THEOREM 1. $\Omega$ is a sequence of length $On$. Then:

(i) there is a hypothesis $h \in \{t,f\}^D$ that is cofinal in $\Omega$

(ii) there is an ordinal $\alpha$ s.t. for all $\beta \geq \alpha$, $\Omega@\beta$ is confinal in $\Omega$; the least such ordinal is the **initial ordinal** for $\Omega$.

(iii) for all ordinals $\alpha$ there is an ordinal $\beta > \alpha$ satisfying the condition that for all hypotheses $h$ cofinal in $\Omega$ there is an ordinal $\gamma$ s.t. $\alpha \leq \gamma < \beta$ and $\Omega@\gamma = h$; such an ordinal is a **completion ordinal** for $\Omega$ above $\alpha$.

THEOREM 2. For all $d \in D$ and $x \in \{t,f\}$,

(i) if $d$ is stably $x$ in $\Omega$ then the value of $d$ is $x$ in all hypotheses cofinal in $\Omega$

(ii) if $lh(\Omega) = On$, then the converse of (i) is true.

An ordinal $\alpha$ is a **reflection ordinal** for $\Omega$ if and only if $\alpha$ is a limit ordinal $< lh(\Omega)$ s.t.

(i) $\alpha \geq$ the initial ordinal for $\Omega$, and

(ii) for all $d \in D$ and $x \in \{t,f\}$, $d$ is stably $x$ in $\Omega \mid \alpha$ if and only if $d$ is stably $x$ in $\Omega$.

THEOREM 3. Let $\Omega$ be a revision sequence for $\rho$ and $\alpha < lh(\Omega)$. If $\Omega@\alpha$ is a fixed point of $\rho$ then for all $\beta$ s.t. $\alpha+\beta < lh(\Omega)$ we have $\Omega@\alpha+\beta = \Omega@\alpha$; furthermore, an object $d \in D$ is stably $x$ in $\Omega$ if and only if $\Omega@\alpha(d) = x$.

A hypothesis $h$ is **recurring** for $\rho$ if and only if $h$ is cofinal in some revision sequence $\Omega$ of length $On$ for $\rho$.

THEOREM 4. All and only recurring hypotheses are reflexive[11]

So if $\alpha$ is a reflection ordinal, then $\Omega \mid \alpha$ reflects all the stabilities and instabilities in $\Omega$.

---

[11] For proofs, see [1] chapter 5.

With these definitions and results in hand, we are ready to show that $\Omega_1$ reaches a fixed point. I use the convention '$P_\alpha(\phi)$' for $\{w \in W : \langle \Omega_1 @\alpha, w \rangle \models \phi\}$.

Let $\zeta$ be the initial ordinal for $\Omega_1$, and let $\xi$ be a reflection ordinal for $\Omega_1$ s.t. $\xi > \zeta$, and let $\mathfrak{M}_2 = \Omega_1 @\xi$. Thus, $\mathfrak{M}_2$ is a reflexive hypothesis for $\Omega_1$; it follows that $\xi$ is a limit ordinal.

I rely on the following lemmas:

LEMMA 5. $\phi$ is stable in $\Omega_1 \mid \xi$ if and only if $\phi$ is stable in $\Omega_1$. [$\xi$ is a reflection ordinal for $\Omega$, so by definition, $\phi$ is stable in $\Omega \mid \xi$ if and only if $\phi$ is stable in $\Omega$.]

LEMMA 6. For any ordinal $\alpha$, if $\exists w \in C$ s.t. $\langle \Omega_1 @\alpha, w \rangle \models \phi$, then $\forall w \in C, \langle \Omega_1 @\alpha, w \rangle \models \phi$. [By induction—the extension of $D(x)$ is the same at all classical worlds in $\mathfrak{M}_1$, and if the extension of $D(x)$ is the same at all classical worlds in $\Omega @\alpha$, then the extension of $D(x)$ is the same at all classical worlds in $\Omega @\alpha+1$.]

To show that $\mathfrak{M}_2$ is a fixed point for $\rho_1$, we prove that the extension of $D(x)$ does not change from $\mathfrak{M}_2$ to $\rho_1(\mathfrak{M}_2)$ on traditional worlds.

Let $u \in C$. Let $Q \in L$. Assume that $\langle \mathfrak{M}_2, u \rangle \models D\langle Q \rangle$. Assume for reductio that $\langle \rho_1(\mathfrak{M}_2), u \rangle \models\sim D\langle Q \rangle$. $D\langle Q \rangle$ is stably true at $u$ in $\Omega_1 \mid \xi$ [else $\langle \mathfrak{M}_2, u \rangle \models\sim D\langle Q \rangle$, since $\xi$ is a limit ordinal]. $D\langle Q \rangle$ is stably true at $u$ in $\Omega_1$ [by Lemma 5]. $\langle \rho_1(\mathfrak{M}_2), u \rangle \models D\langle Q \rangle$. $\bot$. This result shows that the extension of $D(x)$ does not decrease from $\mathfrak{M}_2$ to $\rho_1(\mathfrak{M}_2)$. Now for the other direction.

Assume that $\langle \mathfrak{M}_2, u \rangle \models\sim D\langle Q \rangle$. Assume for reductio 1 that $\langle \rho_1(\mathfrak{M}_2), u \rangle \models D\langle Q \rangle$. $\forall w \in C, R_Q uw \to P_\xi(Q) \in N(w)$. It follows that $R_Q uu$. Hence, $P_\xi(Q) \in N(u)$. Thus, $\forall X \in N(u), u \in X$. Therefore, $u \in P_\xi(Q)$, and it follows that $\langle \mathfrak{M}_2, u \rangle \models Q$. Either $D\langle Q \rangle$ is stably false at $u$ in $\Omega_1 \mid \xi$ or $D\langle Q \rangle$ is unstable at $u$ in $\Omega_1 \mid \xi$ [else $\langle \mathfrak{M}_2, u \rangle \models D\langle Q \rangle$]. Assume for reductio 2 that $D\langle Q \rangle$ is stably false at $u$ in $\Omega_1 \mid \xi$. It follows that $D\langle Q \rangle$ is stably false at $u$ in $\Omega_1$ [by Lemma 5]. Hence, $\langle \rho_1(\mathfrak{M}_2), w \rangle \models\sim D\langle Q \rangle$. $\bot$ (for reductio 2). Now for the other disjunct. Assume for reductio 3 that $D\langle Q \rangle$ is unstable at $u$ in $\Omega_1 \mid \xi$. $D\langle Q \rangle$ is unstable at $u$ in $\Omega_1$ [by Lemma 5]. Assume for conditional proof that $Q$ is stable at $u$ in $\Omega_1$. Hence, $\forall w \in C, Q$ is stable at $w$ in $\Omega_1$ [by Lemma 6]. Let $\beta$ be the stabilization point for $Q$ at $u$ in $\Omega_1$. Then, $\forall \gamma > \beta u \in P_\gamma(Q)$ or $\forall \gamma > \beta u \notin P_\gamma(Q)$. Hence, $\forall w \in C(\forall \gamma > \beta w \in P_\gamma(Q)$ or $\forall \gamma > \beta w \notin P_\gamma(Q))$. Thus, $\forall w \in W(\forall \gamma > \beta w \in P_\gamma(Q)$ or $\forall \gamma > \beta w \notin P_\gamma(Q))$. Hence, $\forall \gamma > \beta P_\gamma(Q) \in N(u)$ or $\forall \gamma > \beta P_\gamma(Q) \notin N(u)$. $R_Q$ is either open or closed; if $R_Q$ is open, then $\sim D\langle Q \rangle$ is stably false at $u$ in $\Omega_1$. Thus, $R_Q$ is closed. Either $\forall \gamma > \beta \langle \Omega_1 @\gamma, u \rangle \models D\langle Q \rangle$ or $\forall \gamma > \beta \langle \Omega_1 @\gamma, u \rangle \models\sim D\langle Q \rangle$. Hence, $D\langle Q \rangle$ is stable at $u$ in $\Omega_1$. By conditional proof, if $Q$ is stable at $u$ in $\Omega_1$ then $D\langle Q\rangle$ is stable at $u$ in $\Omega_1$. So, by contraposition, if $D\langle Q\rangle$ is unstable at $u$ in $\Omega_1$ then $Q$ is unstable at $u$ in $\Omega_1$. Thus, $Q$ is unstable at $u$ in $\Omega_1$. Hence, $Q$ is unstable at $u$ in $\Omega_1 \mid \xi$. Therefore, $\langle M_2, u\rangle \vDash \sim Q$. We have $\bot$ for reductio 3. And we have $\bot$ for reductio 1. Consequently, we have a fixed point, and $\mathfrak{M}_2$ is the intended model for $\mathcal{L}$.

Since $\mathfrak{M}_2$ is a fixed point for $\rho_1$, we know that for $u \in C$:

(D)  $\langle \mathfrak{M}_2, u\rangle \vDash D(\langle \phi\rangle)$ if and only if $\forall w \in WR_\phi uw \to P(\phi) \in N(w)$

So we have a constant domain xeno semantics for $\mathcal{L}$, and it satisfies the intended clause for $D(x)$. We have not said anything about the non-traditional worlds (we have not needed to say anything about them), but to finish the interpretation of $\mathcal{L}$, we can say that if for all $w \in Cw \vDash \phi$, then for all $v \in C'v \vDash \phi$.

Recall that we have been concentrating on $D(x)$ and ignoring $A(x)$ and $S(x)$. If we can define them in terms of $D(x)$, then that would do the trick. We could use the following definitions:

(M1)  $A(\langle \phi\rangle) \leftrightarrow \sim D(\langle \sim \phi\rangle)$

(M2)  $S(\langle \phi\rangle) \leftrightarrow D(\langle \phi\rangle) \vee \sim A(\langle \phi\rangle)$

Let $\mathcal{L}^+$ be the result of adding $A(x)$ and $S(x)$ to $\mathcal{L}$ and let $L^+$ be the set of formulas of $\mathcal{L}^+$. Here is how to interpret the new predicates. Let $\phi \in L^+/L$. Let $\psi$ result from replacing all occurrences of $A\langle \theta\rangle$ in $\phi$ with $\sim D\langle \sim \theta\rangle$ and replacing all occurrences of $S\langle \theta\rangle$ in $\phi$ with $D\langle \theta\rangle \vee D\langle \sim \theta\rangle$. Then $R_\phi = R_\psi$ and $\forall w \in Ww \vDash \phi$ if and only if $w \vDash \psi$.

To summarize the constant domain xeno semantics for the descending truth predicate, '$D(x)$', the ascending truth predicate, '$A(x)$', and the safety predicate, '$S(x)$':

(F)  $\langle \mathfrak{M}, w\rangle \vDash_\nu F(a_1, \ldots, a_n)$ (where $a_i$ is either an individual constant or an individual variable) if and only if $\langle f(a_1), \ldots, f(a_n)\rangle \in I(F, w)$, where if $a_i$ is a variable $x_i$, then $f(a_i) = \nu(x_i)$, and if $a_i$ is an individual constant $c_i$, then $f(a_i) = I(c_i)$ (for each $n$-place predicate $F$).

($\sim$)  $\langle \mathfrak{M}, w\rangle \vDash_\nu \sim \phi$ if and only if it is not the case that $\langle \mathfrak{M}, w\rangle \vDash_\nu \phi$

($\wedge$)  $\langle \mathfrak{M}, w\rangle \vDash_\nu \phi \wedge \psi$ if and only if $\langle \mathfrak{M}, w\rangle \vDash_\nu \phi$ and $\langle M, w\rangle \vDash_\nu \psi$

($\vee$)  $\langle \mathfrak{M}, w\rangle \vDash_\nu \phi \vee \psi$ if and only if $\langle \mathfrak{M}, w\rangle \vDash_\nu \phi$ or $\langle \mathfrak{M}, w\rangle \vDash_\nu \psi$

($\to$)  $\langle \mathfrak{M}, w\rangle \vDash_\nu \phi \to \psi$ if and only if if $\langle \mathfrak{M}, w\rangle \vDash_\nu \phi$, then $\langle \mathfrak{M}, w\rangle \vDash_\nu \psi$

($\leftrightarrow$)    $\langle\mathfrak{M},w\rangle \vDash_\nu \phi \leftrightarrow \psi$ if and only if $\langle\mathfrak{M},w\rangle \vDash_\nu \phi$ iff $\langle\mathfrak{M},w\rangle \vDash_\nu \psi$

($\forall$)    $\langle\mathfrak{M},w\rangle \vDash_\nu \forall x\phi(x)$ if and only if for each $x$-variant $\nu' \langle\mathfrak{M},w\rangle \vDash_{\nu'} \phi(x)$

($\exists$)    $\langle\mathfrak{M},w\rangle \vDash_\nu \exists x\phi(x)$ if and only if there is an $x$-variant $\nu'$ s.t. $\langle\mathfrak{M},w\rangle \vDash_{\nu'} \phi(x)$

For all $u \in C$:

(D)    $\langle\mathfrak{M},u\rangle \vDash_\nu D\langle\phi\rangle$ if and only if $\forall w \in W R_\phi uw \to P_\nu(\phi) \in N(w)$

(A)    $\langle\mathfrak{M},u\rangle \vDash_\nu A\langle\phi\rangle$ if and only if $\exists w \in W R_{\sim\phi} uw \wedge P_\nu(\sim\phi) \notin N(w)$

(S)    $\langle\mathfrak{M},u\rangle \vDash_\nu S\langle\phi\rangle$ if and only if $\forall w \in W(R_\phi uw \to P_\nu(\phi) \in N(w)) \vee \exists u \in W(R_{\sim\phi} uw \wedge P_\nu(\sim\phi) \notin N(w))$

For all $v \in C'$:

(D)    $\langle\mathfrak{M},v\rangle \vDash_\nu D\langle\phi\rangle$ if and only if $\forall u \in C \langle\mathfrak{M},u\rangle \vDash_\nu D\langle\phi\rangle$

(A)    $\langle\mathfrak{M},v\rangle \vDash_\nu A\langle\phi\rangle$ if and only if $\forall u \in C \langle\mathfrak{M},u\rangle \vDash_\nu A\langle\phi\rangle$

(S)    $\langle\mathfrak{M},v\rangle \vDash_\nu S\langle\phi\rangle$ if and only if $\forall u \in C \langle\mathfrak{M},u\rangle \vDash_\nu S\langle\phi\rangle$

We can say $\langle\mathfrak{M},w\rangle \vDash \phi$ if and only if $\phi$ is a closed formula and for all valuations $\nu, \langle\mathfrak{M},w\rangle \vDash_\nu \phi$.

A sentence $\phi$ is *valid* in a xeno model $M$ if and only if $\forall u \in C \langle\mathfrak{M},u\rangle \vDash \phi$

## 4 Soundness

The fixed point theorem from the previous section shows that we have a well-defined notion of truth at a world for an acceptable constant domain xeno model. It follows that we have a well-defined notion of validity for constant domain xeno models. Now, all that is left is to show that ADT is sound with respect to the class of acceptable constant domain xeno models. It follows from this result that ADT is consistent (relative to our background set theory).

In order to prove soundness, we need to go through each of the axioms of ADT and prove that they are valid in any acceptable xeno model. It is a tedious but trivial exercise to demonstrate this, and I do not give the details here.[12] The result is, if $\phi$ is an axiom of ADT, then $\phi$ is valid in any acceptable constant domain xeno model.

---

[12] For example, to show that all instances of axiom schema D1 are valid, assume for an acceptable constant domain xeno model that $u \in C$ $u \vDash D\langle\phi\rangle$. It follows that $\forall w \in W R_\phi uw \to P(\phi) \in N(w)$. By clause 2 of the definition of an acceptable xeno model, $R_\phi$ is reflexive. Thus, $Ruu$. Hence, $P(\phi) \in N(u)$. By clause 4 of the definition of an acceptable xeno frame, $N$ is inclusive. Thus, $u \in P(\phi)$. Hence, $u \vDash \phi$. Therefore, $\vDash D\langle\phi\rangle \to \phi$. The proofs for the other axioms are similar.

ADT is sound with respect to constant domain xeno semantics if and only if for all acceptable constant domain xeno models $\mu$, any set of sentences $\Gamma$ and any sentence $\phi$, if $\phi$ is provable from $\Gamma$, then the argument from $\Gamma$ to $\phi$ is valid. Argument validity is defined in the usual way: for all acceptable constant domain xeno models $\mu$ if all the members of $\Gamma$ are true in $\mu$, then $\phi$ is true in $\mu$. We know that all the classical logical truths are valid and all classical inference rules are valid. So our proof that all axioms of ADT are valid in any acceptable constant domain xeno model completes our soundness proof. ADT is sound with respect to xeno semantics.

## BIBLIOGRAPHY

[1] Gupta, Anil and Belnap Nuel. (1994) *The Revision Theory of Truth*. Cambridge: MIT Press.
[2] Halbach, Volker, Leitgeb, Hannes, and Welch, Philip. (2003) "Possible-Worlds Semantics for Modal Notions Conceived as Predicates," Journal of Philosophical Logic 32: 179-223.
[3] Heck, Richard. (2007). "Self-Reference and the Languages of Arithmetic," *Philosophia Mathematica* 15: 1-29.
[4] Scharp, Kevin. (2007). "Replacing Truth," *Inquiry* 50: 606-621.
[5] Scharp, Kevin. (2008). "Aletheic Vengeance," in *The Revenge of the Liar: New Essays on the Paradox*. Edited by Jc Beall. Oxford University Press.

# Pragmatic Platonism
MARTIN DAVIS

Although I have never thought of myself as a philosopher, Harvey Friedman has told me that I am "an extreme Platonist". Well, extremism in defense of truth may be no vice, but I do feel the need to defend myself from that description.

## Gödel's Platonism

When one thinks of Platonism in mathematics, one naturally thinks of Gödel. In a letter to Gotthard Günther in 1954, he wrote:

> When I say that one can ... develop a theory of classes as objectively existing entities, I do indeed mean by that existence in the sense of ontological metaphysics, by which, however, I do not want to say that abstract entities are present in nature. They seem rather to form a second plane of reality, which confronts us just as objectively and independently of our thinking as nature.[1]

If indeed that's extreme Platonism, it's not what I believe. I don't find myself confronted by such a "second plane of reality".

In his Gibbs lecture of 1951, Gödel made it clear that he rejected any mechanistic account of mind, claiming (with no citations) that

> ... some of the leading men in brain and nerve physiology ... very decidedly deny the possibility of a purely mechanistic explanation of psychical and nervous processes.[2]

In a 1974 letter evidently meant to help comfort Abraham Robinson who was dying of cancer, he was even more emphatic:

> The assertion that our ego consists of protein molecules seems to me one of the most ridiculous ever made.[3]

---
[1] [5], vol IV, pp. 502-505.
[2] [5] vol. III, p.312.
[3] [5] vol. V, p.204.

Alas, I'm stuck with precisely this ridiculous belief. Although I wouldn't mind at all having the transcendental mind Gödel suggests, I'm aware of no evidence that our mental activity is anything but the work of our physical brains.

In his Gibbs lecture Gödel suggests another possibility:

> If mathematics describes an objective world just like physics, there is no reason why inductive methods should not be applied in mathematics just the same as in physics. The fact is that in mathematics we still have the same attitude today that in former times one had toward all science, namely we try to derive everything by cogent proofs from the definitions (that is, in ontological terminology, from the essences of things). Perhaps this method, if it claims monopoly, is as wrong in mathematics as it was in physics.[4]

I will claim that mathematicians have been using inductive methods, appropriately understood, all along. There is a simplistic view that induction simply means the acceptance of a general proposition on the basis of its having been verified in a large number of cases, so that for example we should regard the Riemann Hypothesis as having been established on the basis of the numerical evidence that has been obtained. But this is unacceptable: no matter how much computation has been carried out, it will have verified only an infinitesimal portion of the infinitude of the cases that need to be considered. But inductive methods (even those used in physics) need to be understood in a much more comprehensive sense.

## Gödel Incompleteness and the Metaphysics of Arithmetic

Gödel has claimed that it was his philosophical stance that made his revolutionary discoveries possible and that his Platonism had begun in his youth. However, an examination of the record shows something quite different, namely a gradual and initially reluctant embrace of Platonism as Gödel considered the philosophical implications of his mathematical work [3]. It is at least as true that Gödel's philosophy was the result of his mathematics as that the latter derived from the former.

In 1887, in an article surveying transfinite numbers from mathematical, philosophical, and theological viewpoints, Georg Cantor made a point of attacking a little pamphlet on counting and measuring written by the great scientist Hermann von Helmholtz. Cantor complained that the pamphlet

---

[4] [5], vol. III, p. 313.

expressed an "extreme empirical-psychological point of view with a dogmatism one would not have thought possible ... " He continued:

> Thus, in today's Germany we see, as a reaction against the overblown Kant-Fichte-Hegel-Schelling Idealism, an *academic-positivistic skepticism* that powerfully dominates the scene. This skepticism has inevitably extended its reach even to *arithmetic*, in which domain it has led to its most fateful conclusions. Ultimately, this may turn out most damaging to this positivistic skepticism itself.

In reviewing a collection of Cantor's papers dealing with the transfinite, Frege chose to emphasize the remark just quoted, writing [6]:

> Yes indeed! This is the very reef on which this doctrine will founder. For ultimately, the role of the infinite in arithmetic is not to be denied; yet, on the other hand, there is no way it can coexist with this epistemological tendency. Thus we can foresee that this issue will provide the setting for a momentous and decisive battle.

In a 1933 lecture, Gödel, considering the consequences of his incompleteness theorems, and perhaps not having entirely shaken off the positivism of the Vienna Circle, showed that the "battle" Frege had predicted was taking place in his own mind:

> The result of our previous discussion is that our axioms, if interpreted as meaningful statements, necessarily presuppose a kind of Platonism, which cannot satisfy any critical mind and which does not even produce the conviction that they are consistent.[5]

The axioms to which Gödel referred were an unending sequence produced by permitting variables for ever higher "types" (in contemporary terminology, sets of ever higher rank) and including axioms appropriate to each level. He pointed out that to each of these levels there corresponds an assertion of a particularly simple arithmetic form, what we now would call a $\Pi_1^0$ sentence, which is not provable from the axioms of that level, but which becomes provable at the next level. In the light of later work[6], a $\Pi_1^0$ sentence can be seen as simply asserting that some particular equation

$$p(x_1, x_2, \ldots, x_n) = 0,$$

---

[5] [5] vol. III, p.50.
[6] [4] pp.331-339.

where $p$ is a polynomial with integer coefficients, has no solutions in natural numbers. To say that such a proposition is *true* is just to say that for each choice of natural number values $a_1, a_2, \ldots, a_n$ for the unknowns,

$$p(a_1, a_2, \ldots, a_n) \neq 0.$$

Moreover a proof for each such special case consists of nothing more than the sequence of additions and multiplications needed to compute the value of the polynomial together with the observation that that value is not 0. So in the situation to which Gödel is calling attention, at a given level there is no single proof that subsumes this infinite collection of special cases, while at the next level there is such a proof.

This powerful way of expressing Gödel incompleteness is not available to one who holds to a purely formalist foundation for mathematics. For a formalist, there is no "truth" above and beyond provability in a particular formal system. Post had reacted to this situation by insisting that Gödel's work requires "at least a partial reversal of the entire axiomatic trend of the late nineteenth and early twentieth centuries, with a return to meaning and truth as being of the essence of mathematics".[7] Frege's reference to the "role of the infinite in arithmetic" is very much to the point here. It is the infinitude of the natural numbers, the infinitude of the sequence of formal systems, and finally, the infinitude of the special cases implied by a $\Pi_1^0$ proposition that point to some form of Platonism.

## Infinity in the Seventeenth Century

Hilbert saw the problem of the infinite as central to resolving foundational issues. Perhaps succumbing a bit to hyperbole, he said:

> The infinite has always stirred the emotions of mankind more deeply than any other question; the infinite has stimulated and fertilized reason as few other ideas have; but also the infinite, more than any other notion is in need of clarification.[8]

People have pronounced and speculated about what is and isn't true about infinity since they began thinking abstractly. Aristotle's views on the subject in particular had a great influence. A discovery made by the Italian mathematician Toricelli in 1641 provides a very revealing example.[9] He found that the volume of a certain solid of infinite extent is finite. The

---

[7] [9] p. 295.
[8] [10] p. 371.
[9] This discussion, including the quotations, is based on Paolo Mancosu's wonderful monograph [7].

solid in question is obtained by rotating about an axis a certain plane figure with infinite area. Specifically, in modern terminology, it is the figure bounded by the hyperbola whose equation is $y = 1/x$, the line $x = 1$ and the horizontal asymptote of the hyperbola, namely the $X$-axis. Toricelli's solid is formed by rotating this figure about the $X$-axis. Although showing that this solid of revolution has a finite volume is a routine "homework" problem in a beginning calculus course,

$$\pi \int_1^\infty \frac{1}{x^2} dx = \pi,$$

at the time it created a sensation because it contradicted prevalent views about the infinite. Toricelli himself remarked "... if one proposes to consider a solid, or a plane figure, infinitely extended, everybody immediately thinks that such a figure must be of infinite size." In 1649, Petri Gassendi wrote,

> Mathematicians ... weave those famous demonstrations, some so extraordinary that they even exceed credibility, like what ... Toricelli showed of a certain... solid infinitely long which nevertheless is equal to a finite cylinder.

Writing in 1666, Isaac Barrow found Toricelli's result contradicting what Aristotle had taught. He referred to Aristotle's dictum, "there is no proportion between the finite and the infinite":

> The truth of which statement, a very usual and well known axiom, has been in part broken by ... modern geometricians [who] demonstrate ... equality of ... solids protracted to infinity with other finite ... solids which prodigy ... Toricelli exhibited first.

Much can be learned from this example about the way in which mathematicians expand the applicability of existing methods to new problems and with how they deal with the philosophical problems that may arise. Toricelli used a technique called the method of *indivisibles*, a method pioneered by Cavalieri that provided a short-cut for solving area and volume problems. Toricelli used this technique to prove that his infinite body had the same volume as a certain finite cylinder. The method conceived of each of the two bodies being compared as constituted of a continuum of plane figures. Although there was no rigorous foundation for this, Cavalieri and later Toricelli showed how effective it could be in easily obtaining interesting results. They were well aware of the Eudoxes-Archimedes method of exhaustion (which they called "the method of the ancients"), and used it to confirm their results and/or to convince skeptics.[10] But, Toricelli insisted on the validity of the new method.

What can we say about Toricelli's methodology? He was certainly not seeking to obtain results by "cogent proofs from the definitions" or "in ontological terms, from the essences of things". He was *experimenting* with a mathematical technique that he had learned, and was attempting to see whether it would work in an uncharted realm. In the process, something new about the infinite was discovered. I insist that this was induction from a body of mathematical experience.

## Robustness of Formalism

An interesting example is provided by the development of complex numbers. The fact that the square of any non-zero real number is positive had been generally accepted as implying that there could be no number whose square is negative. Sixteenth century algebra brought this into question. The quadratic formula, essentially known since antiquity, did seem to lead to solutions which did involve square roots of negative quantities. But those were simply regarded as impossible. But the analogous formula for cubic equations, discovered by Tartaglia and published in Cardano's book of 1545, forced a rethinking of the matter. In the case of a cubic equation with real coefficients and three real roots, the formula led to square roots of negative numbers as intermediary steps in the computation. Bombelli discussed this in his book of 1572. In particular, he noted that although the equation $x^3 - 15x - 4 = 0$ had the three roots $4, -2 + \sqrt{3}, -2 - \sqrt{3}$, the Tartaglia formula forced one to consider $\sqrt{-109}$. Soon mathematicians were working freely with complex numbers without questioning whether they really exist in some "second plane of reality". What this experience illustrates is the robustness of mathematical formalisms. These formalisms often point the way to expansions of the subject matter of mathematics before any kind of convincing justification can be supplied. This is again a case of induction in mathematical practice.

Leibniz referred to this very experience when asked to justify the use of infinitesimals. As Mancosu explains

> ...the problem for Leibniz was not, Do infinitely small articles exist? but, Is the use of infinitely small quantities in calculus reliable? [11]

---

[10] The method of exhaustion typically required one to have the answer at hand, whereas with indivisibles the answer could be computed.

[11] [7] p.172.

In justifying his use of infinitesimals in calculus, Leibniz compared this with the use of complex numbers which had become generally accepted although at the time, there was no rigorous justification.

In another example, the rules of algebra, including the manipulation of infinite series was applied to operators with scant justification. This can be seen in Boole's [2] massive tract on differential equations in which marvelous manipulative dexterity is deployed with not a theorem in sight.

## The Ontology of Mathematics

If the objects of mathematics are not in nature and not in a "second plane of reality," then where are they? Perhaps we can learn something from the physicists. Consider for example, the discussion of the "Anthropic Principle" [1]. The advocates of this principle note that the values of certain critical constants are finely tuned to our very existence. Given even minor deviations, the consequence would be: no human race. It is not relevant here whether this principle is regarded as profound or merely tautological. What I find interesting in this discussion of alternate universes whose properties exclude the existence of us, is that no one worries about their ontology. There is simply a blithe confidence that the same reasoning faculty that serves physicists so well in studying the world that we actually do inhabit, will work just as well in deducing the properties of a somewhat different hypothetical world. A more mundane example is the ubiquitous use of idealization. When Newton calculated the motions of the planets assuming that each of the heavenly bodies is a perfect sphere of uniform density or even a mass particle, no one complained that the ontology of his idealized worlds was obscure. The evidence that our minds are up to the challenge of discovering the properties of alternative worlds is simply that we have successfully done so. Induction indeed! This reassurance is not at all absolute. Like all empirical knowledge it comes without a guarantee that it is certain.

My claim is that what mathematicians do is very much the same. We explore simple austere worlds that differ from the one we inhabit both by their stark simplicity and by their openness to the infinite. It is simply an empirical fact that we are able to obtain apparently reliable and objective information about such worlds. And, because of this, any illusion that this knowledge is certain must be abandoned. If, on a neo-Humean morning, I were to awaken to the skies splitting open, hearing a loud voice bellowing, "This ends Phase 1; Phase 2 now begins," I would of course be astonished. But I will not say that I *know* that this will not happen. If presented with a proof that PA is inconsistent or even that some huge natural number is not the sum of four squares, I would be very very skeptical. But I will not say that I *know* that such a proof must be wrong.

## Infinity today.

Mathematical practice obtains information about what it would be like if there were infinitely many things. It is not at all evident a priori that we can do that. But mathematicians have shown us that we can. Our steps are tentative, but as confidence is acquired we move forward. Our theorems are proved in many different ways, and the results are always the same. Our formalisms are robust and yield information beyond the original intent. To doubt the significance of the concrete evidence for the objectivity of mathematical knowledge is like anti-evolutionists doubting the evidence of paleontology by suggesting that those fossils were part of creation. As was discussed above, Gödel's work has left us with a transfinite sequence of formal systems involving larger and larger sets. Models of these systems can be obtained from initial segments of the famous hierarchy obtained by iterating transfinitely the power set operation $\mathcal{P}$:

$$V_0 = \emptyset; \quad V_{\alpha+1} = \mathcal{P}V_\alpha; \quad V_\lambda = \bigcup_{\alpha < \lambda} V_\alpha, \ \lambda \text{ a limit ordinal}$$

Thus, $V_{\omega 2}$ is a model of the original Zermelo axioms. To obtain a model of the more comprehensive Zermelo-Fraenkel (ZF) axioms, no ordinal whose existence is provable in ZF will do.[12] To continue the transfinite sequence of formal systems, it is necessary to enter the realm of large cardinals in which there has been intensive research. Workers in this realm are pioneers on dangerous ground: although we know that no proof of the consistency with ZF of the existence of these enormous sets is possible, it is always conceivable that a proof in ZF of the inconsistency of one of them will emerge thereby destroying a huge body of work. But the empirical evidence is encouraging. Although the defining characteristics of the various large cardinal types that have been studied seeming quite disparate, they line themselves up neatly in order of increasing consistency strength. Moreover, they have shown themselves to be the correct tool for resolving open questions in descriptive set theory.

So far Gödel incompleteness has had only a negligible effect on mathematical practice. Cantor's continuum hypothesis remains a challenge: although the Gödel-Cohen results prove its undecidability from ZF, if the iterative hierarchy is taken seriously, it does have a truth value whether we can ever

---

[12]Because otherwise the consistency of ZF would be provable in ZF contradicting Gödel's second incompleteness theorem. For that matter the set $V_{\omega 2}$ cannot be proved to exist from the Zermelo axioms alone; in ZF its existence follows using Replacement.

find it or not. In the realm of arithmetic many important unsolved problems, including the Riemann Hypothesis and the Goldbach Conjecture, are equivalent to $\Pi_1^0$ sentences. However, so far no undecidable $\Pi_1^0$ sentences have been found that are provably equivalent to questions previously posed (as has been done for uncomputability). However, Harvey Friedman has produced a remarkable collection of $\Pi_1^0$ and $\Pi_2^0$ arithmetic sentences with clear combinatorial content that can only be resolved in the context of large cardinals.

## The Chimerical Effort to Seek Certainty

Mark Twain suggested the lovely notion of a "Sunday truth": something fervently believed in church on Sunday but having no effect on behavior in the rest of the week. Many mathematicians will profess a belief in formalism when foundational matters are discussed. But in their day-to-day work as mathematicians, they remain thoroughgoing Platonists. The "crisis" in foundations from the turn of the 20th century to the 1920s has quietly dissipated. Set theory as a foundation is evident in the initial chapter of many graduate-level textbooks. The obligation to always point out a use of the axiom of choice is a thing of the past. I haven't heard of anyone calling the proof of Fermat's Last Theorem into question because of the large infinities implicit in Grothendiek universes.[13] But there are those who wish to draw a line between safe and unsafe proof methods. The line is drawn by some who insist on some variety of constructivity. Others demand predicativity. Contemporary foundational research makes such notions precise and obtains theorems on the relative strengths of different methods. But there is no pointless attempt to restrict mathematicians. History suggests that they will use whatever methods work including the higher realms of the infinite.

## BIBLIOGRAPHY

[1] Barrow, John D and Frank J Tipler, *The Anthropic Cosmological Principle,* Oxford University Press, 1986.

[2] Boole, George, *A Treatise on Differential Equations,* Macmillan and Co., London 1865.

[3] Davis, Martin, "What Did Gödel Believe and When Did He Believe It?" Bull. Symb. Logic, vol. 11, (2005), pp. 194-206 .

[4] Davis, Martin, Yuri Matiyasevich, and Julia Robinson, "Hilbert's Tenth Problem. Diophantine Equations: Positive Aspects of a Negative Solution," *Proc. Symp. Pure Math., vol XXVIII(1976):* Positive Aspects of a Negative Solution. pp. 323-378.

[5] Feferman, Sol et al, *Kurt Gödel Collected Works, vol. I–V,* Oxford University Press, 1986-2003.

---

[13]Number theorists regard the use of Grothendiek universes as a mere convenience. See [8] for a careful discussion.

[6] Frege, Gottlob, "Rezension von: Georg Cantor. Zum Lehre vom Transfiniten," *Zeitschrift fr Philosophie und philosophische Kritik,* new series, vol. 100(1892), pp. 269-272.

[7] Mancosu, Paolo, *Philosophy of Mathematics & Mathematical Practice in the Seventeenth Century,* Oxford University Press 1996.

[8] McLarty, Colin, "What Does it Take tp Prove Fermat's Last Theorem? Grothendiek and the Logic of Number Theory," *Bull. Symbolic Logic,* vol. 16(2010), pp. 359-377.

[9] Post, E.L., "Recursively Enumerable Sets of Positive Integers and Their Decision Problems," *Bull. Amer. Math. Soc.,* vol. 50(1944), pp. 284-316.
REPRINTED: *The Undecidable,* Martin Davis, editor, Raven Press, New York 1965; Dover New York 2004.
REPRINTED: *Solvability, Provability, Definability: The Collected Works of Emil L. Post,* Martin Davis, editor, Birkhäuser 1994.

[10] van Heijenoort, Jean ed., *From Frege to Gödel: A Source Book in Mathematical Logic, 1879-1931,* Harvard University Press 1967.

# $\omega$-Models and Well-ordering Principles
MICHAEL RATHJEN

ABSTRACT. The purpose of the paper is to present a general methodology which in many cases allows one to establish an equivalence between two types of statements. The first type is concerned with the existence of $\omega$ models of a theory whereas the second type asserts that a certain (usually well-known) elementary operation on orderings preserves the property of being well-ordered. These results have their roots in a theorem of Harvey Friedman (see [11]) which characterizes the theory **ATR**$_0$ by means of a $\Pi^1_2$ sentence of the form *"if $X$ is well ordered then $f(X)$ is well ordered"*, where $f$ is a standard proof theoretic function from ordinals to ordinals. The approach taken here, however, is rather different. The proofs are entirely different from the ones in Friedman's work. The methods used are purely proof-theoretic and crucially involve cut elimination theorems in infinitary logic with ordinal bounds.

The main result presented in this paper is that the following two statements are equivalent over **RCA**$_0$:

(i) Every set is contained in an $\omega$-model of **ATR**;

(ii) If $\mathfrak{X}$ is a well-ordering, then so is $\Gamma_{\mathfrak{X}}$.

Albeit this result is just an example, it may stand for many others as the methodology exemplified in its proof lends itself to a wide range of applications. One could say that every cut elimination theorem in ordinal-theoretic proof theory encapsulates a theorem of this type. Moreover, the technique has the potential for generalization in that it can be lifted up to $\beta$-models and functors acting on ordinal functions.
MSC 03B30 03F05 03F15 03F35 03F35

## 1 Introduction

The present paper can be viewed as a continuation of [2, 25]. Its aim is to present a general proof-theoretic machinery for investigating statements about well-orderings from a reverse mathematics point of view. These statements are of the form

$$\mathbf{WOP}(f) \qquad \textit{"if } X \textit{ is well ordered then } f(X) \textit{ is well ordered"} \qquad (1)$$

where $f$ is a standard proof theoretic function from ordinals to ordinals. There are by now several examples of functions $f$ where the statement **WOP**($f$) has turned out to be equivalent to one of the theories of reverse mathematics over a weak base theory (usually **RCA**$_0$). The first example is due to Girard [13].

**Theorem 1.1.** *(Girard 1987)* Let **WO**($\mathfrak{X}$) express that $\mathfrak{X}$ is a well ordering. Over **RCA**$_0$ the following are equivalent:

(i) Arithmetic Comprehension

(ii) $\forall \mathfrak{X}\, [\mathbf{WO}(\mathfrak{X}) \to \mathbf{WO}(2^{\mathfrak{X}})]$.

More recently two new theorems appeared in preprints [18, 11] and were finally published in [19]. These results give characterizations of the form (1) for the theories **ACA**$_0^+$ and **ATR**$_0$, respectively, in terms of familiar proof-theoretic functions. **ACA**$_0^+$ denotes the theory **ACA**$_0$ augmented by an axiom asserting that for any set $X$ the $\omega$-th jump in $X$ exists while **ATR**$_0$ asserts the existence of sets constructed by transfinite iterations of arithmetical comprehension. $\alpha \mapsto \varepsilon_\alpha$ denotes the usual $\varepsilon$ function while $\varphi$ stands for the two-place Veblen function familiar from predicative proof theory (cf. [27]). More detailed descriptions of **ATR**$_0$ and the function $\mathfrak{X} \mapsto \varphi \mathfrak{X} 0$ will be given shortly. Definitions of the familiar subsystems of reverse mathematics can be found in [34].

**Theorem 1.2.** *(Montalban, Marcone)* Over **RCA**$_0$ the following are equivalent:

(i) **ACA**$_0^+$

(ii) $\forall \mathfrak{X}\, [\mathbf{WO}(\mathfrak{X}) \to \mathbf{WO}(\varepsilon_{\mathfrak{X}})]$.

A proof-theoretic proof for Theorem 1.2 was given by Afshari and Rathjen [2].

**Theorem 1.3.** *(Friedman, unpublished)* Over **RCA**$_0$ the following are equivalent:

(i) **ATR**$_0$

(ii) $\forall \mathfrak{X}\, [\mathbf{WO}(\mathfrak{X}) \to \mathbf{WO}(\varphi \mathfrak{X} 0)]$.

There is a proof of this result in [18] and again there is a proof using proof theory which is due to Rathjen and Weiermann [26]. The original proofs of Theorem 1.3 and 1.2 used recursion-theoretic and combinatorial results about linear orderings. They build on a result from [9] to the effect that there is no arithmetic sequence of degrees descending by $\omega$-jumps. The latter result was then improved by Steel [35] to descent by Turing jumps: If $Q \subseteq \text{Pow}(\omega) \times \text{Pow}(\omega)$ is arithmetic, then there is no sequence $\{A_n \mid n \in \omega\}$ such that (a) for every $n$, $A_{n+1}$ is the unique set such that $Q(A_n, A_{n+1})$, (b) for every $n$, $A'_{n+1} \leq_T A_n$.

For a proof theorist, theorems 1.2 and 1.3 bear a striking resemblance to cut elimination theorems for infinitary logics. Hearing the statements, but not the proofs, the author was prompted to look for proof-theoretic ways of obtaining these results. The hope was that this would also unearth a common pattern behind them and possibly lead to generalizations. The project commenced in [2] in collaboration with Bahareh Afshari, where a purely proof-theoretic proof of Theorem 1.2 was presented. Joint work with Andreas Weiermann led to [25], giving a new (and again proof-theoretic) proof of 1.3. The main result I want to prove in this paper is the following:

**Theorem 1.4.** Over **RCA**$_0$ the following are equivalent:

(i) $\forall \mathfrak{X} \, [\mathbf{WO}(\mathfrak{X}) \to \mathbf{WO}(\Gamma_\mathfrak{X})]$

(ii) Every set is contained in a countable $\omega$-model of **ATR**.

At this point it might be useful to state precisely what a countable coded $\omega$-model is.

**Definition 1.5.** *Let $T$ be a theory in the language of second order arithmetic, $L_2$. A countable coded $\omega$-model of $T$ is a set $W \subseteq \mathbb{N}$, viewed as encoding the $L_2$-model*

$$\mathbb{M} = (\mathbb{N}, \mathcal{S}, +, \cdot, 0, 1, <)$$

*with $\mathcal{S} = \{(W)_n \mid n \in \mathbb{N}\}$ such that $\mathbb{M} \models T$ (where $(W)_n = \{m \mid \langle n, m \rangle \in W\}$; $\langle , \rangle$ some coding function).*

This definition can be made in **RCA**$_0$ (see [34], Definition VII.2). We write $X \in W$ if $\exists n \, X = (W)_n$.

Another result in the same vein as Theorem 1.4 is from impredicative proof theory. Here we turn to the ordinal representation system used for the ordinal analysis of the theory **ID**$_1$ of non-iterated inductive definitions, which can be expressed in terms of the $\theta$-function (cf. [6]). **ID**$_1$ has the same strength as the subsystem of second order arithmetic based on bar induction, **BI** (cf. [6, 7, 26]). In Simpson's book the acronym used for **BI** is $\Pi^1_\infty$-**TI**$_0$ (cf. [34, §VII.2]). In place of the function $\theta$ we prefer to work with simpler ordinal representations based on the $\psi$-function introduced in [5] or the $\vartheta$-function of [26]. For definiteness we refer to [26]. Given a well-ordering $\mathfrak{X}$, the relativized versions $\vartheta_\mathfrak{X}$ and $\psi_\mathfrak{X}$ of the $\vartheta$-function and the $\psi$-function, respectively, are obtained by adding all the ordinals from $\mathfrak{X}$ to the sets $C_n(\alpha, \beta)$ of [26, §1] and $C_n(\alpha)$ of [26, Definition 3.1] as initial segments, respectively. The resulting well-orderings $\vartheta_{\mathfrak{X}}(\varepsilon_{\Omega+1})$ and $\psi_{\mathfrak{X}}(\varepsilon_{\Omega+1})$ are equivalent owing to [26, Corollary 3.2].

The next Theorem is obtained by the same methodology as 1.4, but it will not be proved in this paper as its proof is too long to be incorporated.

**Theorem 1.6.** Over **RCA**$_0$ the following are equivalent:

(i) *Every set is contained in a countable coded $\omega$-model of* **BI**.

(ii) $\forall \mathfrak{X} \, [\mathbf{WO}(\mathfrak{X}) \to \mathbf{WO}(\psi_{\mathfrak{X}}(\varepsilon_{\Omega+1}))]$.

At first glance, Theorems 1.4 and 1.6 appear to be of a different type than Theorems 1.2 and 1.3. But the similarity becomes more apparent owing to the next result.

**Theorem 1.7.** (**RCA**$_0$)

(i) **ACA**$_0^+$ *is equivalent to the statement that every set is contained in a countable coded $\omega$-model of* **ACA**.

(ii) **ATR**$_0$ *is equivalent to the statement that every set is contained in a countable coded $\omega$-model of* $\Delta_1^1$-**CA** (*or* $\Sigma_1^1$-**DC**).

**Proof:** (i) follows from [2, Lemma 3.4]. (ii) follows from [34, VIII.4.19]. □

As a consequence of Theorems 1.2, 1.3, and 1.7 we have:

**Corollary 1.8.** (**RCA**$_0$)

(i) $\forall \mathfrak{X} \, [\mathbf{WO}(\mathfrak{X}) \to \mathbf{WO}(\varepsilon_{\mathfrak{X}})]$ *is equivalent to the statement that every set is contained in a countable coded $\omega$-model of* **ACA**.

(ii) $\forall \mathfrak{X} \, [\mathbf{WO}(\mathfrak{X}) \to \mathbf{WO}(\varphi \mathfrak{X} 0)]$ *is equivalent to the statement that every set is contained in a countable coded $\omega$-model of* $\Delta_1^1$-**CA** (*or* $\Sigma_1^1$-**DC**).

Taking any of the Theorems 1.2, 1.3, 1.7, 1.6 the proof of the direction $(i) \Rightarrow (ii)$ can be directly inferred or gleaned from a result or proof in the proof-theoretic literature. The harder part is always the implication $(ii) \Rightarrow (i)$. For a theory $T$ let $\mathrm{Mod}_\omega(T)$ be the statement that every set is contained in a countable coded $\omega$-model of $T$. There exists, however, an Ansatz which given a theory $T$ (a subsystem of second order arithmetic) can help one to find a function $f$ on orderings such that the appertaining statements $\mathrm{Mod}_\omega(T)$ and $\mathbf{WOP}(f)$ are equivalent. The first step consists in an attempt to find an $\omega$-model of $T$ via the method of search trees in

$\omega$-logic. This gives rise to a tree $\mathcal{D}$. In case one finds an infinite path $\mathbb{P}$ on $\mathcal{D}$ it can be used to define an $\omega$-model of $T$. The less desirable outcome would be that all paths in $\mathcal{D}$ are finite, ending in a simple axiom. To get out of this predicament one can scour the proof-theoretic literature to find an infinitary proof system $T_\infty$ which enjoys cut elimination such that $\mathcal{D}$ can be viewed as a skeleton of an infinitary proof in $T_\infty$ of the empty sequent. If $T_\infty$ is chosen optimal, the desired function $f$ is the one which measures the cost of cut elimination in $T_\infty$. As there is no cut free proof of the empty sequent, **WOP**$(f)$ implies that $\mathcal{D}$ must possess an infinite path after all, and hence there is an $\omega$-model of $T$ as desired.

## Acknowledgements

This research was supported by a Royal Society International Joint Projects award (2006/R3) and the *Swedish Collegium for Advanced Study*. I would like to thank the *Swedish Collegium for Advanced Study* in Uppsala for providing an excellent research environment. I am grateful for very inspiring discussions with Antonio Montalbán during his visits to Leeds in November 2010.

## 2 The ordering $\Gamma_{\mathfrak{x}}$

In this paper we use ordinal functions stemming from the early days of ordinal representation systems. Before we give a formal definition of $\Gamma_{\mathfrak{x}}$ it might be useful to recall some of the historical background.

### 2.1 A brief history of early ordinal representation systems

In 1904, Hardy [14] wanted to "construct" a subset of $\mathbb{R}$ of size $\aleph_1$. His method was to represent countable ordinals via increasing sequence of natural numbers and then to correlate a decimal expansion with each such sequence. Hardy used two processes on sequences which led to explicit representations for all ordinals $< \omega^2$. Veblen [37] in 1908 extended the initial segment of the countable for which fundamental sequences can be given effectively. The new tools he devised were the operations of **derivation** and **transfinite iteration** applied to **continuous increasing functions** on ordinals.

**Definition 2.1.** *Let* **ON** *be the class of ordinals. A (class) function* $f:$ **ON** $\to$ **ON** *is said to be* **increasing** *if* $\alpha < \beta$ *implies* $f(\alpha) < f(\beta)$ *and* **continuous** *(in the order topology on* **ON***) if*

$$f(\lim_{\xi<\lambda} \alpha_\xi) = \lim_{\xi<\lambda} f(\alpha_\xi)$$

*holds for every limit ordinal* $\lambda$ *and increasing sequence* $(\alpha_\xi)_{\xi<\lambda}$. $f$ *is called* **normal** *if it is increasing and continuous.*

The function $\beta \mapsto \omega + \beta$ is normal while $\beta \mapsto \beta + \omega$ is not continuous at $\omega$ since $\lim_{\xi<\omega}(\xi+\omega) = \omega$ but $(\lim_{\xi<\omega}\xi) + \omega = \omega+\omega$.

**Definition 2.2.** *The* **derivative** *$f'$ of a function $f : \mathbf{ON} \to \mathbf{ON}$ is the function which enumerates in increasing order the solutions of the equation $f(\alpha) = \alpha$, also called the* **fixed points** *of $f$.*

If $f$ is a normal function, $\{\alpha : f(\alpha) = \alpha\}$ is a proper class and $f'$ will be a normal function, too.

**Definition 2.3.** *Now, given a normal function $f : \mathbf{ON} \to \mathbf{ON}$, define a hierarchy of normal functions as follows:*

$$f_0 = f \qquad f_{\alpha+1} = f'_\alpha$$
$$f_\lambda(\xi) = \xi^{th} \text{ element of } \bigcap_{\alpha<\lambda} (\text{Range of } f_\alpha) \qquad \text{for } \lambda \text{ a limit ordinal.}$$

In this way, from the normal function $f$ we get a two-place function, $\varphi_f(\alpha,\beta) := f_\alpha(\beta)$. Veblen then discusses the hierarchy when $f(\alpha) = 1+\alpha$. We shall use the starting function $\ell(\alpha) = \omega^\alpha$. Instead of $\varphi_\ell(\alpha,\beta)$ it is customary to simply write $\varphi\alpha\beta$.

The least ordinal $\gamma > 0$ closed under $\varphi = \varphi_\ell$, i.e. the least ordinal $> 0$ satisfying $(\forall \alpha, \beta < \gamma) \varphi\alpha\beta < \gamma$ is the famous ordinal $\Gamma_0$ which Feferman [8] and Schütte [28, 29] determined to be the least ordinal 'unreachable' by predicative means.

In general, $\Gamma_\alpha$ denotes the $\alpha^{th}$ ordinal closed under $\varphi$.

## 2.2 Definition of $\Gamma_{\bar{x}}$

Via simple coding procedures, countable well-orderings and functions on them can be expressed in the language of second order arithmetic, $L_2$. Variables $X, Y, Z, \ldots$ are supposed to range over subsets of $\mathbb{N}$. Using an elementary injective pairing function $\langle,\rangle$ (e.g. $\langle n,m\rangle := (n+m)^2 + n + 1$), every set $X$ encodes a sequence of sets $(X)_i$, where $(X)_i := \{m \mid \langle i,m\rangle \in X\}$. We also adopt from [34], II.2 the method of encoding a finite sequence $(n_0, \ldots, n_{k-1})$ of natural numbers as a single number $\langle n_0, \ldots, n_{k-1}\rangle$.

**Definition 2.4.** *Every set of natural numbers $Q$ can be viewed as encoding a binary relation $<_Q$ on $\mathbb{N}$ via $n <_Q m$ iff $\langle n,m\rangle \in Q$. The* **field** *of $Q$, $\mathrm{fld}(Q)$ is the set $\{n \mid \exists m [n <_Q m \lor m <_Q n]\}$.*

*We say that $Q$ is a* **well-ordering** *if $<_Q$ is a well-ordering, that is $<_Q$ is a linear ordering of its field and every non-empty subset $U$ of $\mathrm{fld}(Q)$ has a $<_Q$-least element.*

**Definition 2.5.** *Let $Q$ be a linear ordering. Let $\Gamma_u := \langle 0, u\rangle$, $\varphi u a := \langle 1, \langle u, a\rangle\rangle$ and $\alpha_1 + \ldots + \alpha_n := \langle 2, \langle \alpha_1, \ldots, \alpha_n\rangle\rangle$ if $n > 1$.* **SC** *is the set*

$\{\Gamma_u \mid u \in \mathit{fld}(Q)\}$.

We introduce the ordering $\Gamma_Q$ by inductively defining its field $\mathit{fld}(\Gamma_Q)$, the ordering $<_{\Gamma_Q}$, the set $\mathbf{H}$ of additive principal members of $\mathit{fld}(\Gamma_Q)$ and the critical level function $\mathbf{h}(\alpha)$ for $\alpha \in \mathit{fld}(\Gamma_Q)$:

1. $0 \in \mathit{fld}(\Gamma_Q)$ and $\mathbf{h}(0) = 0$.

2. $0 <_{\Gamma_Q} \alpha$ if $\alpha \in \mathit{fld}(\Gamma_Q)$ and $\alpha \neq 0$.

3. If $u \in \mathit{fld}(Q)$ then $\Gamma_u \in \mathit{fld}(\Gamma_Q)$, $\Gamma_u \in \mathbf{H}$ and $\mathbf{h}(\Gamma_u) = \Gamma_u$.

4. $\Gamma_u <_{\Gamma_Q} \Gamma_v$ iff $u <_Q v$.

5. If $\alpha, \beta \in \mathit{fld}(\Gamma_Q)$, $\alpha \notin \mathbf{SC}$ and $\mathbf{h}(\beta) \leq_{\Gamma_Q} \alpha$ then $\varphi\alpha\beta \in \mathit{fld}(\Gamma_Q)$, $\varphi\alpha\beta \in \mathbf{H}$, and $\mathbf{h}(\varphi\alpha\beta) = \alpha$.

6. If $\alpha, \beta \in \mathit{fld}(\Gamma_Q)$, $\alpha \in \mathbf{SC}$, $\mathbf{h}(\beta) \leq_{\Gamma_Q} \alpha$ and $\beta \neq 0$ then $\varphi\alpha\beta \in \mathit{fld}(\Gamma_Q)$, $\varphi\alpha\beta \in \mathbf{H}$, and $\mathbf{h}(\varphi\alpha\beta) = \alpha$.

7. If $\varphi\alpha\beta, \Gamma_u \in \mathit{fld}(\Gamma_Q)$ then

$$\varphi\alpha\beta <_{\Gamma_Q} \Gamma_u \quad \textit{iff} \quad \alpha, \beta < \Gamma_u,$$
$$\Gamma_u <_{\Gamma_Q} \varphi\alpha\beta \quad \textit{iff} \quad \Gamma_u \leq_{\Gamma_Q} \alpha \vee \Gamma_u \leq_{\Gamma_Q} \beta.$$

8. If $\alpha_1, \ldots, \alpha_n \in \mathit{fld}(\Gamma_Q)$, $n > 1$, $\alpha_1, \ldots, \alpha_n \in \mathbf{H}$, and $\alpha_n \leq_{\Gamma_Q} \ldots \leq_{\Gamma_Q} \alpha_1$, then

$$\alpha_1 + \ldots + \alpha_n \in \mathit{fld}(\Gamma_Q)$$

and $\mathbf{h}(\alpha_1 + \ldots + \alpha_n) = 0$.

9. If $\alpha_1 + \ldots + \alpha_n, \beta_1 + \ldots + \beta_m \in \mathit{fld}(\Gamma_Q)$, then

$$\alpha_1 + \ldots + \alpha_n <_{\Gamma_Q} \beta_1 + \ldots + \beta_m \textit{ iff}$$
$$n < m \wedge \forall i \leq n\, \alpha_i = \beta_i \quad \text{or}$$
$$\exists i \leq \min(n,m)[\alpha_i <_{\Gamma_Q} \beta_i \wedge \forall j < i\, \alpha_j = \beta_j].$$

10. If $\alpha_1 + \ldots + \alpha_n \in \mathit{fld}(\Gamma_Q)$ and $\beta \in \mathbf{H}$ then

$$\beta <_{\Gamma_Q} \alpha_1 + \ldots + \alpha_n \quad \textit{iff} \quad \beta \leq_{\Gamma_Q} \alpha_1$$
$$\alpha_1 + \ldots + \alpha_n <_{\Gamma_Q} \beta \quad \textit{iff} \quad \alpha_1 <_{\Gamma_Q} \beta.$$

11. If $\varphi\xi\alpha, \varphi\zeta\beta \in \mathit{fld}(\Gamma_Q)$, then

$$\varphi\xi\alpha <_{\Gamma_Q} \varphi\zeta\beta \quad \mathit{iff} \quad \begin{aligned} &\xi <_{\Gamma_Q} \zeta \wedge \alpha <_{\Gamma_Q} \varphi\zeta\beta \quad \text{or} \\ &\xi = \zeta \wedge \alpha <_{\Gamma_Q} \beta \quad \text{or} \\ &\zeta <_{\Gamma_Q} \xi \wedge \varphi\xi\alpha <_{\Gamma_Q} \beta. \end{aligned}$$

**Lemma 2.6.** (**RCA$_0$**)

(i) If $Q$ is a linear ordering then so is $\Gamma_Q$.

(ii) $\Gamma_Q$ is elementary recursive in $Q$.

# 3 Proof of the Main Theorem: The easy direction

The implication $(ii) \Rightarrow (i)$ of Theorem 1.4 even holds on the basis of intuitionistic logic. Assume (ii) and suppose $\mathbf{WO}(\mathfrak{X})$. Let $U$ be an arbitrary set of natural numbers. By (2) we can pick an $\omega$-model $\mathbb{A}$ of **ATR** which contains $\mathfrak{X}$ and $U$. Inside $\mathbb{A}$ we have transfinite induction on $\mathfrak{X}$ for arbitrary formulae with parameters from $\mathbb{A}$. It therefore follows from [23, Lemma 4.13,4.16] that $\mathbb{A} \models \mathbf{WO}(\Gamma_{\mathfrak{X}})$. Since $U \in \mathbb{A}$ it follows that $U$ has a $\Gamma_{\mathfrak{X}}$-least element unless $U = \emptyset$. Consequently $\mathbf{WO}(\Gamma_{\mathfrak{X}})$ holds as $U$ was an arbitrary set of naturals. □

# 4 Proof of the Main Theorem: The hard direction part 1

Given a set $Q \subseteq \mathbb{N}$ we are to find an $\omega$-model $\mathbb{M}$ of **ATR** containing $Q$. To find $\mathbb{M}$ we follow Schütte's method of proof search (deduction chains) from [27, II§4] which he used to prove the completeness theorem for first order logic (cf. [27, Theorem 5.7]). The method has to be extended to $\omega$-logic, though. Rather than working in the Schütte calculus of positive and negative forms we work in a Gentzen sequent calculus with finite sets of formulas, called sequents. Before we embark on the technical details let's recall the history of this method.

## The method of search trees in $\omega$-logic

An extremely elegant and efficient proof procedure for first order logic consists in producing the search or decomposition tree (in German "Stammbaum") of a given formula. It proceeds by decomposing the formula according to its logical structure and amounts to applying logical rules backwards. This decomposition method has been employed by Schütte [31, 30] to prove the completeness theorem. It is closely related to the method of "semantic tableaux" of Beth [4] and the tableaux of Hintikka [17]. Ultimately, the whole idea derives from Gentzen [12].

The decomposition tree method can also be extended to prove the $\omega$-completeness theorem due to Henkin [15] and Orey [22]. Schütte [32] used it to prove $\omega$-completeness in the arithmetical case.

$\omega$-logic is obtained from first-order logic by adding the rule

$$(\omega) \quad \frac{F(0), F(1), \ldots, F(m), \ldots}{\forall x \, F(x)}$$

with infinitely many premises. The $\omega$-rule is usual attributed to Hilbert [16], though Tarski [36] says that he introduced the rule in 1927 in an unpublished talk to the Polish Philosophical Society at Warsaw. The restriction of the rule, with the premises being enumerated by a recursive function, is sometimes referred to as *Novikov's rule* who in [21] introduced calculi with "constructive" infinite conjunctions and disjunctions.

## 4.1 Deduction chains in $\omega$-logic

For what follows it is convenient (but by no means essential) that **ATR**$_0$ can be axiomatized via a single sentence.

**Lemma 4.1.** **ATR**$_0$ can be axiomatized via a single $\Pi_2^1$ sentence $\forall X \, C(X)$.

**Proof:** **ATR**$_0$ is equivalent over **ACA**$_0$ to the statement that every two well-orderings are comparable (see [34, Theorem V.6.8]). This statement can be expressed via a $\Pi_2^1$ sentence. Moreover, **ACA**$_0$ can be axiomatized via a single $\Pi_2^1$ sentence (see [34, Lemma VIII.1.5]). □

Our formalization of the language of second order arithmetic, L$_2$, will slightly deviate from standard procedures in that it will not have any function symbols. Instead it has a constant $\bar{n}$ for each natural number $n$ and symbols for primitive recursive relations (though we usually omit the bar on top of 0).

**Definition 4.2.**

*(i) Let $U_0, U_1, U_2, \ldots$ be an enumeration of the free set variables of* L$_2$. *We shall assume that all predicate symbols of the language* L$_2$ *are symbols for primitive recursive relations.* L$_2$ *contains predicate symbols for the primitive recursive relations of equality and inequality and possibly more (or all) primitive recursive relations. If $R$ is a symbol in* L$_2$ *for a primitive recursive relation we denote by $R^{\mathbb{N}}$ the primitive recursive relation it stands for. The formula $R(\bar{k}_1, \ldots, \bar{k}_r)$ ($\neg R(\bar{k}_1, \ldots, \bar{k}_r)$) is said to be* true *if $R^{\mathbb{N}}(k_1, \ldots, k_r)$ is true (is false).*

*(ii)* Henceforth a **sequent** will be a finite set of $L_2$-formulas without free number variables.

*(iii)* A sequent $\Gamma$ is **axiomatic** if it satisfies at least one of the following conditions:

1. $\Gamma$ contains a true **literal**, i.e. a true formula of either form $R(\bar{k}_1,\ldots,\bar{k}_r)$ or $\neg R(\bar{k}_1,\ldots,\bar{k}_r)$, where $R$ is a predicate symbol in $L_2$ for a primitive recursive relation.
2. $\Gamma$ contains formulae $\bar{k} \in U$ and $\bar{k} \notin U$ for some set variable $U$ and number $k$.

*(iv)* A sequent is **reducible** or a **redex** if it is not axiomatic and contains a formula which is not a literal.

**Definition 4.3.** For $Q \subseteq \mathbb{N}$ define
$$\bar{Q}(n) = \begin{cases} \bar{n} \in U_0 & \text{if } n \in Q \\ \bar{n} \notin U_0 & \text{otherwise} \end{cases}$$

**Definition 4.4.** Fix $Q \subseteq \mathbb{N}$. Let $\forall X\, C(X)$ be the sentence of Lemma 4.1 which axiomatizes $\mathbf{ATR}_0$.

A **$Q$-deduction chain** is a finite string
$$\Gamma_0, \Gamma_1, \ldots, \Gamma_k$$
of sequents $\Gamma_i$ constructed according to the following rules:

*(i)* $\Gamma_0 = \neg\bar{Q}(0), \neg C(U_0)$.

*(ii)* $\Gamma_i$ is not axiomatic for $i < k$.

*(iii)* If $i < k$ and $\Gamma_i$ is not reducible then
$$\Gamma_{i+1} = \Gamma_i, \neg\bar{Q}(i+1), \neg C(U_{i+1}).$$

*(iv)* Every reducible $\Gamma_i$ with $i < k$ is of the form
$$\Gamma_i', E, \Gamma_i''$$
where $E$ is not a literal and $\Gamma_i'$ contains only literals. $E$ is said to be the **redex** of $\Gamma_i$.

Let $i < k$ and $\Gamma_i$ be reducible. $\Gamma_{i+1}$ is obtained from $\Gamma_i = \Gamma_i', E, \Gamma_i''$ as follows:

1. If $E \equiv E_0 \vee E_1$ then
$$\Gamma_{i+1} = \Gamma'_i, E_0, E_1, \Gamma''_i, \neg \bar{Q}(i+1), \neg C(U_{i+1}).$$

2. If $E \equiv E_0 \wedge E_1$ then
$$\Gamma_{i+1} = \Gamma'_i, E_j, \Gamma''_i, \neg \bar{Q}(i+1), \neg C(U_{i+1})$$
where $j = 0$ or $j = 1$.

3. If $E \equiv \exists x\, F(x)$ then
$$\Gamma_{i+1} = \Gamma'_i, F(\bar{m}), \Gamma''_i, \neg \bar{Q}(i+1), \neg C(U_{i+1}), E$$
where $m$ is the first number such that $F(\bar{m})$ does not occur in $\Gamma_0, \ldots, \Gamma_i$, providing $x$ occurs free in $F(x)$, and $m = 0$ if $x$ does not occur free in $F(x)$.

4. If $E \equiv \forall x\, F(x)$ then
$$\Gamma_{i+1} = \Gamma'_i, F(\bar{m}), \Gamma''_i, \neg \bar{Q}(i+1), \neg C(U_{i+1})$$
for some $m$.

5. If $E \equiv \exists X\, F(X)$ then
$$\Gamma_{i+1} = \Gamma'_i, F(U_m), \Gamma''_i, \neg \bar{Q}(i+1), \neg C(U_{i+1}), E$$
where $m$ is the first number such that $F(U_m)$ does not occur in $\Gamma_0, \ldots, \Gamma_i$, providing $X$ occurs free in $F(X)$, and $m = 0$ if $X$ does not occur free in $F(X)$.

6. If $E \equiv \forall X\, F(X)$ then
$$\Gamma_{i+1} = \Gamma'_i, F(U_m), \Gamma''_i, \neg \bar{Q}(i+1), \neg C(U_{i+1})$$
where $m$ is the first number such that $m \neq i+1$ and $U_m$ does not occur in $\Gamma_i$.

The set of $Q$-deduction chains forms a tree $\mathcal{D}_Q$ labeled with strings of sequents. We will now consider two cases.

**Case I:** $\mathcal{D}_Q$ is not well-founded. Then $\mathcal{D}_Q$ contains an infinite path $\mathbb{P}$. Now define a set $M$ via
$$(M)_i = \{k \mid \bar{k} \notin U_i \text{ occurs in } \mathbb{P}\}.$$
Set $\mathbb{M} = (\mathbb{N}; \{(M)_i \mid i \in \mathbb{N}\}, +, \cdot, 0, 1, <)$.

For a formula $F$, let $F \in \mathbb{P}$ mean that $F$ occurs in $\mathbb{P}$, i.e. $F \in \Gamma$ for some $\Gamma \in \mathbb{P}$.

**Claim:** Under the assignment $U_i \mapsto (M)_i$ we have

$$F \in \mathbb{P} \quad \Rightarrow \quad \mathbb{M} \models \neg F. \tag{2}$$

The Claim will imply that $\mathbb{M}$ is an $\omega$-model of **ATR**. Also note that $(M)_0 = Q$, thus $Q$ is in $\mathbb{M}$. The proof of (2) follows by induction on $F$ using Lemma 4.5 below. The upshot of the foregoing is that we can prove Theorem 1.4 under the assumption that $\mathcal{D}_Q$ is ill-founded for all sets $Q \subseteq \mathbb{N}$.

**Lemma 4.5.** Let $Q$ be an arbitrary subset of $\mathbb{N}$ and $\mathcal{D}_Q$ be the corresponding deduction tree. Moreover, suppose $\mathcal{D}_Q$ is not well-founded. Then $\mathcal{D}_Q$ has an infinite path $\mathbb{P}$. $\mathbb{P}$ has the following properties:

1. $\mathbb{P}$ does not contain literals which are true in $\mathbb{N}$.

2. $\mathbb{P}$ does not contain formulas $s \in U_i$ and $t \notin U_i$ for constant terms $s$ and $t$ such that $s^\mathbb{N} = t^\mathbb{N}$.

3. If $\mathbb{P}$ contains $E_0 \vee E_1$ then $\mathbb{P}$ contains $E_0$ and $E_1$.

4. If $\mathbb{P}$ contains $E_0 \wedge E_1$ then $\mathbb{P}$ contains $E_0$ or $E_1$.

5. If $\mathbb{P}$ contains $\exists x F(x)$ then $\mathbb{P}$ contains $F(\bar{n})$ for all $n$.

6. If $\mathbb{P}$ contains $\forall x F(x)$ then $\mathbb{P}$ contains $F(\bar{n})$ for some $n$.

7. If $\mathbb{P}$ contains $\exists X F(X)$ then $\mathbb{P}$ contains $F(U_m)$ for all $m$.

8. If $\mathbb{P}$ contains $\forall X F(X)$ then $\mathbb{P}$ contains $F(U_m)$ for some $m$.

9. $\mathbb{P}$ contains $\neg C(U_m)$ for all $m$.

10. $\mathbb{P}$ contains $\neg \bar{Q}(m)$ for all $m$.

**Proof:** Standard. □

**Corollary 4.6.** If $\mathcal{D}_Q$ is ill-founded then there exists a countable coded $\omega$-model of **ATR** which contains $Q$.

## 5 Proof of the Main Theorem: The hard direction part 2

The remainder of the paper will be devoted to ruling out the possibility that for some $Q$, $\mathcal{D}_Q$ could be a well-founded tree. This is the place where the principle $\forall \mathfrak{X}\, [\mathbf{WO}(\mathfrak{X}) \to \mathbf{WO}(\Gamma_{\mathfrak{X}})]$ in the guise of cut elimination for an infinitary proof system enters the stage. Aiming at a contradiction, suppose that $\mathcal{D}_Q$ is a well-founded tree. Let $\mathfrak{X}_0$ be the Kleene-Brouwer ordering on $\mathcal{D}_Q$ (see [34, Definition V.1.2]). Then $\mathfrak{X}_0$ is a well-ordering. In a nutshell, the idea is that a well-founded $\mathcal{D}_Q$ gives rise to a derivation of the empty sequent (contradiction) in the infinitary proof systems $T_Q^\infty$ from [24, Section 3]. To make this step more transparent we introduce two intermediate systems $\mathbf{KPl}_0$ and $\mathbf{KPl}_Q^\infty$. $\mathbf{KPl}_0$ is a formal set theory with the natural numbers as urelements. It has a constant $\mathbb{N}$ for the set of natural numbers and a unary predicate symbol $\mathbf{Ad}$ to convey that a set is an admissible set. The axioms of $\mathbf{KPl}_0$ are the usual axioms of Peano arithmetic for the urelements plus the schema of induction on the naturals for arbitrary formulae, extensionality for sets, an axiom saying that the natural numbers (urelements) form a set, an axiom saying that every set is contained in an admissible set and axioms saying that every admissible set is transitive and satisfies the axioms of Kripke-Platek set theory (see [3]), $\mathbf{KP}$, but with the axiom of foundation omitted. It is easy to show that $\mathbf{ATR}$ can be viewed as a subtheory of $\mathbf{KPl}_0$ wherein the second order quantifiers of $\mathbf{ATR}$ are interpreted as set quantifiers ranging over subsets of $\mathbb{N}$.

**Lemma 5.1.** $\mathbf{ATR}$ is a subtheory of $\mathbf{KPl}_0$.

**Proof:** We argue informally in $\mathbf{KPl}_0$. Suppose $\prec$ is a well-ordering on a subset of $\mathbb{N}$, $u, v$ are the free variables of a bounded formula $B(u, v)$, i.e. all quantifiers in $B(u, v)$ are restricted. According to the axioms of $\mathbf{KPl}_0$ we can find an admissible set $\mathbb{A}$ such that $\mathbb{N}, \prec \in \mathbb{A}$ and such that all parameters occurring in $B(u, v)$ are also elements of $\mathbb{A}$. We use induction on $\prec$ to show that for every $n \in \mathbb{N}$ the following statement $C(n)$ holds: there exists a function $f_n \in \mathbb{A}$ with domain $\{i \in \mathbb{N} \mid i \preceq n\}$ such that

$$(\forall i \preceq n)\; f_n(i) = \{\langle j, m \rangle \in \mathbb{N} \times \mathbb{N} \mid j \prec i \wedge B(m, \bigcup_{l \prec j} f_n(l))\}. \quad (3)$$

Note that each function $f_n$ is uniquely determined by (3). Inductively assume that we have a function $f_n$ of this form for all $n \prec k$. By $\Sigma$ collection, bounded separation and union in $\mathbb{A}$, $g_k := \bigcup_{n \prec k} f_n$ is a set in $\mathbb{A}$. Thus the function $f_k$ with domain $\{i \mid i \preceq k\}$ defined by

$$f_k(i) = \begin{cases} g_k(i) & \text{if } i \prec k \\ \{\langle k, m\rangle \mid m \in \mathbb{N} \wedge B(m, \bigcup_{n \prec k} f_n(n))\} & \text{if } i = k \end{cases}$$

is also an element of $\mathbb{A}$. Moreover, $f_k$ satisfies (3) (when we replace $f_n$ by $f_k$), whence $C(k)$ holds.

In view of the foregoing, in order to verify that transfinite arithmetical recursion is provable in $\mathbf{KPl_0}$ we only need to ensure that the above employment of transfinite induction is legitimate. To this end pick an admissible set $\mathbb{B}$ such that $\mathbb{A} \in \mathbb{B}$. Then $\{n \in \mathbb{N} \mid C(n)\}$ is a set by bounded separation in $\mathbb{B}$. □

## 5.1 A sequent calculus for $\mathbf{KPl_0}$

The *language* of $\mathbf{KPl_0}$, $\mathcal{L}$, consists of: *free variables* $a_1, a_2, a_3, \ldots$, *bound variables* $x_1, x_2, x_3, \ldots$, constants, predicate symbols, the *logical symbols* $\neg, \wedge, \vee, \forall, \exists$; and parentheses.

The *constants* are $\mathbb{N}$ for the set of natural numbers and for each natural number $n$ a constant $\bar{n}$. The *terms* are the constants and free variables and will be denoted by letters $s, t, s_0, t_0, \ldots$.

The *predicate symbols* are $\in$ for elementhood, $\mathbf{Ad}$ for the class of admissible sets, a unary predicate $\mathbf{Set}$ to signify that an object is a set, a unary predicate P to denote an arbitrary set of natural numbers, two binary predicates $\mathbf{SUC}, =_\mathbb{N}$ for the successor relation and the identity on natural numbers, respectively. Further, there are two ternary relations $\mathbf{ADD}, \mathbf{MULT}$ for the graphs of addition and multiplication on natural numbers, respectively.

*Formulae* are built from atomic and *negated* atomic formulae by means of the connectives $\wedge, \vee$ and the following construction steps: If $t$ is a term, $a$ is a free variable and $F(a)$ is a formula in which the bound variable $x$ does not occur, then $(\forall x \in t)F(x), (\exists x \in t)F(x), \forall x F(x), \exists x F(x)$ are formulae.

A formula which contains only bounded quantifiers, i.e. quantifiers of the form $(\forall x \in t), (\exists x \in s)$, is said to be a $\Delta_0$-*formula*.

The *negation*, $\neg A$, of a non–atomic formula $A$ is defined to be the formula obtained from $A$ by (i) putting $\neg$ in front any atomic subformula, (ii) replacing $\wedge, \vee, (\forall x \in t), (\exists x \in t), \forall x, \exists x$ by $\vee, \wedge, (\exists x \in t), (\forall x \in t), \exists x, \forall x$, respectively, and (iii) dropping double negations.

*Equality* is defined by

$$a = b \;:\Leftrightarrow\; \big(\neg\mathbf{Set}(a) \wedge \neg\mathbf{Set}(b) \wedge a =_\mathbb{N} b\big) \vee$$
$$\big(\mathbf{Set}(a) \wedge \mathbf{Set}(b) \wedge (\forall x \in a)(x \in b) \wedge (\forall x \in b)(x \in a)\big).$$

As a result of this, we will have to state the Axiom of Extensionality in a different way than usually.

We use $A, B, C, ..., F(a), G(a), ..$ as meta–variables for formulae. Upper case Greek letters $\Delta, \Gamma, \Lambda, ...$ range over finite sets of formulae. The meaning of $\{A_1, ..., A_n\}$ is the disjunction $A_1 \vee \cdots \vee A_n$. $\Gamma, A$ stands for $\Gamma \cup \{A\}$. As usual, $A \to B$ abbreviates $\neg A \vee B$. We shall write $s = \{y \in t : F(y)\}$ for $(\forall y \in s)[y \in t \wedge F(y)] \wedge (\forall y \in t)[F(y) \to y \in t]$. We use $(\forall x_1, ..., x_n \in \mathbb{N})$ as an abbreviation for $(\forall x_1 \in \mathbb{N}) ... (\forall x_n \in \mathbb{N})$.

The axioms of $\mathbf{KPl}_0$ fall into several groups.

**Logical axioms**

1. $\Gamma, A, \neg A$ for each atomic formula $A$.

**Ontological axioms.**

1. $\Gamma, \mathbf{Set}(s) \leftrightarrow s \notin \mathbb{N}$.

2. $\Gamma, \bar{n} \in \mathbb{N}$
   for every number constant $\bar{n}$.

3. $\Gamma, t \in s \to \mathbf{Set}(s)$.

4. $\Gamma, J(s_1, ..., s_n) \to s_1 \in \mathbb{N} \wedge ... \wedge s_n \in \mathbb{N}$
   when $J$ is one of the symbols $\mathrm{P}, \mathbf{SUC}, =_{\mathbb{N}}, \mathbf{ADD}, \mathbf{MULT}$.

**Number-theoretic axioms.**

1. $(\forall x \in \mathbb{N}) \neg \mathbf{SUC}(x, \bar{0})$.

2. $(\forall x \in \mathbb{N})[x \neq_{\mathbb{N}} \bar{0} \to (\exists y \in \mathbb{N}) \mathbf{SUC}(y, x)]$.

3. $(\forall x \in \mathbb{N}) (\exists y \in \mathbb{N}) \mathbf{SUC}(x, y)$.

4. $\Gamma, \mathbf{SUC}(\bar{n}, \overline{n+1})$ for all numbers $n$.

5. $(\forall x, y, z \in \mathbb{N}) [\mathbf{SUC}(x, y) \wedge \mathbf{SUC}(x, z) \to y =_{\mathbb{N}} z]$.

6. $(\forall x, y, z \in \mathbb{N}) [\mathbf{SUC}(y, x) \wedge \mathbf{SUC}(z, x) \to y =_{\mathbb{N}} z]$.

7. $(\forall x, y, z \in \mathbb{N}) (\forall v \in \mathbb{N}) [\mathbf{ADD}(y, x, z) \wedge \mathbf{ADD}(y, x, v) \to z =_{\mathbb{N}} v]$.

8. $(\forall x, y \in \mathbb{N}) (\exists z \in \mathbb{N}) \mathbf{ADD}(x, y, z)$.

9. $(\forall x \in \mathbb{N}) \mathbf{ADD}(x, \bar{0}, x)$.

10. $(\forall x, y, z, v, w \in \mathbb{N}) [\mathbf{ADD}(x, y, z) \wedge \mathbf{SUC}(y, v) \wedge \mathbf{SUC}(z, w) \to \mathbf{ADD}(x, v, w)]$.

11. $(\forall x, y, z, v \in \mathbb{N}) [\mathbf{MULT}(y, x, z) \wedge \mathbf{MULT}(y, x, v) \to z =_\mathbb{N} v]$.
12. $(\forall x, y \in \mathbb{N}) (\exists z \in \mathbb{N}) \mathbf{MULT}(x, y, z)$.
13. $(\forall x \in \mathbb{N}) \mathbf{MULT}(x, \bar{0}, \bar{0})$.
14. $(\forall x, y, z, v, w \in \mathbb{N}) [\mathbf{MULT}(x, y, z) \wedge \mathbf{SUC}(y, v) \wedge \mathbf{ADD}(z, x, w) \to \mathbf{MULT}(x, v, w)]$.

**Equality and Extensionality axioms.**

1. $\Gamma, s = s$.
2. $\Gamma, s = t \wedge A(s) \to A(t)$ for all atomic formulae $A$.

**$\mathbb{N}$-induction.**

$\bar{0} \in s \wedge (\forall x, y \in \mathbb{N}) [x \in s \wedge \mathbf{SUC}(x, y) \to y \in s] \to (\forall x \in \mathbb{N}) (x \in s)$

**Set-theoretic axioms**
**Ad**-*Limit:*
$$\Gamma, \exists y \, (t \in y \wedge \mathbf{Ad}(y)).$$

**Ad**-*Linearity:*
$$\Gamma, \mathbf{Ad}(s) \wedge \mathbf{Ad}(t) \to s \in t \vee s = t \vee t \in s.$$

**Ad-Axioms**
(**Ad**1): $\Gamma, \mathbf{Ad}(s) \to \mathbb{N} \in s \wedge (\forall x \in s) (\forall z \in x) z \in s$
(**Ad**2): $\Gamma, \mathbf{Ad}(s) \to A^s$

where $A^s$ is the relativization of $A$ to $s$ and $A$ is a universal closure of one of the following axioms:
*Pairing:* $\exists x \, (x = \{a, b\})$

*Union:* $\exists x \, (x = \bigcup a)$

$\Delta_0$-*Separation:*
$$\exists x \, (x = \{y \in a : F(y)\})$$

for all $\Delta_0$–formulae $F(b)$

$\Delta_0$-*Collection:*
$$(\forall x \in a) \exists y G(x, y) \to \exists z (\forall x \in a)(\exists y \in z) G(x, y)$$

for all $\Delta_0$–formulae $G(b, c)$.
The *logical rules of inference* are:

($\wedge$) $\dfrac{\Gamma, A \quad \Gamma, A'}{\Gamma, A \wedge A'}$  ($\vee$) $\dfrac{\Gamma, A_i}{\Gamma, A_0 \vee A_1}$ if $i \in \{0,1\}$

($b\forall$) $\dfrac{\Gamma, a \in s \to F(a)}{\Gamma, (\forall x \in s) F(x)}$  ($\forall$) $\dfrac{\Gamma, F(a)}{\Gamma, \forall x F(x)}$

($b\exists$) $\dfrac{\Gamma, t \in s \wedge F(t)}{\Gamma, (\exists x \in s) F(x)}$  ($\exists$) $\dfrac{\Gamma, F(t)}{\Gamma, \exists x F(x)}$

(Cut) $\dfrac{\Gamma, A \quad \Gamma, \neg A}{\Gamma}$

where in ($\forall$) and ($b\forall$) the free variable $a$ is not to occur in the conclusion of the inference.

## 5.2 The infinitary calculus $\mathbf{KPl}_Q^\infty$

In what follows we fix $Q \subseteq \mathbb{N}$. In the main, the infinitary version of $\mathbf{KPl}_0$, denoted $\mathbf{KPl}_Q^\infty$, is obtained from $\mathbf{KPl}_0$ by adding the $\omega$-rule and the basic diagram of $Q$. $\mathbf{KPl}_0$ and $\mathbf{KPl}_Q^\infty$ have the same language. Again, we will be working in a Tait-style formalization of set theory with formulae in negation normal form, i.e. negations only in front of atomic formulas.

$\mathbf{KPl}_Q^\infty$, has the following axioms and rules.

### Basic Axioms

1. **Logical axioms:**

   $\Gamma, A, \neg A$ for each atomic formula $A$.

2. **Ontological** and **number-theoretic axioms:**

   (A1) $\Gamma, s \notin t, s \in t$.

   (A2) $\Gamma, P(\bar{n})$ if $n \in Q$.
   
   $\Gamma, \neg P(\bar{n})$ if $n \notin Q$.

   (A3) $\Gamma, R(\bar{n}_1, \ldots, \bar{n}_k)$ if $R(n_1, \ldots, n_k)$ is true, where $R$ is one of the symbols $\mathbf{SUC}, \mathbf{ADD}, \mathbf{MULT}, =_\mathbb{N}$.

   (A4) $\Gamma, \neg R(\bar{n}_1, \ldots, \bar{n}_k)$ if $R(n_1, \ldots, n_k)$ is false, where $R$ is one of the symbols $\mathbf{SUC}, \mathbf{ADD}, \mathbf{MULT}, =_\mathbb{N}$.

   (A5) $\Gamma, \bar{n} \in \mathbb{N}$.

   (A6) $\Gamma, s \in \mathbb{N}, \mathbf{Set}(s)$.

   (A7) $\Gamma, \neg \mathbf{Set}(s), s \notin \mathbb{N}$.

   (A8) $\Gamma, \neg \mathbf{Set}(\bar{n})$.

(A9) $\Gamma, s \in \mathbb{N}, \neg P(s)$.

(A10) $\Gamma, s_i \in \mathbb{N}, \neg J(s_1, \ldots, s_k)$ if $1 \leq i \leq k$ and $J$ is one of the symbols **SUC, ADD, MULT,** $=_\mathbb{N}$.

(A11) $\Gamma, s \notin \bar{n}$.

3. **Equality** and (at the same time) **extensionality axioms**:
$\Gamma, s = t \wedge A(s) \to A(t)$ for all atomic formulae $A$.

4. **Set-theoretic axioms**:

 **Ad**-*Limit*:   $\Gamma, \exists y\, (s \in y \wedge \mathbf{Ad}(y))$.
 **Ad**-*Linearity*:  $\Gamma, \mathbf{Ad}(s) \wedge \mathbf{Ad}(t) \to s \in t \vee s = t \vee t \in s$.
 (**Ad1**):     $\Gamma, \mathbf{Ad}(s) \to \mathbb{N} \in s \wedge (\forall x \in s)(\forall z \in x)\, z \in s$.
 (**Ad2**):     $\Gamma, \mathbf{Ad}(s) \to A^s$,
        where $A$ is a universal closure of one of the following axioms:
 *Pairing*:    $\exists x\, (x = \{a,b\})$.
 *Union*:     $\exists x\, (x = \bigcup a)$.
 $\Delta_0$-*Separation*: $\exists x\, (x = \{y \in a : F(y)\})$ for all $\Delta_0$-formulae $F(b)$
 $\Delta_0$-*Collection*: $(\forall x \in a)\exists y\, G(x,y) \to \exists z (\forall x \in a)(\exists y \in z) G(x,y)$
        for all $\Delta_0$-formulae $G(b,c)$.

Below $a, b$ always denote free variables. The rules of $\mathbf{KPl}_Q^\infty$ are:

$$(\wedge) \quad \frac{\Gamma, A \quad \Gamma, A'}{\Gamma, A \wedge A'}$$

$$(\vee) \quad \frac{\Gamma, A_i}{\Gamma, A_0 \vee A_1} \quad \text{if } i = 0 \text{ or } i = 1$$

$$(\forall_{\mathbb{N}}) \quad \frac{\Gamma, F(\bar{n}) \text{ for all number constants } \bar{n}}{\Gamma, (\forall x \in \mathbb{N}) F(x)}$$

$$(\exists_{\mathbb{N}}) \quad \frac{\Gamma, F(\bar{n}) \text{ for some number constant } \bar{n}}{\Gamma, (\exists x \in \mathbb{N}) F(x)}$$

$$(b\forall^\infty) \quad \frac{\Gamma, b \in s \to F(b) \text{ for all } b}{\Gamma, (\forall x \in s) F(x)}$$

$$(b\exists^\infty) \quad \frac{\Gamma, t \in s \wedge F(t) \text{ for some } t}{\Gamma, (\exists x \in s) F(x)}$$

$$(\forall^\infty) \quad \frac{\Gamma, F(b) \text{ for all } b}{\Gamma, \forall x\, F(x)}$$

$$(\exists^\infty) \quad \frac{\Gamma, F(t) \text{ for some } t}{\Gamma, \exists x\, F(x)}$$

$$(\text{Cut}) \quad \frac{\Gamma, A \quad \Gamma, \neg A}{\Gamma}$$

The *degree* of a formula $A$ of $\mathcal{L}(\mathbf{KPl}_Q^\infty)$, $\deg(A)$, is defined as follows:

1. $\deg(A) = 0$ if $A$ is $\Delta_0$.

2. $\deg((\exists x \in t) F(x)) := \deg((\forall x \in t) F(x)) := \deg(F(\bar{0})) + 2$ if $F(\bar{0})$ is not $\Delta_0$.

3. $\deg(\exists x\, F(x)) := \deg(\forall x\, F(x)) := \deg(F(\bar{0})) + 1$.

4. $\deg(A \wedge B) := \deg(A \vee B) := \max\{\deg(A), \deg(B)\} + 1$ if $A \wedge B$ is not $\Delta_0$.

The relation $\mathbf{KPl}_Q^\infty \vdash^\beta_k \Gamma$ is inductively defined as follows:

1. If $\Gamma$ is an axiom of $\mathbf{KPl}_Q^\infty$, then $\mathbf{KPl}_Q^\infty \vdash^\beta_k \Gamma$ for all $\beta$ and $k$.

2. If $\mathbf{KPl}_Q^\infty \vdash_k^{\beta_i} \Gamma_i$ and $\beta_i < \beta$ hold for every premise $\Gamma_i$ of a rule other than (Cut), then $\mathbf{KPl}_Q^\infty \vdash_k^\beta \Gamma$ if $\Gamma$ is the conclusion of that rule.

3. If $\mathbf{KPl}_Q^\infty \vdash_k^{\beta_0} \Gamma, A$, $\mathbf{KPl}_Q^\infty \vdash_k^{\beta_1} \Gamma, \neg A$, $\beta_0, \beta_1 < \beta$ and $\deg(A) < k$, then $\mathbf{KPl}_Q^\infty \vdash_k^\beta \Gamma$.

If $\Gamma$ is a set of formulae we use the notation $\Gamma[a_1, \ldots, a_r]$ to convey that all free variables of formulae in $\Gamma$ are contained in the set $\{a_1, \ldots, a_r\}$. We use $F[a_1, \ldots, a_r]$ to convey the analogous thing for a formula $F$.

**Theorem 5.2.** If $\mathbf{KPl}_0$ proves a sequent $\Gamma[a_1, \ldots, a_r]$, then there exist $k < \omega$ and $\alpha < \omega + \omega$ such that for all terms $s_1, \ldots, s_n$,

$$\mathbf{KPl}_Q^\infty \vdash_k^\alpha \Gamma[s_1, \ldots, s_r].$$

**Proof:** This is routine. For induction one has to use the $\omega$-rule $(\forall_\mathbb{N})$. □

**Theorem 5.3.** Let $k < \omega$ and $\Theta$ be a finite set of arithmetical sentences. Then we have:

$$\mathbf{KPl}_Q^\infty \vdash_k^\alpha \Theta \quad \Rightarrow \quad \mathbf{KPl}_Q^\infty \vdash_0^{\Gamma_{\epsilon_\omega \alpha + 2}} \Theta.$$

**Proof:** This follows from Corollary 7.6 at the very end of the paper. □

There are different ways of formalizing infinite deductions in theories like **PA**. We just mention [33] and [10].

### 5.3 Finishing the proof of the main Theorem

Recall that in order to finish the proof of Theorem 1.4 we want to show that $\mathcal{D}_Q$ is not well-founded. Aiming at a contradiction, suppose that $\mathcal{D}_Q$ is a well-founded tree, i.e. all paths in $\mathcal{D}_Q$ are finite, and thus every maximal path ends in a sequent which contains a basic axiom. It is then possible to conceive of $\mathcal{D}_Q$ as a skeleton of a proof in $\mathbf{KPl}_Q^\infty$. Each formula $A$ of $L_2$ can be identified with a formula of $\mathbf{KPl}_Q^\infty$ arising by the following two steps:

1. Replace all second order quantifiers $\forall X \ldots$ and $\exists X \ldots$ by $\forall X(X \subseteq \mathbb{N} \to \ldots)$ and $\exists X(X \subseteq \mathbb{N} \wedge \ldots)$, respectively. (Here we adopt the convention that variables of $L_2$ other than $U_0$ are also variables of the language of $\mathbf{KPl}_Q^\infty$ and $X \subseteq \mathbb{N}$ is an abbreviation for $(\forall v \in X)(v \in \mathbb{N})$.)

2. Replace each subformula of the form $t \in U_0$ by $\mathrm{P}(t)$.

As for (2), note that the variable $U_0$ is axiomatically linked in deduction chains to the set $Q$. In $\mathbf{KPl}_Q^\infty$ this role is taken over by the predicate symbol P.

Now $\mathcal{D}_Q$ can be conceived of as a derivation of the empty sequent $\emptyset$ with **hidden cuts** involving cut formulae of the shape $\neg\bar{Q}(i)$ and $\neg C(U_i)$. Note that by Theorem 5.2 and Lemma 5.1, we have the following result:

**Lemma 5.4.** There exist fixed $k_0, k_1 < \omega$ such that for all $i < \omega$:

(i) $\mathbf{KPl}_Q^\infty \vdash_{k_1}^{\omega+k_0} \neg U_i \subseteq \mathbb{N},\ C(U_i)$.

(ii) $\mathbf{KPl}_Q^\infty \vdash_0^0 \bar{Q}(i)$.

Thus if $\Lambda$ is the sequent attached to a node $\tau$ of $\mathcal{D}_Q$ and $(\Lambda_i)_{i \in I}$ is an enumeration of the sequents attached to the immediate successor nodes of $\tau$ in $\mathcal{D}_Q$ then the transition

$$\frac{(\Lambda_i)_{i \in I}}{\Lambda}$$

can be viewed as a combination of three inferences in $\mathbf{KPl}_Q^\infty$, the first one being logical inferences and the other two being cuts. To make this formally precise, let $\mathfrak{X}_0$ be the Kleene-Brouwer ordering of this tree (see [34, Definition V.1.2]). Note that $\mathfrak{X}_0$ has a maximal element which is the bottom node $\langle\rangle$ of the tree. Next let $\mathfrak{X}_1$ be the well-ordering $\mathfrak{X}_0 \cdot \omega$ defined in [25, Definition 2.2]. At each node $\tau$ of $\mathcal{D}_Q$ the pertaining sequent is of the form $\Gamma_\tau, \neg\bar{Q}(j), \neg C(U_j)$, where $j$ is the highest number such that the pair $\neg\bar{Q}(j), \neg C(U_j)$ belongs to the sequent. We shall convey this by writing:

$$\mathcal{D}_Q \vdash^\tau \Gamma_\tau, \neg\bar{Q}(j), \neg C(U_j).$$

We then have the following result.

**Lemma 5.5.** Let $\tau \in \mathfrak{X}_0$ and suppose the free set variables of $\Gamma_\tau$ with indices $> 0$ are among $U_{i_1}, \ldots, U_{i_r}$. Then there exists a fixed $k$ such that

$$\mathcal{D}_Q \vdash^\tau \Gamma_\tau, \neg\bar{Q}(j), \neg C(U_j)$$

implies

$$\mathbf{KPl}_Q^\infty \vdash_k^{\omega+\tau\cdot\omega} \neg U_{i_1} \subseteq \mathbb{N}, \ldots, \neg U_{i_r} \subseteq \mathbb{N}, \Gamma_\tau. \qquad (4)$$

Here the ordinal $\omega + \tau \cdot \omega$ is a member of $\mathfrak{X}_1$.

**Proof:** Below we use the notation $\tau * 0$ to denote the node of the tree obtained by appending $0$ to the string $\tau$, i.e. what is usually denoted by $\tau^\frown\langle 0\rangle$.

Note that $\deg(C(U_i)) = 3$. So we may put $k := \max(k_1, 4)$ with $k_1$ taken from Lemma 5.4.

The proof proceeds by induction on $\tau$ with respect to the well-ordering $\mathfrak{X}_1$. If $\Gamma_\tau, \neg \bar{Q}(j), \neg C(U_j)$ is axiomatic (in the sense of Definition 4.2 (iii)), then $\Gamma_\tau, \neg \bar{Q}(j)$ an axiom of $\mathbf{KPl}_Q^\infty$ according to (A1), (A3) or (A4). Thence, by Lemma 5.4, (Cut) and weakening:

$$\mathbf{KPl}_Q^\infty \vdash_k^{\omega + \tau \cdot \omega} \neg U_{i_1} \subseteq \mathbb{N}, \ldots, \neg U_{i_r} \subseteq \mathbb{N}, \Gamma_\tau.$$

Now suppose $\Gamma_\tau$ has a redex $E$ of the form $E \equiv \exists X\, F(X)$. Then $\Gamma_\tau = \Gamma'_\tau, E, \Gamma''_\tau$ for some $\Gamma'_\tau, \Gamma''_\tau$, and, moreover,

$$\mathcal{D}_Q \vdash^{\tau * 0} \Gamma'_\tau, F(U_m), \Gamma''_\tau, E, \neg \bar{Q}(j), \neg C(U_j), \neg \bar{Q}(i+1), \neg C(U_{i+1})$$

for some $i, m$. Since $\tau * 0$ is smaller than $\tau$ in the Kleene-Brouwer ordering, by the induction hypothesis (and a little help from weakening):

$$\mathbf{KPl}_Q^\infty \vdash_k^{\omega + (\tau * 0) \cdot \omega} \neg U_{i_1} \subseteq \mathbb{N}, \ldots, \neg U_{i_r} \subseteq \mathbb{N}, \neg U_m \subseteq \mathbb{N}, \neg U_j \subseteq \mathbb{N},$$
$$\neg U_{i+1} \subseteq \mathbb{N} \Gamma'_\tau, F(U_m), \Gamma''_\tau, E, \neg \bar{Q}(j), \neg C(U_j),$$
$$\neg \bar{Q}(i+1), \neg C(U_{i+1}).$$

Owing to Lemma 5.4 we get for some $k_0$, using several cuts:

$$\mathbf{KPl}_Q^\infty \vdash_k^{\omega + (\tau * 0) \cdot \omega + k_0} \neg U_{i_1} \subseteq \mathbb{N}, \ldots, \neg U_{i_r} \subseteq \mathbb{N}, \neg U_m \subseteq \mathbb{N}, \neg U_j \subseteq \mathbb{N},$$
$$\neg U_{i+1} \subseteq \mathbb{N}, \Gamma'_\tau, F(U_m), \Gamma''_\tau, E.$$

Since $\mathbf{KPl}_Q^\infty \vdash_0^{k'} U_m \subseteq \mathbb{N}, \neg U_m \subseteq \mathbb{N}$ for some $k'$ we get

$$\mathbf{KPl}_Q^\infty \vdash_k^{\omega + (\tau * 0) \cdot \omega + k_0 + 1} \neg U_{i_1} \subseteq \mathbb{N}, \ldots, \neg U_{i_r} \subseteq \mathbb{N}, \neg U_m \subseteq \mathbb{N}, \neg U_j \subseteq \mathbb{N},$$
$$\neg U_{i+1} \subseteq \mathbb{N}, \Gamma'_\tau, U_m \subseteq \mathbb{N} \wedge F(U_m), \Gamma''_\tau, E$$

via $(\wedge)$ and thus

$$\mathbf{KPl}_Q^\infty \vdash_k^{\omega + (\tau * 0) \cdot \omega + k_0 + 2} \neg U_{i_1} \subseteq \mathbb{N}, \ldots, \neg U_{i_r} \subseteq \mathbb{N}, \neg U_m \subseteq \mathbb{N}, \neg U_j \subseteq \mathbb{N},$$
$$\neg U_{i+1} \subseteq \mathbb{N}, \Gamma'_\tau, \Gamma''_\tau, E$$

since $E \equiv \exists X (X \subseteq \mathbb{N} \wedge F(X))$. If $i+1, j, m \in \{i_1, \ldots, i_r\}$ we are done. If this is not the case we can substitute the set constant $\mathbb{N}$ for any of the variables whose index does not belong to $\{i_1, \ldots, i_r\}$ everywhere in the derivation. This does not change the length of the derivation. As a result we have:

$$\mathbf{KPl}_Q^\infty \vdash_k^{\omega+(\tau*0)\cdot\omega+k_0+2} \neg U_{i_1} \subseteq \mathbb{N}, \ldots, \neg U_{i_r} \subseteq \mathbb{N}, \neg \mathbb{N} \subseteq \mathbb{N}, \Gamma'_\tau, \Gamma''_\tau, E\,.$$

Since $\mathbf{KPl}_Q^\infty \vdash_0^{k''} \mathbb{N} \subseteq \mathbb{N}$, a cut yields

$$\mathbf{KPl}_Q^\infty \vdash_k^{\omega+(\tau*0)\cdot\omega+k_0+3} \neg U_{i_1} \subseteq \mathbb{N}, \ldots, \neg U_{i_r} \subseteq \mathbb{N}, \Gamma'_\tau, \Gamma''_\tau, E$$

and hence

$$\mathbf{KPl}_Q^\infty \vdash_k^{\omega+\tau\cdot\omega} \neg U_{i_1} \subseteq \mathbb{N}, \ldots, \neg U_{i_r} \subseteq \mathbb{N}, \Gamma_\tau\,.$$

Next suppose $\Gamma_\tau$ has a redex $E$ of the form $E \equiv \forall X\, F(X)$. Then $\Gamma_\tau = \Gamma'_\tau, E, \Gamma''_\tau$ for some $\Gamma'_\tau, \Gamma''_\tau$, and, moreover,

$$\mathcal{D}_Q \vdash^{\tau*0} \Gamma'_\tau, F(U_m), \Gamma''_\tau, E, \neg \bar{Q}(j), \neg C(U_j), \neg \bar{Q}(i+1), \neg C(U_{i+1})$$

for some $i, m$ with the proviso that $U_m$ occurs only in $F(U_m)$, $i+1 \neq m$ and $j \neq m$. By the induction hypothesis and weakening we have:

$$\mathbf{KPl}_Q^\infty \vdash_k^{\omega+(\tau*0)\cdot\omega} \neg U_{i_1} \subseteq \mathbb{N}, \ldots, \neg U_{i_r} \subseteq \mathbb{N}, \neg U_m \subseteq \mathbb{N}, \neg U_j \subseteq \mathbb{N},$$
$$\neg U_{i+1} \subseteq \mathbb{N}, \Gamma'_\tau, F(U_m), \Gamma''_\tau, E, \neg \bar{Q}(j), \neg C(U_j),$$
$$\neg \bar{Q}(i+1), \neg C(U_{i+1}).$$

Owing to Lemma 5.4 we get

$$\mathbf{KPl}_Q^\infty \vdash_k^{\omega+(\tau*0)\cdot\omega+k_0} \neg U_{i_1} \subseteq \mathbb{N}, \ldots, \neg U_{i_r} \subseteq \mathbb{N}, \neg U_j \subseteq \mathbb{N}, \neg U_{i+1} \subseteq \mathbb{N}$$
$$\Gamma'_\tau, U_m \subseteq \mathbb{N} \to F(U_m), \Gamma''_\tau, E,$$

for some $k_0$, using several cuts and $(\vee)$ twice. As $U_m$ is an eigenvariable we infer (via $(\forall^\infty)$) that

$$\mathbf{KPl}_Q^\infty \vdash_k^{\omega+(\tau*0)\cdot\omega+k_0+1} \neg U_{i_1} \subseteq \mathbb{N}, \ldots, \neg U_{i_r} \subseteq \mathbb{N}, \neg U_j \subseteq \mathbb{N}, \neg U_{i+1} \subseteq \mathbb{N},$$
$$\Gamma'_\tau, \Gamma''_\tau, E$$

since $E \equiv \forall X(X \subseteq \mathbb{N} \to F(X))$. If $i+1, j \in \{i_1, \ldots, i_r\}$ we are done. If this is not the case we can substitute everywhere in the derivation the set constant $\mathbb{N}$ for any of the variables whose index does not belong to $\{i_1, \ldots, i_r\}$. This does not change the length of the derivation. As a result we have:

$$\mathbf{KPl}_Q^\infty \vdash_k^{\omega+(\tau*0)\cdot\omega+k_0+1} \neg U_{i_1} \subseteq \mathbb{N}, \ldots, \neg U_{i_r} \subseteq \mathbb{N}, \neg \mathbb{N} \subseteq \mathbb{N}, \Gamma'_\tau, \Gamma''_\tau, E\,.$$

Since $\mathbf{KPl}_Q^\infty \vdash_0^{k''} \mathbb{N} \subseteq \mathbb{N}$, a cut yields

$$\mathbf{KPl}_Q^\infty \vdash_k^{\omega+(\tau*0)\cdot\omega+k_0+2} \neg U_{i_1} \subseteq \mathbb{N}, \ldots, \neg U_{i_r} \subseteq \mathbb{N}, \Gamma'_\tau, \Gamma''_\tau, E$$

and hence

$$\mathbf{KPl}_Q^\infty \vdash_k^{\omega+\tau\cdot\omega} \neg U_{i_1} \subseteq \mathbb{N}, \ldots, \neg U_{i_r} \subseteq \mathbb{N}, \Gamma_\tau.$$

Finally, if the redex $E$ is of the form $E_0 \vee E_1$ or $E_0 \wedge E_1$ the desired assertion follows by similar (simpler) considerations. □

We can now finish the proof of Theorem 1.4. Let $\tau_0$ be the bottom node of the tree $\mathcal{D}_Q$. By Lemma 5.5 we have

$$\mathbf{KPl}_Q^\infty \vdash_k^{\omega+\tau_0\cdot\omega} \emptyset.$$

Going to the well-ordering $\Gamma_{\aleph_1}$ we can employ Theorem 5.3, arriving at

$$\mathbf{KPl}_Q^\infty \vdash_0^{\Gamma_{\varepsilon_\omega \tau_0+2}} \emptyset.$$

However, this is impossible since a cut free derivation in $\mathbf{KPl}_Q^\infty$ cannot produce the empty sequent as any derivation starts from axioms and formulae can only disappear via cuts.

## 6 Prospectus

A statement of the form $\mathbf{WOP}(f)$ is $\Pi_2^1$ and therefore cannot be equivalent to a theory whose axioms have a higher complexity, like for instance $\Pi_1^1$-comprehension. After $\omega$-models come $\beta$-models and the theory $\Pi_1^1$-$\mathbf{CA}$ has a characterization in terms of countable coded $\beta$-models (see [34, VII]), namely via the statement "every set belongs to a countably coded $\beta$-model". An $\omega$-model $\mathfrak{A}$ is a $\beta$-model if the concept of well ordering is absolute with respect to $\mathfrak{A}$.

The question arises whether the methodology of this paper can be extended to more complex axiom systems, in particular to those characterizable via $\beta$-models? The answer will be couched as a conjecture. First of all, to get equivalences one has to climb up in the type structure. Given a functor

$$F : (\mathbb{LO} \to \mathbb{LO}) \to (\mathbb{LO} \to \mathbb{LO}),$$

where $\mathbb{LO}$ is the class of linear orderings, we consider the statement:

$\mathbf{WOPP}(F):$ $\quad \forall f \in (\mathbb{LO} \to \mathbb{LO}) \, [\mathbf{WOP}(f) \to \mathbf{WOP}(F(f))].$

There is also a variant of **WOPP**($F$) which should basically encapsulate the same "power". Given a functor

$$G : (\mathbb{LO} \to \mathbb{LO}) \to \mathbb{LO}$$

consider the statement:

**WOPP**$_1(G)$ :  $\quad \forall f \in (\mathbb{LO} \to \mathbb{LO}) \, [\textbf{WOP}(f) \to \textbf{WOP}(G(f))]$.

**Conjecture 6.1.** Statements of the form **WOPP**($F$) (or **WOPP**$_1(F)$), where $F$ comes from some ordinal ordinal representation system used for an ordinal analysis of a theory $T_F$, are equivalent to statements of the form "every set belongs to a countable coded $\beta$-model of $T_F$".

The conjecture may be a bit vague, but it has been corroborated in some cases (around $\Pi^1_1$-**CA**), and, what is perhaps more important, the proof technology exhibited in this paper seems to be sufficiently malleable as to be applicable to the extended scenario of $\beta$-models, too.

At this point I'd like to point out that Antonio Montálban has made several (precise) conjectures about statements of the type 6.1 being equivalent to $\Pi^1_1$-**CA** in [20].

# 7 Appendix: The infinitary calculus $\mathcal{T}_Q^\infty$

We still do not have a complete proof of Theorem 1.4 because we haven't proved the cut elimination Theorem 5.3 for **KPl**$_Q^\infty$. This appendix is devoted to this task. It requires the introduction of yet another proof calculus, the system $\mathcal{T}_Q^\infty$ from [24].

In addition to the constants and relation symbols of **KPl**$_Q^\infty$, the language of $\mathcal{T}_Q^\infty$, $\mathcal{L}(\mathcal{T}_Q^\infty)$, has the following symbols:

- constants $\mathbf{M}_\alpha$ for all $1 \leq \alpha \leq \omega$.

- free variables $a^i, b^i, c^i, \ldots$ for all $i < \omega$.

Letting $\mathbf{M}_0 := \mathbb{N}$, the intended meaning of $\mathbf{M}_{n+1}$ is the least admissible set above $\mathbf{M}_n$ while $\mathbf{M}_\omega = \bigcup_{n<\omega} \mathbf{M}_n$.

Variables $a^i$ are supposed to range over elements of $\mathbf{M}_{i+1}$ which are not numbers, i.e. sets.

The *terms of* $\mathcal{L}(\mathcal{T}_Q^\infty)$ consist of the constants and variables. Each term $t$ possesses a *level*, $|t|$, which is defined as follows:

$$\begin{aligned} |\bar{n}| &:= 0 \\ |\mathbb{N}| &:= 0 \\ |\mathbf{M}_\alpha| &:= \alpha \\ |a^i| &:= i. \end{aligned}$$

The *atomic formulae* of $\mathcal{L}(\mathcal{T}_Q^\infty)$ are obtained from atomic formulae of $\mathcal{L}(\mathbf{KPl}_Q^\infty)$ by replacing all its free variables with terms of $\mathcal{L}(\mathcal{T}_Q^\infty)$ having levels $< \omega$ (hence $\mathbf{M}_\omega$ does not appear in atomic formulae of $\mathcal{L}(\mathcal{T}_Q^\infty)$).

*Formulae* are built from atomic and *negated* atomic formulae by means of the connectives $\wedge, \vee$ and the following construction step: If $s$ is a term and $a^\alpha$ is a free variable of $\mathcal{L}(\mathcal{T}_Q^\infty)$ and $\mathcal{F}(a^\alpha)$ is a formula in which the bound variable $x$ does not occur, then $(\forall x \in s)\mathcal{F}(x)$ and $(\exists x \in s)\mathcal{F}(x)$ are formulae.

Notice that formally $\mathcal{L}(\mathcal{T}_Q^\infty)$ formulae do not have unbounded quantifiers, albeit the quantifiers $(\forall x \in \mathbf{M}_\omega)$ and $(\exists x \in \mathbf{M}_\omega)$ can be viewed as unbounded as they range over the entire universe of discourse of $\mathcal{T}_Q^\infty$.

Below we use the relation $\equiv$ to mean syntactical identity. For terms $s, t$ we set

$$s \triangleleft t :\Leftrightarrow \begin{cases} [\, s \text{ is a numeral and } t \equiv \mathbb{N}\,] \\ \text{or} \quad [\,|s| < |t| \text{ and } t \equiv \mathbf{M}_\alpha \text{ for some } \alpha > 0\,] \\ \text{or} \quad [\,|s| \leq |t| \text{ and } t \equiv a^i \text{ for some } i\,]. \end{cases}$$

For terms $s, t$ with $s \triangleleft t$ we set

$$s \overset{\circ}{\in} t :\equiv \begin{cases} \bar{0} =_\mathbb{N} \bar{0} & \text{if } t \equiv \mathbb{N} \\ \bar{0} =_\mathbb{N} \bar{0} & \text{if } t \equiv \mathbf{M}_\alpha \text{ for some } \alpha > 0 \\ s \in t & \text{if } t \equiv a^i \text{ for some } i. \end{cases}$$

The *rank* of formulae and terms is determined as follows.

1. $rk(t) := \omega \cdot |t|$.

2. $rk(s \overset{\circ}{\in} t) := rk(s \overset{\circ}{\notin} t) := \max(rk(s)+6, rk(t)+1)$.

3. $rk(\mathbf{Ad}(s)) := rk(\neg \mathbf{Ad}(s)) := rk(s) + 9$.

4. $rk(J(s_1,\ldots,s_n)) = rk(\neg J(s_1,\ldots,s_n)) = \max(rk(s_1),\ldots,rk(s_n)) + 1$ if $J$ is a predicate symbol other than $\in$ and $\mathbf{Ad}$.

5. $rk((\exists x \in t)F(x)) := rk((\forall x \in t)F(x)) := max(rk(t), rk(F(\bar{0})) + 2)$ provided that $t$ is not a variable.

6. $rk((\exists x \in a^i)F(x)) := rk((\forall x \in a^i)F(x)) := \max(rk(a^i)+6, rk(F(a^i)))+2$.

7. $rk(A \wedge B) := rk(A \vee B) := \max(rk(A), rk(B)) + 1$.

Let $0 < k < \omega$. A formula of $\mathcal{T}_Q^\infty$ is $\Delta_0(k)$ if all the terms occurring in it have levels $< k$.

A formula is $\Sigma(k)$ if it is in the smallest class of formulae containing the $\Delta_0(k)$ formulae which is closed under $\wedge, \vee$ and bounded quantifiers $(\exists x \in t)$, $(\forall x \in s)$, $(\exists x \in a^i)$, $(\forall x \in a^i)$, providing $|s|, i < k$ and $|t| \leq k$, where $s, t$ are closed terms.

A formula is $\Sigma_\infty(k)$ if it is in the smallest class of formulae containing the $\Delta_0(k)$ formulae which is closed under $\wedge, \vee$ and bounded quantifiers $(\exists x \in t)$, $(\forall x \in s)$, $(\exists x \in a^i)$, $(\forall x \in a^i)$, providing $i < k$ and $|s|, |t| \leq k$, where $s, t$ are closed terms.

Observe that if $A$ is $\Delta_0(k)$, then $rk(A) < \omega \cdot k$. If $B$ is $\Sigma_\infty(k)$, then $rk(B) < \omega \cdot k + \omega$. If $C$ is of the form $(\exists x \in t)F(x)$, where $t$ is a closed term with $|t| = k$ and $F(\bar{0})$ is $\Delta_0(k)$, then $rk(C) = \omega \cdot k$.

The *logical, ontological and arithmetic axioms* of $T_Q^\infty$ are:

(A1) $\Gamma, s \notin t, s \in t$.

(A2) $\Gamma, \mathrm{P}(\bar{n})$ if $n \in Q$.
$\Gamma, \neg \mathrm{P}(\bar{n})$ if $n \notin Q$.

(A3) $\Gamma, R(\bar{n}_1, \ldots, \bar{n}_k)$ if $R(n_1, \ldots, n_k)$ is true, where $R$ is one of the symbols $\mathbf{SUC}, \mathbf{ADD}, \mathbf{MULT}, =_\mathbb{N}$.

(A4) $\Gamma, \neg R(\bar{n}_1, \ldots, \bar{n}_k)$ if $R(n_1, \ldots, n_k)$ is false, where $R$ is one of the symbols $\mathbf{SUC}, \mathbf{ADD}, \mathbf{MULT}, =_\mathbb{N}$.

(A5) $\Gamma, \bar{n} \in \mathbb{N}$.

(A6) $\Gamma, s \notin \mathbb{N}$ if $s$ is not a numeral.

(A7) $\Gamma, \mathbf{Set}(s)$ if $s$ is not a numeral.

(A8) $\Gamma, \neg \mathbf{Set}(\bar{n})$.

(A9) $\Gamma, \neg \mathrm{P}(s)$ if $s$ is not a numeral.

(A10) $\Gamma, \neg J(s_1, \ldots, s_k)$ if, for some $1 \leq i \leq k$, $s_i$ is not a numeral and $J$ is one of the symbols $\mathbf{SUC}, \mathbf{ADD}, \mathbf{MULT}, =_\mathbb{N}$.

(A11) $\Gamma, a^i \notin a^i$.

Let $0 < \kappa < \omega$. The *set-theoretical axioms* of $T_Q^\infty$ are:

(**Extens.**) $\Gamma, r \neq s, t \neq t', s \notin t, r \in t'$    $t$ not a numeral.
(**Pair**) $\Gamma, (\exists x \in \mathbf{M}_\kappa)(s \in x \wedge t \in x)$    if $s, t \triangleleft \mathbf{M}_\kappa$.
(**Union**) $\Gamma, (\exists x \in \mathbf{M}_\kappa)(\forall y \in s)(\forall z \in y)(z \in x)$    if $s \triangleleft \mathbf{M}_\kappa$.
($\Delta_0$-**Sep**) $\Gamma, (\exists x \in \mathbf{M}_\kappa)(x = \{y \in s : \mathcal{F}[y, \vec{t}]\})$    if $\mathcal{F}$ is $\Delta_0$ and $s, \vec{t} \triangleleft \mathbf{M}_\kappa$.

The *rules* of $T_Q^\infty$ are:

$(\wedge)$ $\quad\dfrac{\Gamma, A \quad \Gamma, A'}{\Gamma, A \wedge A'}$

$(\vee)$ $\quad\dfrac{\Gamma, A_i}{\Gamma, A_0 \vee A_1}\quad$ if $i = 0$ or $i = 1$

$(\forall)$ $\quad\dfrac{\cdots \Gamma, s\mathring{\in}t \to F(s) \cdots \text{ (for all } s \triangleleft t)}{\Gamma, (\forall x \mathbin{\varepsilon} t)F(x)}$

$(\exists)$ $\quad\dfrac{\Gamma, s\mathring{\in}t \wedge F(s)}{\Gamma, (\exists x \mathbin{\varepsilon} t)F(x)}\quad$ if $s \triangleleft t$

$(\notin)$ $\quad\dfrac{\cdots \Gamma, s\mathring{\in}t \to r \neq s \cdots \text{ (for all } s \triangleleft t)}{\Gamma, r \notin t}$

$(\in)$ $\quad\dfrac{\Gamma, s\mathring{\in}t \wedge r = s}{\Gamma, r \in t}\quad$ if $s \triangleleft t$.

$(\neg\mathbf{Ad})$ $\quad\dfrac{\cdots \Gamma, \mathbf{M}_\kappa \neq t \cdots (\kappa \leq |t|)}{\Gamma, \neg\mathbf{Ad}(t)}$

$(\mathbf{Ad})$ $\quad\dfrac{\Gamma, \mathbf{M}_\kappa = t}{\Gamma, \mathbf{Ad}(t)}\quad$ if $\kappa \leq |t|$

$(\text{Cut})$ $\quad\dfrac{\Gamma, A \quad \Gamma, \neg A}{\Gamma}$

$(\Delta_0(\kappa)\text{-Col})\quad\dfrac{\Gamma, (\forall x \in s)(\exists y \in \mathbf{M}_\kappa)\mathcal{F}(x, y)}{\Gamma, (\exists z \in \mathbf{M}_\kappa)(\forall x \in s)(\exists y \in z)\mathcal{F}(x, y)}\quad \mathcal{F}(\bar{0}, \bar{0})\ \Delta_0(\kappa).$

The relation $T_Q^\infty \vdash_\rho^\beta \Gamma$ is inductively defined as follows:

1. If $\Gamma$ is an axiom of $T_Q^\infty$, then $T_Q^\infty \vdash_\rho^\beta \Gamma$ for all $\beta$ and $\rho$.

2. If $T_Q^\infty \vdash_\rho^{\beta_i} \Gamma_i$ and $\beta_i < \beta$ hold for every premise $\Gamma_i$ of a rule other than (Cut), then $T_Q^\infty \vdash_\rho^\beta \Gamma$ if $\Gamma$ is the conclusion of that rule.

3. If $T_Q^\infty \vdash_\rho^{\beta_0} \Gamma, A$, $T_Q^\infty \vdash_\rho^{\beta_1} \Gamma, \neg A$, $\beta_0, \beta_1 < \beta$ and $rk(A) < \rho$, then $T_Q^\infty \vdash_\rho^\beta \Gamma$.

**Theorem 7.1.** Let $k, m < \omega$. For every finite set $\Gamma[a_1, \ldots, a_r]$ of $\mathbf{KPl}_Q^\infty$ formulae and $\mathcal{T}_Q^\infty$ terms $s_1, \ldots, s_r$ of levels $< \omega$,

$$\mathbf{KPl}_Q^\infty \vdash_k^\alpha \Gamma[a_1, \ldots, a_r] \quad \Rightarrow \quad \mathcal{T}_Q^\infty \vdash_{\omega^2+k}^{\omega+\omega^\alpha} \Gamma[s_1, \ldots, s_r]^{\mathbf{M}_\omega},$$

where $\Gamma[s_1, \ldots, s_r]^{\mathbf{M}_\omega}$ arises from $\Gamma[a_1, \ldots, a_r]$ by substituting $s_i$ for $a_i$ and replacing unbounded quantifiers $\forall x$ and $\exists x$ by $(\forall x \in \mathbf{M}_\omega)$ and $(\exists x \in \mathbf{M}_\omega)$, respectively.

**Proof:** Note that the highest rank a term of $\mathcal{T}_Q^\infty$ can have is $\omega^2$. One easily computes that whenever a formula $F[\vec{a}]$ of $\mathbf{KPl}_Q^\infty$ has degree $m$, then $rk(F[\vec{s}]) \leq \omega^2 + m$ holds for all $\mathcal{T}_Q^\infty$ terms $\vec{s}$ of levels $< \omega$.

The proof of this theorem, which proceeds by induction on $\alpha$, is a simplification of the proof of [24, Theorem 3.17] for $\mathcal{T}^\infty$ as the rule ($\Delta_0$-Coll) does not exist in $\mathcal{T}_Q^\infty$ and in the case of (Cut) one can just apply the induction hypothesis and the above observation about the rank of $F[\vec{s}]$. □

Below we use the function $\varepsilon$ which is defined by $\varepsilon_\alpha := \varphi 1 \alpha$.

**Theorem 7.2.** *(Cut elimination I)* Let $\Gamma$ be a set of formulas of rank $< \rho + \omega$, where $\rho := \omega \cdot \alpha$. Furthermore, we will assume that all the derivations considered below neither contain variables of level $\alpha$ nor Extensionality axioms with terms of level $\alpha$.

(i) If $\mathcal{T}_Q^\infty \vdash_{\rho+m+2}^{\delta} \Gamma$, then $\mathcal{T}_Q^\infty \vdash_{\rho+m+1}^{\omega^\delta} \Gamma$.

(ii) If $\alpha$ is a limit or 0 and $\mathcal{T}_Q^\infty \vdash_{\rho+m+1}^{\delta} \Gamma$, then $\mathcal{T}_Q^\infty \vdash_{\rho+m}^{\omega^\delta} \Gamma$.

(iii) If $\mathcal{T}_Q^\infty \vdash_{\rho+n+1}^{\beta} \Gamma$, then $\mathcal{T}_Q^\infty \vdash_{\rho+1}^{\omega_n(\beta)} \Gamma$.

(iv) If $\alpha$ is a limit or 0 and $\mathcal{T}_Q^\infty \vdash_{\rho+n}^{\beta} \Gamma$, then $\mathcal{T}_Q^\infty \vdash_{\rho}^{\omega_n(\beta)} \Gamma$.

(v) If $\mathcal{T}_Q^\infty \vdash_{\rho+\omega}^{\beta} \Gamma$, then $\mathcal{T}_Q^\infty \vdash_{\rho+1}^{\varepsilon_\beta} \Gamma$.

(vi) If $\alpha$ is a limit or 0 and $\mathcal{T}_Q^\infty \vdash_{\rho+\omega}^{\beta} \Gamma$, then $\mathcal{T}_Q^\infty \vdash_{\rho}^{\varepsilon_\beta} \Gamma$.

**Proof:** [24, Theorem 3.21]. □

**Theorem 7.3.** *(Cut elimination II)* Let $n = k+1$ and $\Gamma$ be a set of $\Sigma_\infty(k)$ formulae. Let $\rho := \omega \cdot n$.

If $\mathcal{T}_Q^\infty \vdash_{\rho+1}^{\beta} \Gamma$, then $\mathcal{T}_Q^\infty \vdash_{\rho}^{\varphi \varepsilon_{\beta+n+1} 0} \Gamma$.

**Proof:** [24, Theorem 3.22]. □

Define $\omega_0(\beta) := \beta$ and $\omega_{n+1}(\beta) := \omega^{\omega_n(\beta)}$.

**Theorem 7.4.** *(Cut elimination III)* Let $k < \omega$ and $\Lambda$ be a set of $\Sigma_\infty(k)$ formulae. Then:

$$\mathcal{T}_Q^\infty \vdash^{\alpha}_{\omega^2} \Lambda \;\Rightarrow\; \mathcal{T}_Q^\infty \vdash^{\Gamma_\alpha}_{\omega \cdot k + 1} \Lambda.$$

**Proof:** The proof proceeds by induction on $\alpha$.

If the last inference was not (Cut), then the desired assertion follows easily from the induction hypotheses applied to the premises, using the same inference rule. Now suppose that the last inference was (Cut). Then

$$\mathcal{T}_Q^\infty \vdash^{\alpha_0}_{\omega^2} \Lambda, A \quad \text{and} \quad \mathcal{T}_Q^\infty \vdash^{\alpha_0}_{\omega^2} \Lambda, \neg A \tag{5}$$

for some $\alpha_0 < \alpha$. Let $m_0$ be minimal such that $A, \neg A$ are $\Sigma_\infty(m_0)$ formulae.

**Case 1:** $m_0 < k$. The induction hypothesis then yields

$$\mathcal{T}_Q^\infty \vdash^{\Gamma_{\alpha_0}}_{\omega \cdot k + 1} \Lambda, A \quad \text{and} \quad \mathcal{T}_Q^\infty \vdash^{\Gamma_{\alpha_0}}_{\omega \cdot k + 1} \Lambda, \neg A.$$

Thus since $rk(A) < \omega \cdot k$, via (Cut) we get

$$\mathcal{T}_Q^\infty \vdash^{\Gamma_\alpha}_{\omega \cdot k + 1} \Lambda.$$

**Case 2:** $k \leq m_0$. The induction hypothesis then yields

$$\mathcal{T}_Q^\infty \vdash^{\Gamma_{\alpha_0}}_{\omega \cdot m_0 + 1} \Lambda, A \quad \text{and} \quad \mathcal{T}_Q^\infty \vdash^{\Gamma_{\alpha_0}}_{\omega \cdot m_0 + 1} \Lambda, \neg A.$$

Thus, by (Cut),

$$\mathcal{T}_Q^\infty \vdash^{\Gamma_{\alpha_0}+1}_{\omega \cdot k + l} \Lambda \tag{6}$$

for some $l < \omega$.

Let $\mu := \omega^{rk(\mathbf{M}_{m_0} \in \mathbf{M}_{m_0})}$. By substituting $\mathbf{M}_{m_0}$ for variables of level $m_0$ occurring in derivation (6) and subsequently replacing Extensionality axioms with occurrences of $\mathbf{M}_{m_0}$ by derivations according to [24, Lemma 3.12], we arrive at $\mathcal{T}_Q^\infty \vdash^{\mu+\mu+\Gamma_{\alpha_0}+1}_{\omega \cdot m_0 + l} \Lambda$. To the latter we may apply cut elimination I (Theorem 7.2, (iii)) to obtain

$$\mathcal{T}_Q^\infty \vdash^{\omega_l(\mu+\mu+\Gamma_{\alpha_0}+1)}_{\omega \cdot m_0 + 1} \Lambda. \tag{7}$$

As $\mu, \Gamma_{\alpha_0} < \Gamma_\alpha$, we have $\omega_l(\mu + \mu + \Gamma_{\alpha_0} + 1) < \Gamma_\alpha$; thus $\mathcal{T}_Q^\infty \vdash^{\Gamma_\alpha}_{\omega \cdot m_0 + 1} \Lambda$. So we are done if $m_0 = k$.

Suppose $m_0 > k$. Let $\beta_0 := \omega_l(\mu + \mu + \Gamma_{\alpha_0} + 1)$. By applying cut elimination II (Theorem 7.3) to (7), we obtain

$$\mathcal{T}_Q^\infty \vdash^{\varphi(\varepsilon_{\beta_0 + m_0 + 1})0}_{\omega \cdot (m_0 - 1) + \omega} \Lambda. \tag{8}$$

Let $\eta := \omega^{rk(\mathbf{M}_{m_0-1} \in \mathbf{M}_{m_0-1})}$. By substituting $\mathbf{M}_{m_0-1}$ for variables of level $m_0 - 1$ occurring in the derivation (8) and subsequently replacing Extensionality axioms with occurrences of $\mathbf{M}_{m_0-1}$ by derivations according to [24, Lemma 3.12], we arrive at

$$\mathcal{T}_Q^\infty \vdash^{\eta + \eta + \varphi(\varepsilon_{\beta_0 + m_0 + 1})0}_{\omega \cdot (m_0 - 1) + \omega} \Lambda.$$

Hence, letting $\beta_1 := \eta + \eta + \varphi(\varepsilon_{\beta_0 + m_0 + 1})0$, cut elimination I (Theorem 7.2,(v)) yields

$$\mathcal{T}_Q^\infty \vdash^{\varepsilon_{\beta_1}}_{\omega \cdot (m_0 - 1) + 1} \Lambda. \tag{9}$$

If $k = m_0 - 1$ we are done. If $k < m_0 - 1$, we have to repeat the above procedure again and again until we arrive after finitely many steps at an ordinal $\beta_r$ such that

$$\mathcal{T}_Q^\infty \vdash^{\varepsilon_{\beta_r}}_{\omega \cdot k + 1} \Lambda. \tag{10}$$

Since we started out with ordinals $< \Gamma_\alpha$ and applied the functions $+$, $\cdot$ and $\varphi$ finitely many times to these ordinals we arrive at an ordinal $\varepsilon_{\beta_r}$ which is still smaller than $\Gamma_\alpha$. Thus from (10) we conclude that

$$\mathcal{T}_Q^\infty \vdash^{\Gamma_\alpha}_{\omega \cdot k + 1} \Lambda.$$

□

**Corollary 7.5.** If $k < \omega$ and $\Theta$ is a finite set of arithmetical sentences, then:

$$\mathcal{T}_Q^\infty \vdash^{\beta}_{\omega^2 + k} \Theta \quad \Rightarrow \quad \mathcal{T}_Q^\infty \vdash^{\Gamma_{\varepsilon_{\beta+1}}}_{0} \Theta.$$

**Proof:** First we use cut elimination I (Theorem 7.2(ii)) to get:

$$\mathcal{T}_Q^\infty \vdash^{\omega_k(\beta)}_{\omega^2} \Theta.$$

By cut elimination III (Theorem 7.4) we obtain

$$\mathcal{T}_Q^\infty \mid\!\frac{\Gamma_{\omega_k(\beta)}}{\omega+1}\; \Theta. \tag{11}$$

Let $\beta_0 := \Gamma_{\omega_k(\beta)}$. Applying cut elimination II (Theorem 7.3) to (11) we have

$$\mathcal{T}_Q^\infty \mid\!\frac{\varphi\varepsilon_{\beta_0+2}0}{\omega}\; \Theta. \tag{12}$$

The derivation of (12) may contain free variables of level 0. We can get rid of them by substituting $\mathbb{N} = \mathbf{M_0}$ for these variables and subsequently we can replace extensionality axioms with occurrences of $\mathbf{M_0}$ in this derivation using [24, Lemma 3.12]. As a result, since $\omega^{rk(\mathbf{M_0}\in\mathbf{M_0})} = \omega^6$ we also have

$$\mathcal{T}_Q^\infty \mid\!\frac{\varphi\varepsilon_{\beta_0+2}0}{\omega}\; \Theta, \tag{13}$$

where the derivation no longer contains free variables nor extensionality axioms. Let $\beta_1 := \varphi\varepsilon_{\beta_0+2}0$. Via a final application cut elimination I (Theorem 7.2(v)) we therefore get:

$$\mathcal{T}_Q^\infty \mid\!\frac{\varepsilon_{\beta_1}}{0}\; \Theta.$$

One easily computes that $\varepsilon_{\beta_1} < \Gamma_{\varepsilon_{\beta+1}}$; whence $\mathcal{T}_Q^\infty \mid\!\frac{\Gamma_{\varepsilon_{\beta+1}}}{0}\; \Theta$. □

**Corollary 7.6.** Let $k < \omega$ and $\Theta$ be a finite set of arithmetical sentences.

(i) $\mathbf{KPl}_Q^\infty \mid\!\frac{\alpha}{k}\; \Theta \;\Rightarrow\; \mathcal{T}_Q^\infty \mid\!\frac{\Gamma_{\varepsilon_{\omega\alpha+2}}}{0}\; \Theta.$

(ii) $\mathbf{KPl}_Q^\infty \mid\!\frac{\alpha}{k}\; \Theta \;\Rightarrow\; \mathbf{KPl}_Q^\infty \mid\!\frac{\Gamma_{\varepsilon_{\omega\alpha+2}}}{0}\; \Theta.$

**Proof:** (i) follows from Theorem 7.1 and Corollary 7.5.

(ii) follows from (i) and the observation that a cut free derivation of $\Theta$ in $\mathcal{T}_Q^\infty$ consists entirely of arithmetical sentences and all the inferences and axioms used therein are inferences and axioms of $\mathbf{KPl}_Q^\infty$ too. Thus such a derivation is a derivation of $\mathbf{KPl}_Q^\infty$ as well. □

AFFILIATION: Department of Pure Mathematics, University of Leeds, Leeds LS2 9JT, England, rathjen@maths.leeds.ac.uk.

# BIBLIOGRAPHY

[1] B. Afshari: *Proof-Theoretic Strengths of Hierarchies of Theories*, PhD thesis, University of Leeds, U.K., 2008.
[2] B. Afshari and M. Rathjen: *Reverse Mathematics and Well-ordering Principles: A pilot study*, Annals of Pure and Applied Logic 160 (2009) 231-237.

[3] J Barwise: *Admissible Sets and Structures*, Springer, Berlin 1975.
[4] E.W. Beth: *The Foundations of Mathematics*, (North Holland, Amsterdam, 1959)
[5] W. Buchholz: *A new system of proof–theoretic ordinal functions*, Ann. Pure Appl. Logic **32** (1986) 195–207.
[6] W. Buchholz, S. Feferman, W. Pohlers, W. Sieg: *Iterated inductive definitions and subsystems of analysis* (Springer, Berlin, 1981).
[7] W. Buchholz and K. Schütte: *Proof theory of impredicative subsystems of analysis* (Bibliopolis, Naples, 1988).
[8] S. Feferman: *Systems of predicative analysis*, Journal of Symbolic Logic 29 (1964) 1–30.
[9] H. Friedman: *Uniformly Defined Descending Sequences of Degrees*, Journal of Symbolic Logic 41 (1976) 363–367.
[10] H. Friedman and S. Sheard: *Elementary descent recursion and proof theory*, Annals of Pure and Applied Logic 71 (1995) 1–45.
[11] H. Friedman, A. Montalban, A. Weiermann: *A characterization of* $ATR_0$ *in terms of Kruskal-like tree theorems* unpublished draft.
[12] G. Gentzen: *Untersuchungen über das logische Schließen*, Mathematische Zeitschrift 39 (1935) 176–210, 405–431.
[13] J.-Y. Girard: *Proof Theory and logical complexity, vol 1* (Bibliopolis, Napoli, 1987).
[14] G.H. Hardy: *A theorem concerning the infinite cardinal numbers*. Quarterly Journal of Mathematics 35 (1904) 87–94.
[15] L. Henkin: *A generalization of the concept of* $\omega$*-consistency*, Journal of Symbolic Logic 19 (1954) 183–196.
[16] D. Hilbert: *Die Grundlegung der elementaren Zahlentheorie*, Mathematische Annalen 104 (1930/31) 485–494.
[17] K.J.J. Hintikka: *Form and content in quantification theory*, Acta Philosophica Fennica 8 (1955) 7–55.
[18] A. Marcone, A. Montalbán: *The epsilon function for computability theorists*, draft, 2007.
[19] A. Marcone, A. Montalbán: *The Veblen functions for computability theorists*, Journal of Symbolic Logic 76 (2011) 575–602.
[20] A. Montalbán: *Ordinal functors and* $\Pi_1^1$-$CA_0$, draft December 2009.
[21] P.S. Novikov: *On the consistency of certain logical calculi*, Math. Sbornik 12 (1943) 231–261.
[22] S. Orey: *On* $\omega$*-consistency and related properties*, Journal of Symbolic Logic 21 (1956) 246–252.
[23] M. Rathjen: *The strength of Martin-Löf type theory with a superuniverse. Part I.* Archive for Mathematical Logic 39 (2000) 1-39.
[24] M. Rathjen: *The strength of Martin-Löf type theory with a superuniverse. Part II.* Archive for Mathematical Logic 40 (2001) 207-233.
[25] M. Rathjen and A. Weiermann: *Reverse Mathematics and Well-ordering Principles.* In: S. Cooper, A. Sorbi (eds.): *Computability in Context: Computation and Logic in the Real World* (Imperial College Press, 2011) 351-370.
[26] M. Rathjen and A. Weiermann: *Proof-theoretic investigations on Kruskal's theorem,* Annals of Pure and Applied Logic 60 (1993) 49–88.
[27] K. Schütte: *Proof Theory*, Springer-Verlag, Berlin, Heidelberg, 1977.
[28] K. Schütte: *Eine Grenze für die Beweisbarkeit der transfiniten Induktion in der verzweigten Typenlogik*, Archiv für Mathematische Logik und Grundlagenforschung 67 (1964) 45–60.
[29] K. Schütte: *Predicative well-orderings*, in: Crossley, Dummet (eds.), Formal systems and recursive functions (North Holland, 1965) 176–184.
[30] K. Schütte: *Beweistheorie*, (Springer, Berlin, 1960).
[31] K. Schütte: *Ein System des verknüpfenden Schließens*, Archiv für mathematische Logik und Grundlagenforschung 2 (1956) 55–67.

[32] K. Schütte: *Beweistheoretische Erfassung der unendlichen Induktion in der Zahlentheorie*, Mathematische Annalen 122 (1951) 369–389.
[33] H. Schwichtenberg: *Proof Theory: Some applications of cut-elimination*. In: Handbook of Mathematical Logic (J. Barwise ed.) North Holland 1977, pp. 867-895.
[34] S.G. Simpson: *Subsystems of Second Order Arithmetic*, Springer-Verlag, Berlin, Heidelberg, 1999.
[35] J. Steel: *Descending sequences of degrees*, Journal of Symbolic Logic 40 (1975) 59–61.
[36] A. Tarski: *Some observations on the concept of $\omega$-consistency and $\omega$-completeness*, in: *Logic, Semantics, Metamathematics*, (Clarendon, Oxford, 1956) 279–295.
[37] O. Veblen: *Continuous increasing functions of finite and transfinite ordinals,* Trans. Amer. Math. Soc. 9 (1908) 280–292.

# Ordering Free Products in Reverse Mathematics

REED SOLOMON

## 1 Introduction

This paper is a contribution to Harvey Friedman's project of reverse mathematics in which one tries to classify theorems of ordinary mathematics by the set theoretic axioms required to prove them.[1] All of the result in this paper are formalized in $\mathsf{RCA}_0$, the weakest of the standard systems in reverse mathematics, and concern classical theorems about countable ordered groups. The reader who is not familiar with reverse mathematics is referred to Simpson [8] for the background definitions and to Solomon [9] and [10] for specific information concerning reverse mathematics and ordered groups. We begin with some background definitions before stating our main results.

DEFINITION 1 ($\mathsf{RCA}_0$). A *partially ordered (p.o.) group* is a group $G$ together with a partial order $\leq_G$ on the elements of $G$ such that for any $a, b, c \in G$, if $a \leq_G b$ then $a \cdot_G c \leq_G b \cdot_G c$ and $c \cdot_G a \leq_G c \cdot_G b$. If the order is linear, then $(G, \leq_G)$ is called a *fully ordered (f.o.) group*.

We frequently drop the subscripts on $\cdot_G$ and $\leq_G$ when the ambient group is clear. An order preserving homomorphism between p.o. groups is called an *o-homomorphism*. If an o-homomorphism is onto, it is an *o-epimorphism*, and if it is a bijection, then it is an *o-isomorphism*.

If a subgroup $H$ of $G$ is normal, then the quotient $G/H$ inherits a natural group structure. We form the quotient group $G/H$ in $\mathsf{RCA}_0$ by picking the $\leq_\mathbb{N}$-least representative of each coset. Since $aH = bH \leftrightarrow a^{-1}b \in H$, $\mathsf{RCA}_0$ suffices to define the quotient group $G/H$.

If $G$ is a p.o. group, then we need an additional condition on a normal subgroup $H$ for the order on $G$ to induce an order on $G/H$. $H$ is *convex* if for any $a, b \in H$ and $g \in G$, we have that $a \leq g \leq b$ implies $g \in H$. If $H$ is a convex normal subgroup, then the *induced order* $\leq_{G/H}$ on $G/H$ is defined for $a, b \in G/H$ by $a \leq_{G/H} b \leftrightarrow \exists h \in H(a \leq_G bh)$. In the general case for a

---
[1]The author thanks two anonymous referees for many helpful suggestions.

p.o. group $G$ and a convex normal subgroup $H$, $\mathsf{ACA}_0$ is required to prove the existence of the induced order. (See [9] for a proof.) However, if $G$ is an f.o. group, then $\mathsf{RCA}_0$ is strong enough to form the induced order.

THEOREM 2 ($\mathsf{RCA}_0$). *Let $(G, \leq_G)$ be an f.o. group and $H$ a convex normal subgroup. The induced order $\leq_{G/H}$ on $G/H$ exists.*

**Proof.** Let $a, b \in G/H$ and $a \neq b$. Because $a$ and $b$ are representatives of different cosets, $ab^{-1} \notin H$.

CLAIM 3. $\exists h \in H\, (a \leq_G bh)$ if and only if $a \leq_G b$.

If $a \leq_G b$ then, because $1_G \in H$, it follows that $\exists h \in H\, (a \leq_G bh)$. For the other direction, suppose for a contradiction that $\exists h \in H(a \leq_G bh)$ and $b <_G a$. Then $b <_G a \leq_G bh$ and so $1_G <_G b^{-1}a \leq_G h$. Since $H$ is convex, $b^{-1}a \in H$. Because $H$ is normal, $bb^{-1}ab^{-1} \in H$, and hence $ab^{-1} \in H$ which is a contradiction. The induced order can now be given by a $\Sigma^0_0$ condition: $aH \leq_{G/H} bH$ if and only if $aH = bH$ or $a <_G b$. □

It is well known that every group $G$ is isomorphic to a quotient group $F/N$ of a free group $F$ by a normal subgroup $N$. A similar result holds for ordered groups. Every f.o. group $(G, \leq)$ is o-isomorphic to a quotient group $(F/N, \leq_{F/N})$ of a fully ordered free group $(F, \leq_F)$ by a convex normal subgroup $N$. Classical proofs of this fact can be found in Fuchs [4] and Kokorin and Kopytov [5]. The main result of this paper is to give a proof of this classical theorem in $\mathsf{RCA}_0$. (This result was first announced in Solomon [10].)

THEOREM 4 ($\mathsf{RCA}_0$). *Every countable f.o. group is the o-epimorphic image of a fully ordered free group. Hence, for every f.o. group $(G, \leq_G)$, there is a fully ordered free group $(F, \leq_F)$ and a convex normal subgroup $N$ of $F$ such that $(G, \leq_G)$ is o-isomorphic to $(F/N, \leq_{F/N})$.*

Because Theorem 4 is provable in $\mathsf{RCA}_0$, it follows that for every computably fully ordered computable group $(G, \leq_G)$, there is a computably ordered free group $(F, \leq_F)$ and a computable convex normal subgroup $N$ such that $(G, \leq_G) \cong (F/N, \leq_{F/N})$, which answers an open question from Downey and Kurtz [2]. In this paper, we obtain a proof of Theorem 4 after first proving the following theorem in $\mathsf{RCA}_0$.

THEOREM 5 ($\mathsf{RCA}_0$). *The free product of f.o. groups $A$ and $B$ is fully orderable.*

There are also a number of different classical proofs of Theorem 5. We modify and formalize the classical proof of this theorem contained in Kokorin and Kopytov [5]. Before proving these theorems, we give additional background information about ordered groups in this section and explain how to formalize free groups and free products in $\mathsf{RCA}_0$ in Section 2. In Section 3, we prove Theorem 4 in $\mathsf{RCA}_0$ using Theorem 5 and an idea from Revesz [7]. In Sections 4 and 5, we prove Theorem 5 in $\mathsf{RCA}_0$. In Section 6, we present proofs of some technical formulas used in previous sections.

It follows from Theorem 5 that $\mathsf{RCA}_0$ proves that the free group on infinitely many generators is fully orderable, and hence that the natural computable presentation of this free group admits a computable order. The standard classical proof that free groups are fully orderable uses the fact that the lower central series of a free group has length $\omega$. In Dabkowska, Dabkowski, Harizanov and Tongha [1], this fact and the Fox calculus (for determining where elements of the free group sit in the lower central series) were used to show that free groups have orders of every Turing degree (including computable orders). Rather than developing properties of the lower central series of free groups in $\mathsf{RCA}_0$, we obtain our effective proof that free groups are fully orderable as a corollary of Theorem 5 using the facts (shown in $\mathsf{RCA}_0$ in Section 2) that the free group on two generators is isomorphic to the free product $\mathbb{Z} * \mathbb{Z}$ and the free group on countably many generators embeds into the free group on two generators.

When working with ordered groups, it is often easier to work with the set of positive elements in the group rather than the order $\leq_G$. Proofs of the following results (which are straightforward formalizations of the standard classical proofs) can be found in Solomon [9].

DEFINITION 6 ($\mathsf{RCA}_0$). The *positive cone* of a p.o. group $(G, \leq_G)$ is the set $P(G) = \{g \in G \mid 1_G \leq_G g\}$.

Because $ab^{-1} \in P(G) \leftrightarrow b \leq_G a$, $\mathsf{RCA}_0$ suffices to define the positive cone from the order and vice versa. There is a classical set of algebraic conditions that determines if an arbitrary subset of a group is the positive cone of some full or partial order. If $G$ is a group and $X \subseteq G$, then $X^{-1} = \{g^{-1} \mid g \in X\}$. $X$ is a *full subset of* $G$ if $X \cup X^{-1} = G$ and $X$ is a *pure subset of* $G$ if $X \cap X^{-1} \subseteq \{1_G\}$.

THEOREM 7 ($\mathsf{RCA}_0$). *A subset $P$ of a group $G$ is the positive cone of some partial order on $G$ if and only if $P$ is a normal pure semigroup with identity. Furthermore, $P$ is the positive cone of a full order if and only if in addition $P$ is full.*

If $(G_i, \leq_{G_i})$, $i \in \mathbb{N}$, is a uniform sequence of f.o. groups, then $\mathsf{RCA}_0$ suffices to lexicographically order the (restricted) direct product $\prod_{i \in \mathbb{N}} G_i$. The elements of the product are represented by finite sequences $\sigma$ such that $\sigma(i) \in G_i$ and the last element of $\sigma$ is not the identity. The sequences are multiplied componentwise (removing any trailing identity elements). We order distinct sequences $\sigma$ and $\tau$ in the product by $\sigma < \tau$ if and only if

$\sigma(i) <_{G_i} \tau(i)$ for the $\leq_\mathbb{N}$-least $i$ such that $\sigma(i) \neq \tau(i)$. (If $\sigma$ is an initial segment of $\tau$, then $\sigma < \tau$ if and only if $1_{G_j} <_{G_j} \tau(j)$ for the least $j \geq \text{lh}(\sigma)$ such that $\tau(j) \neq 1_{G_j}$, where $\text{lh}(\sigma)$ denotes the length of $\sigma$.) When ordering a direct product in $\mathsf{RCA}_0$, it is important that the sequence of orders is given uniformly. In general, the fact that an infinite direct product of fully orderable groups is fully orderable is equivalent to $\mathsf{WKL}_0$. (See Solomon [9].)

## 2 Free Groups and Free Products

In this section, we give a brief presentation of free groups in second order arithmetic, indicating informally how to define these groups in $\mathsf{RCA}_0$. For a more formal presentation, including proofs of the stated properties, see Downey, Hirschfedlt, Lempp and Solomon [3]. After handling free groups, we give a more formal presentation of free products in $\mathsf{RCA}_0$, although we often omit proofs which are straightforward formalizations of their classical counterparts as contained in [6]. [3] contains the formal versions of these properties for free groups and the modifications required for free products are minimal. Throughout this section, we work in $\mathsf{RCA}_0$. We let $\text{Fin}_X$ denote the set of codes for finite sequences of elements of a set $X$, $\text{lh}(\sigma)$ denote the length of $\sigma \in \text{Fin}_X$, and $x \frown y$ denote the concatenation of $x, y \in \text{Fin}_X$.

Let $A \subseteq \mathbb{N}$. When defining the free group on the generators $A$, it is convenient to think of the elements of $A$ as distinct symbols in an alphabet. Let $a^1$ stand for the pair $\langle a, 1 \rangle$ and $a^{-1}$ stand for the pair $\langle a, -1 \rangle$. In this section, $\epsilon$ will always denote either $1$ or $-1$. The set of *words over* $A$ is $\text{Word}_A = \text{Fin}_{\tilde{A}}$ where $\tilde{A} = \{a^\epsilon \mid a \in A \text{ and } \epsilon = \pm 1\}$. The empty sequence is denoted by $1_A$. In keeping with standard mathematical notation, we write $a_0^{\epsilon_0} \cdots a_{k-1}^{\epsilon_{k-1}}$ for the sequence $\sigma \in \text{Word}_A$ with $\text{lh}(\sigma) = k$ and $\sigma(i) = a_i^{\epsilon_i}$.

An element $x \in \text{Word}_A$ is *reduced* if there is no place in the sequence where $a^1$ and $a^{-1}$ appear next to each other for any $a \in A$. More formally, $x \in \text{Red}_A$ if and only if $x \in \text{Word}_A$ and

$$\forall i < (\text{lh}(x) - 1) \left( \pi_1(x(i)) \neq \pi_1(x(i+1)) \vee \pi_2(x(i)) = \pi_2(x(i+1)) \right)$$

where $\pi_1$ and $\pi_2$ are the standard projection functions on pairs. Both $\text{Word}_A$ and $\text{Red}_A$ have $\Sigma_0^0$ definitions, so $\mathsf{RCA}_0$ proves they exist.

To define the group structure on $\text{Red}_A$, we need a function that maps words to reduced words by repeatedly removing symbols $a^1$ and $a^{-1}$ which occur next to each other. Two words $x, y \in \text{Word}_A$ are *1 step equivalent*, denoted $x \sim_1 y$, if either they are the same sequence or one results from the other by deleting a pair $a^1, a^{-1}$ that appear next to each other. This concept can be formalized by a $\Sigma_0^0$ formula, so $\mathsf{RCA}_0$ proves the existence of the set of pairs $\langle x, y \rangle$ such that $x \sim_1 y$. Two words $x, y \in \text{Word}_A$ are *freely equivalent*, denoted $x \sim y$, if there is a finite sequence $\sigma$ of elements of $\text{Word}_A$ such that $\sigma(0) = x$, $\sigma(\text{lh}(\sigma) - 1) = y$, and $\sigma(i) \sim_1 \sigma(i+1)$ for $0 \le i < \text{lh}(\sigma) - 1$. Although formalizing this concepts uses a $\Sigma_1^0$ definition, we can capture free equivalence by defining a function $\rho$ by primitive recursion as follows.

$$\rho : \text{Word}_A \to \text{Red}_A$$
$$\rho(1_A) = 1_A$$
$$\rho(a^\epsilon) = a^\epsilon \text{ for } a \in A, \epsilon = \pm 1$$

If $\rho(U) = a_0^{\epsilon_0} \cdots a_{k-1}^{\epsilon_{k-1}}$ then

$$\rho(U^\frown a^\epsilon) = \begin{cases} a_0^{\epsilon_0} \cdots a_{k-2}^{\epsilon_{k-2}} & \text{if } a = a_{k-1} \text{ and } \epsilon_{k-1} + \epsilon = 0 \\ a_0^{\epsilon_0} \cdots a_{k-1}^{\epsilon_{k-1}} a^\epsilon & \text{otherwise} \end{cases}$$

It follows that $x \sim \rho(x)$, $\rho(x) \in \text{Red}_A$, $x \sim y$ if and only if $\rho(x) = \rho(y)$, and

$$\{\langle x, y \rangle \mid x \sim y\} = \{\langle x, y \rangle \mid \rho(x) = \rho(y)\}$$

DEFINITION 8 (RCA$_0$). Let $A \subseteq \mathbb{N}$. The *free group on the generators* $A$ consists of the elements $\text{Red}_A$ with the identity element $1_A$ and the multiplication $x \cdot y = \rho(x^\frown y)$.

The definition of the free product of two groups $A * B$ is similar. Instead of using sequences of generators and inverses as elements, we will use sequences whose elements alternate between $A$ and $B$. For example, if $a_i \in A$ and $b_i \in B$ then strings such as

$$\langle a_1, b_3, a_2 \rangle \text{ and } \langle b_2, a_1 \rangle$$

are in $A * B$. To form the free product, we start with the set of finite strings over $A \cup B$. Strings are reduced by removing occurrences of $1_A$ and $1_B$ and by multiplying elements of the same group which appear next to each other in the string. For example,

$$\langle a_1, 1_A, b_2, b_3 \rangle \mapsto \langle a_1, b_2 \cdot_B b_3 \rangle.$$

The definitions and lemmas for free products parallel those given for free groups.

DEFINITION 9 (RCA$_0$). For groups $A$ and $B$, we define $\text{Word}_{A*B} = \text{Fin}_{A \cup B}$ and we let $1_{A*B}$ denote the empty sequence.

DEFINITION 10 (RCA$_0$). The set of *reduced words* is defined by $x \in \text{Red}_{A*B}$ if and only if $x \in \text{Word}_{A*B}$ and one of the following conditions holds:

1. $x = 1_{A*B}$

2. For all $i < \mathrm{lh}(x)$, $x(i)$ is not $1_A$ or $1_B$, and for all $i < (\mathrm{lh}(x) - 1)$, if $x(i) \in A$ then $x(i+1) \in B$ and if $x(i) \in B$ then $x(i+1) \in A$.

DEFINITION 11 (RCA$_0$). Two words $x, y \in \mathrm{Word}_{A*B}$ are *1 step equivalent*, $x \sim_1 y$, if and only if $x = y$ or one of the following conditions holds:

1. $\mathrm{lh}(x) = \mathrm{lh}(y) + 1$ and the sequence $y$ is the same as $x$ except one occurrence of $1_A$ or $1_B$ is removed (or the symmetric condition with the roles of $x$ and $y$ reversed).

2. $\mathrm{lh}(x) = \mathrm{lh}(y) + 1$ and there is an $i < \mathrm{lh}(y)$ such that $y(j) = x(j)$ for all $j < i$, $y(j) = x(j+1)$ for all $j > i$, and either $x(i), x(i+1) \in A$ and $y(i) = x(i) \cdot_A x(i+1)$, or $x(i), x(i+1) \in B$ and $y(i) = x(i) \cdot_B x(i+1)$ (or the symmetric condition with the roles of $x$ and $y$ reversed).

DEFINITION 12 (RCA$_0$). Two words $x, y \in \mathrm{Word}_{A*B}$ are *freely equivalent*, $x \sim y$, if there exists a finite sequence $\sigma$ of elements of $\mathrm{Word}_{A*B}$ such that $\sigma(0) = x$, $\sigma(\mathrm{lh}(\sigma) - 1) = y$, and $\forall i < (\mathrm{lh}(\sigma) - 1)\ (\sigma(i) \sim_1 \sigma(i+1))$.

As in the case of free groups, this condition is $\Sigma_1^0$, so we use a function $\rho : \mathrm{Word}_{A*B} \to \mathrm{Red}_{A*B}$ to form the set of pairs $\langle x, y \rangle$ with $x \sim y$ in RCA$_0$.

$$\rho(1_{A*B}) = 1_{A*B}$$

$$\rho(\langle g \rangle) = \begin{cases} 1_{A*B} & \text{if } g = 1_A \vee g = 1_B \\ \langle g \rangle & \text{otherwise} \end{cases}$$

If $\rho(U) = \langle h_1, h_2, \ldots, h_r \rangle$ then

$$\rho(U^\frown \langle g \rangle) = \begin{cases} \langle h_1, \ldots, h_r \rangle & \text{if } g = 1_A \vee g = 1_B \\ \langle h_1, \ldots, h_{r-1} \rangle & \text{if } g = h_r^{-1} \\ \langle h_1, \ldots, h_r, g \rangle & \text{if } (g \in A \setminus 1_A \wedge h_r \in B) \\ & \quad \vee (g \in B \setminus 1_B \wedge h_r \in A) \\ \langle h_1, \ldots, h_{r-1}, h_r \cdot_A g \rangle & \text{if } g \in A \setminus 1_A \wedge h_r \in A \setminus g^{-1} \\ \langle h_1, \ldots h_{r-1}, h_r \cdot_B g \rangle & \text{if } g \in B \setminus 1_B \wedge h_r \in B \setminus g^{-1} \end{cases}$$

As in the free group case, we use properties of $\rho$ to show that each word in $\mathrm{Word}_{A*B}$ is freely equivalent to a unique reduced word.

LEMMA 13 (RCA$_0$). For all $W_1, W_2, W \in \mathrm{Word}_{A*B}$:

1. $\rho(W) \in \mathrm{Red}_{A*B}$ and $\rho(W) \sim W$

2. $W \in \mathrm{Red}_{A*B} \to \rho(W) = W$

3. $\rho(W_1^\frown W_2) = \rho(\rho(W_1)^\frown W_2)$

4. $\rho(W_1^\frown \langle 1_A \rangle^\frown W_2) = \rho(W_1^\frown \langle 1_B \rangle^\frown W_2) = \rho(W_1^\frown W_2)$

5. If $g, h \in A$ then $\rho(W^\frown \langle g, h \rangle) = \rho(W^\frown \langle g \cdot_A h \rangle)$, and similarly for $g, h \in B$

6. If $g, h \in A$ then $\rho(W_1^\frown \langle g, h \rangle^\frown W_2) = \rho(W_1^\frown \langle g \cdot_A h \rangle^\frown W_2)$, and similarly for $g, h \in B$

The proof of this lemma is a series of inductions formalizing the classical arguments in [6]. (Similar properties in the context of free groups are shown formally in $\mathsf{RCA}_0$ in [3].) It follows from this lemma that $\mathsf{RCA}_0$ can prove that $x \sim y$ if and only if $\rho(x) = \rho(y)$ and that for every $x \in \mathrm{Word}_{A*B}$ there is a unique $y \in \mathrm{Red}_{A*B}$ such that $x \sim y$. Thus, we obtain

$$\{\langle x, y \rangle \mid x \sim y\} = \{\langle x, y \rangle \mid \rho(x) = \rho(y)\}.$$

and hence $\mathsf{RCA}_0$ can form the set of pair $\langle x, y \rangle$ such that $x \sim y$.

DEFINITION 14. ($\mathsf{RCA}_0$) Let $A$ and $B$ be groups. The *free product of $A$ and $B$*, denoted $A * B$, consists of the elements $\mathrm{Red}_{A*B}$ with the empty sequence $1_{A*B}$ as the identity and multiplication given by $x \cdot y = \rho(x^\frown y)$.

Unraveling the definitions, we obtain the connection between free products of $\mathbb{Z}$ and free groups.

PROPOSITION 15 ($\mathsf{RCA}_0$). *The free product $\mathbb{Z} * \mathbb{Z}$ is isomorphic to $F_2$, the free group on two generators.*

**Proof.** Let $a, b$ denote the generators of $F_2$. We illustrate the isomorphism with an example. Consider the element $\langle a, a, b^{-1}, a, b, b \rangle$ of $F_2$. For notational convenience, let $a$ also denote 1 in the first copy of $\mathbb{Z}$ and $b$ denote 1 in the second copy of $\mathbb{Z}$. The given element of $F_2$ corresponds to the element $\langle 2a, -b, a, 2b \rangle$ of $\mathbb{Z} * \mathbb{Z}$. The isomorphism from $\mathbb{Z} * \mathbb{Z}$ to $F_2$ is built by expanding elements resembling $na$ in sequences in $\mathbb{Z} * \mathbb{Z}$ to n-tuples $\langle a, a, \ldots, a \rangle$ in which $a$ appears $n$ times. That is,

$$\langle 2a, -b, a, 2b \rangle \mapsto \langle a, a, b^{-1}, a, b, b \rangle.$$

Writing this map formally, we obtain the isomorphism. □

## 3 Main Result

Having formalized the notions of free group and free product in the last section, we use standard mathematical notation for elements of these groups.

Specifically, if $X = \{x_0, x_1, \ldots\}$ is the set of generators of a free group, then we write an arbitrary element of Word$_X$ as $x_{i_1}^{n_1} \cdots x_{i_k}^{n_k}$ with $n_i \in \mathbb{Z}\backslash 0$, where $x_i^{n_i}$ refers to $x_i^1$ or $x_i^{-1}$ repeated $|n_i|$ times. In this section, we prove Theorem 4 (every countable f.o. group is the o-epimorphic image of an f.o. free group) assuming the following theorem (which will be proved in later sections).

**THEOREM 16 (RCA$_0$).** *The free product of two f.o. groups is fully orderable.*

Combining Theorem 16 and Proposition 15, we obtain the following corollary.

**COROLLARY 17 (RCA$_0$).** *The free group on two generators is fully orderable.*

We use Corollary 17 to show that the free group on countably many generators is fully orderable. Let $\mathbb{N}^+$ denote the set of strictly positive natural numbers.

**LEMMA 18 (RCA$_0$).** *Let $F$ be the free group on the two generators $x, y$. For each $i \in \mathbb{N}^+$ let $\alpha_i = x^i y^i$.*

1. *The word $\alpha_{i_1}^{n_1} \cdots \alpha_{i_k}^{n_k}$ with $i_{j+1} \neq i_j$, $n_j \in \mathbb{Z} \setminus \{0\}$, and $k > 0$ freely reduces to a word ending in $x^\epsilon y^{i_k}$ if $n_k > 0$ and $y^\epsilon x^{-i_k}$ if $n_k < 0$.*

2. *No product $\alpha_{i_1}^{n_1} \cdots \alpha_{i_k}^{n_k}$ with the above restrictions is the identity element.*

**Proof.** Assuming the first property holds, there is either an $x$ or a $y$ with a nonzero exponent in the reduced form of $\alpha_{i_1}^{n_1} \cdots \alpha_{i_k}^{n_k}$. Therefore, the second property follows immediately from the first. The first property is proved by induction on $k$. If $k = 1$ then

$$\alpha_{i_1}^{n_1} = (x^{i_1} y^{i_1})^{n_1}$$

which satisfies the first property. If $k > 1$, then split into cases depending on the signs of $n_k$ and $n_{k-1}$.

**CASE 19.** $n_k > 0$ and $n_{k-1} > 0$

By the induction hypothesis, $\alpha_{i_1}^{n_1} \cdots \alpha_{i_{k-1}}^{n_{k-1}}$ reduces to a word ending in $x^\epsilon y^{i_{k-1}}$. Thus $\alpha_{i_1}^{n_1} \cdots \alpha_{i_k}^{n_k}$ reduces to a word ending in $x^\epsilon y^{i_{k-1}}(x^{i_k} y^{i_k})^{n_k}$. Since $n_k > 0$, this word ends in $x^{i_k} y^{i_k}$. The case when $n_k, n_{k-1} < 0$ is similar.

**CASE 20.** $n_k > 0$ and $n_{k-1} < 0$

By the induction hypothesis, $\alpha_{i_1}^{n_1} \cdots \alpha_{i_{k-1}}^{n_{k-1}}$ reduces to a word ending in $y^\epsilon x^{-i_{k-1}}$. Hence $\alpha_{i_1}^{n_1} \cdots \alpha_{i_k}^{n_k}$ reduces to a word ending in $y^\epsilon x^{-i_{k-1}} (x^{i_k} y^{i_k})^{n_k}$. If $n_k = 1$ then we have a word ending in $y^\epsilon x^{i_k - i_{k-1}} y^{i_k}$. By assumption, $i_k - i_{k-1} \neq 0$ so we have satisfied the first property. If $n_k > 1$, then this word ends in $x^{i_k} y^{i_k}$. The case when $n_k < 0$ and $n_{k-1} > 0$ is similar. □

**PROPOSITION 21 (RCA$_0$).** *The free group on a countable number of generators is fully orderable.*

**Proof.** Let $F$ and $\alpha_i$ be as in Lemma 18 and let $P(F)$ be the positive cone for some full order on $F$. Let $G$ be the free group on the generators $x_i$ with $i \in \mathbb{N}^+$. Define the homomorphism $\psi : G \to F$ sending $x_{i_1}^{n_1} \cdots x_{i_k}^{n_k} \mapsto \alpha_{i_1}^{n_1} \cdots \alpha_{i_k}^{n_k}$. If $x_{i_1}^{n_1} \cdots x_{i_k}^{n_k}$ is fully reduced and $x_{i_1}^{n_1} \cdots x_{i_k}^{n_k} \neq 1_G$, then $\alpha_{i_1}^{n_1} \cdots \alpha_{i_k}^{n_k}$ satisfies the hypotheses of Lemma 18. Hence $\alpha_{i_1}^{n_1} \cdots \alpha_{i_k}^{n_k} \neq 1_F$ so $\psi$ is a monomorphism. The set $P(G) = \{x_{i_1}^{n_1} \cdots x_{i_k}^{n_k} \mid \alpha_{i_1}^{n_1} \cdots \alpha_{i_k}^{n_k} \in P(F)\}$ is the positive cone of a full order on $G$. □

We can use a technique from Revesz [7] to give a proof of Theorem 4, which is restated below.

**THEOREM 22 (RCA$_0$).** *Every countable f.o. group is the o-epimorphic image of an f.o. free group.*

**Proof.** Let $(G, \leq_G)$ be an f.o. group, $P(G)$ be the positive cone of $\leq_G$, and $g_0, g_1, \ldots$ be an enumeration of $G \setminus \{1_G\}$. Let $F$ be the free group on the generators $x_0, x_1, \ldots$ and, by Proposition 21, let $P(F)$ be the positive cone of some full order on $F$. Define the epimorphism $\varphi : F \to G$ by $\varphi(x_{i_1}^{n_1} \cdots x_{i_k}^{n_k}) = g_{i_1}^{n_1} \cdots g_{i_k}^{n_k}$. We produce a new order $\tilde{P}(F)$ on $F$ under which $\varphi$ is order preserving.

Define an embedding $\psi : F \to G \times F$ by $\psi(a) = \langle \varphi(a), a \rangle$. Order $G \times F$ lexicographically, i.e. by the positive cone

$$\langle a, b \rangle \in P(G \times F) \leftrightarrow (a \in P(G) \wedge a \neq 1_G) \vee (a = 1_G \wedge b \in P(F)).$$

Since $\psi$ is a monomorphism, we define a new positive cone on $F$ by $\tilde{P}(F) = \{a \mid \psi(a) \in P(G \times F)\}$. It remains to show that $\varphi$ is order preserving under $\tilde{P}(F)$. Rewriting the definition of $\tilde{P}(F)$ we have

$$a \in \tilde{P}(F) \leftrightarrow (\varphi(a) \in P(G) \wedge \varphi(a) \neq 1_G) \vee (\varphi(a) = 1_G \wedge a \in P(F)).$$

Let $\leq_F$ be the order corresponding to $\tilde{P}(F)$. Suppose that $a \leq_F b$. Since $a^{-1}b \in \tilde{P}(F)$, we have $\varphi(a^{-1}b) = \varphi(a)^{-1}\varphi(b) \in P(G)$ by the definition of $\tilde{P}(F)$ and hence $\varphi(a) \leq_G \varphi(b)$. Now, suppose that $c, d \in G$ and $c <_G d$. Since $\varphi$ is onto, there are $a, b \in F$ with $\varphi(a) = c$ and $\varphi(b) = d$. Since $c <_G d$, we have that $c^{-1}d \in P(G)$ and $c^{-1}d \neq 1_G$. Because $\varphi(a^{-1}b) = c^{-1}d$ and $c^{-1}d \neq 1_G$, we have (by the definition of $\tilde{P}(F)$) that $a^{-1}b \in \tilde{P}(F)$ and so $a <_F b$. Therefore, $\varphi$ is an o-epimorphism from $(F, \leq_F)$ onto $(G, \leq_G)$. □

## 4 Ordered Rings and Triangular Matrices

In this section, we lay the framework to prove Theorem 16, which we do in the next section. The proof given here is a modification of the proof given in Kokorin and Kopytov [5]. The idea is to embed the free product into a set of infinite upper triangular matrices over a fully ordered ring. In this section, we introduce ordered rings and formalize the spaces of upper triangular matrices in second order arithmetic.

DEFINITION 23 ($\mathsf{RCA}_0$). A *ring* is a set $R$ together with two functions $+_R$, $\cdot_R$ and two constants $0_R$, $1_R$ which satisfy the usual axioms for a commutative ring with identity. (We often drop the subscripts $R$ when there is no chance of confusion.)

In this paper, rings will always be commutative with identity. As with groups, the subscripts will be dropped when the context is clear.

DEFINITION 24 ($\mathsf{RCA}_0$). A *partially ordered (p.o.) ring* is a ring $R$ together with a partial order $\leq_R$ on $R$ such that $a \leq_R b$ implies $a+c \leq_R b+c$ for all $a, b, c \in R$, and $a \leq_R b$ and $c >_R 0_R$ implies $ca \leq_R cb$ and $ac \leq_R bc$. If $\leq_R$ is linear, then $(R, \leq_R)$ is a *fully ordered (f.o.) ring*.

The positive cone of a p.o. ring $R$ is $P(R) = \{r \in R \mid r \geq_R 0_R\}$ and negative cone is $-P(R) = \{-r \mid r \in P(R)\}$. $\mathsf{RCA}_0$ can verify properties of $P$ similar to those for p.o. groups such as $P \cap -P = \{0\}$ and for fully ordered rings $P \cup -P = R$. Classically, positive cones for rings also satisfy $P + P \subseteq P$ and $PP \subseteq P$. Although $\mathsf{RCA}_0$ is not strong enough to prove that the sets $P + P$ and $PP$ exist, it is strong enough to show

$$\forall x, y \in R \ (x \in P \wedge y \in P \to x + y \in P)$$
$$\forall x, y \in R \ (x \in P \wedge y \in P \to xy \in P).$$

In the context of $\mathsf{RCA}_0$, we take $P + P \subseteq P$ to stand for the top formula and $PP \subseteq P$ to stand for the bottom formula.

THEOREM 25 ($\mathsf{RCA}_0$). *A subset $P$ of a ring $R$ is the positive cone of a partial order on $R$ if and only if $P \cap -P = \{0\}$, $P + P \subseteq P$ and $PP \subseteq P$. $P$ is the positive cone of a full order if and only if $P$ also satisfies $P \cup -P = R$.*

**Proof.** It is straightforward to check that any positive cone satisfies these requirements. Conversely, if $P$ is a set with these properties, then the order can be defined by $a \leq b$ if and only if $b - a \in P$. □

Given a f.o. ring $K$, we define a class of upper triangular matrices with the rows and columns indexed by the elements of $\mathbb{N}^+$ and with positive invertible elements along the main diagonal. Such a matrix resembles:

$$\begin{pmatrix} k_{11} & k_{12} & k_{13} & k_{14} & \cdots \\ 0 & k_{22} & k_{23} & k_{24} & \cdots \\ 0 & 0 & k_{33} & k_{34} & \cdots \\ \vdots & \vdots & \vdots & \ddots & \ddots \end{pmatrix}$$

where each $k_{ii}$ is in $P(K)$ and has a multiplicative inverse. Since there are an uncountable number of such matrices, we represent them in second order arithmetic as a class of functions.

DEFINITION 26 (RCA$_0$). Let $(K, \leq_K)$ be a fully ordered ring with positive cone $P(K)$. The function $f : \mathbb{N}^+ \times \mathbb{N}^+ \to K$ is *in the class* Tri$_K$, denoted $f \in$ Tri$_K$, if and only if it satisfies the following conditions.

1. For all $i > j$, $f(i,j) = 0_K$.

2. For all $i$, $f(i,i) \in P(K)$ and $\exists x \in K(f(i,i) \cdot x = 1_K)$.

We define an order and a group structure on Tri$_K$. Given $f, g \in$ Tri$_K$, we define $f < g$ if and only if for some pair $\langle i, j \rangle \in \mathbb{N}^+ \times \mathbb{N}^+$ with $i \leq j$ the following two conditions hold:

1. $f(i,j) <_K g(i,j)$.

2. $f(k, k+s) = g(k, k+s)$ for all $k, s$ such that $i + s < j$ or $i + s = j$ and $k < i$.

A pair $\langle i, j \rangle$ for which these conditions hold is called a witness for $f < g$. When $f$ and $g$ are viewed as matrices, these conditions say that we compare $f$ and $g$ down the diagonals, starting with the main diagonal, then the diagonal to its right, and so on, until we find the first place that $f$ and $g$ differ. The entries of $f$ and $g$ are compared in the order indicated in this picture:

$$\begin{pmatrix} 1 & \omega & \omega+\omega & \omega+\omega+\omega & \cdots \\ \cdot & 2 & \omega+1 & \omega+\omega+1 & \ddots \\ \cdot & \cdot & 3 & \omega+2 & \ddots \\ \cdot & \cdot & \cdot & 4 & \ddots \\ \vdots & \vdots & \vdots & \vdots & \ddots \end{pmatrix}$$

That is, we compare $f$ and $g$ by comparing two ordered sequences of elements of $K$ with order type $\omega \cdot \omega$. If $f \neq g$, then the relationship between $f$ and $g$ is determined by the relationship between the elements of $K$ at the first place where these ordered sequences differ. While $\mathrm{RCA}_0$ is strong enough to prove that $\omega \cdot \omega$ is well ordered, we will eventually need a stronger comparability result. Our goal is to embed $A * B$ into $\mathrm{Tri}_K$ (for a particular $K$ we construct in the next section) and pull the order back from $\mathrm{Tri}_K$ to $A * B$. For this process, we need to be able to determine the order between elements of $\mathrm{Tri}_K$ (in the image of the embedding) uniformly. $\mathrm{RCA}_0$ is not strong enough to uniformly compare sequences of length $\omega \cdot \omega$.

To be more specific, let $s_i(x, y)$ (for $i \in \mathbb{N}$) be a sequence of functions. For each $i$, think of $s_i(x, y)$ as representing an $\omega \cdot \omega$ length sequence in which the $(n \cdot \omega + m)$-th value is $s_i(n, m)$. Assume that for all $i \neq j$, there is at least one pair $(n, m)$ for which $s_i(n, m) \neq s_j(n, m)$. $\mathrm{RCA}_0$ is not strong enough to prove the existence of a function $g(x, y)$ such that for all $i \neq j$, $g(i, j)$ is the least pair $(n, m)$ (in the $\omega \cdot \omega$ order) at which $s_i$ and $s_j$ differ.

In the next section, we define a countable subgroup of $\mathrm{Tri}_K$ for which $\mathrm{RCA}_0$ can determine the order. (For this subgroup, we will reduce the length of the well order along which we do our comparisons from $\omega \cdot \omega$ to $\omega$.) For now, our goal is to define the group structure and to prove that the elements of $\mathrm{Tri}_K$ satisfy the axioms for a partially ordered group with this order.

Given $f, g \in \mathrm{Tri}_K$, we define the product $f \cdot g$ to be the function

$$f \cdot g : \mathbb{N}^+ \times \mathbb{N}^+ \to K$$

$$(f \cdot g)(i, j) = \sum_{n=1}^{\infty} f(i, n) g(n, j).$$

This definition is the standard definition for matrix multiplication. Since $f(i, n) = 0$ for $n < i$ and $g(n, j) = 0$ for $n > j$, if $n$ is not between $i$ and $j$, then $f(i, n)g(n, j) = 0$. Therefore, if $i > j$, then the sum is 0, and if $i \leq j$, then the infinite sum reduces to the finite sum $\sum_{n=i}^{j} f(i, n)g(n, j)$. Thus, $\mathrm{RCA}_0$ proves $f \cdot g$ well defined. Since $(f \cdot g)(i, i) = f(i, i)g(i, i)$, $(f \cdot g)(i, i)$ is both positive and invertible, and hence $f \cdot g \in \mathrm{Tri}_K$. The matrix $I \in \mathrm{Tri}_K$ defined by $I(i, i) = 1_K$ and $I(i, j) = 0_K$ for $i \neq j$ plays the role of the identity element in $\mathrm{Tri}_K$. The next two lemmas show that $\mathrm{RCA}_0$ proves the associativity of the multiplication and the existence of inverses.

LEMMA 27 ($\mathrm{RCA}_0$). *For all $f, g, h \in \mathrm{Tri}_K$, $(f \cdot g) \cdot h = f \cdot (g \cdot h)$.*

**Proof.** For $j < i$, both of these products have the value 0 on input $(i, j)$.

For $i \leq j$, direct calculation shows that

$$((f \cdot g) \cdot h)(i,j) = (f \cdot (g \cdot h))(i,j) = \sum_{i \leq n \leq m \leq j} f(i,n)g(n,m)h(m,j).$$

$\square$

LEMMA 28 (RCA$_0$). *If $f \in \text{Tri}_K$, then $f$ has an inverse $g \in \text{Tri}_K$, in the sense that $f \cdot g = g \cdot f = I$, given by:*

$$g(i,j) = \begin{cases} 0 & j < i \\ f(i,j)^{-1} & i = j \\ -\frac{f(i,j)}{f(i,i)f(j,j)} + \sum_{i<k_1<j} \frac{f(i,k_1)f(k_1,j)}{f(i,i)f(k_1,k_1)f(j,j)} \\ \quad - \sum_{i<k_1<k_2<j} \frac{f(i,k_1)f(k_1,k_2)f(k_2,j)}{f(i,i)f(k_1,k_1)f(k_2,k_2)f(j,j)} + \cdots & i < j \\ \quad \cdots + (-1)^{j-i} \frac{f(i,i+1)\cdots f(j-1,j)}{f(i,i)f(i+1,i+1)\cdots f(j,j)} \end{cases}$$

Since $f(n,n)$ is invertible, we write it in the denominator of a fraction as shorthand for $f(n,n)^{-1}$. The proof of Lemma 28 is presented in Section 6.

It remains to verify that the order interacts in the appropriate way with the group structure. If $f \in \text{Tri}_K$, we say $f \in P(\text{Tri}_K)$ if and only if $I \leq f$ in the order given above. If $f \neq I$ this is equivalent to either

$$\exists i \; [f(i,i) > 1 \wedge \forall j < i(f(j,j) = 1)]$$

or

$$\forall i \; (f(i,i) = 1) \wedge \exists i,j \left( i < j \wedge f(i,j) > 0 \wedge \right.$$
$$\left. \forall k \; \forall s > 0 \; ((i+s < j \vee (i+s = j \wedge k < i)) \to f(k, k+s) = 0) \right).$$

The proof of the following lemma is delayed until Section 6.

LEMMA 29 (RCA$_0$).

1. *If $f, g \in P(\text{Tri}_K)$ then $f \cdot g \in P(\text{Tri}_K)$.*

2. *If $f \in P(\text{Tri}_K)$ and $f \neq I$ then $f^{-1} \notin P(\text{Tri}_K)$.*

3. *If $f \in P(\text{Tri}_K)$ and $g \in \text{Tri}_K$ then $gfg^{-1} \in P(\text{Tri}_K)$.*

## 5   Free Products of Fully Ordered Groups

In this section we prove Theorem 16. The proof has several steps, so we outline them here. Given fully ordered groups $A$ and $B$, we form a larger group $C$ of which $A$ and $B$ are direct summands. We use the orders on $A$ and $B$ to fully order the group ring $\mathbb{Q}[C]$ and form the ordered matrix group $\text{Tri}_{\mathbb{Q}[C]}$. The free product $A * B$ is embedded in $\text{Tri}_{\mathbb{Q}[C]}$ and we define an order on $A * B$ by pulling back the order from $\text{Tri}_{\mathbb{Q}[C]}$.

Let $A$ and $B$ be fully ordered groups. We first define a larger ordered group $C$. For each pair $\langle i, j \rangle \in \mathbb{N}^+ \times \mathbb{N}^+$, let $x_{ij}$ and $y_{ij}$ generate copies of $\mathbb{Z}$ ordered such that $x_{ij}^n$ and $y_{ij}^n$ are positive if and only if $n \geq 0$. (Note that we are writing these groups multiplicatively.) For each $i \in \mathbb{N}^+$, let $u_i$ and $v_i$ generate copies of $\mathbb{Z}$ ordered in the same way. The notation $\langle x_{ij} \rangle$ is used for the group generated by $x_{ij}$, and similarly for $\langle y_{ij} \rangle$, $\langle u_i \rangle$, and $\langle v_i \rangle$.

The group $C$ is the restricted direct product

$$C = A \times B \times \prod_{i,j=1}^{\infty} \langle x_{ij} \rangle \times \prod_{i,j=1}^{\infty} \langle y_{ij} \rangle \times \prod_{i=1}^{\infty} \langle u_i \rangle \times \prod_{i=1}^{\infty} \langle v_i \rangle.$$

Since there is a uniform sequence of orders on the factors of $C$, $C$ can be lexicographically ordered in $\mathsf{RCA}_0$. As above, note that $C$ is written multiplicatively even though many of the summands are normally written additively. We use $x_{ij}$ to denote the element of $C$ which is the identity in all components of $C$ except the $\langle x_{ij} \rangle$ component and has value $x_{ij}$ in the $\langle x_{ij} \rangle$ component. We abuse notation similarly for $a \in A$, $b \in B$ and the generators $u_i, v_i, y_{ij}$.

Let $\mathbb{Q}[C]$ be the group ring of $C$ over $\mathbb{Q}$. Formally, the elements of $\mathbb{Q}[C]$ are the finite sums $\sum \alpha_i c_i$ with $\alpha_i \in \mathbb{Q} \setminus \{0\}$, $c_i \in C$ and all the $c_i$ distinct. Addition is defined by:

$$\sum_{i \in I} \alpha_i c_i + \sum_{j \in J} \beta_j c_j = \sum_{i \in I \setminus J} \alpha_i c_i + \sum_{j \in J \setminus I} \beta_j c_j + \sum_{i \in I \cap J} (\alpha_i + \beta_i) c_i$$

with the stipulation that any terms in the third sum with $\alpha_i + \beta_i = 0$ are removed. Multiplication is defined by:

$$\left( \sum_{i \in I} \alpha_i c_i \right) \left( \sum_{j \in J} \beta_j c_j \right) = \sum_{i \in I} \sum_{j \in J} (\alpha_i \beta_j) c_i c_j$$

where the terms with the same value from $C$ in this finite sum are collected and any term with coefficient 0 is dropped. The additive identity is the empty sum and the multiplicative identity is the sum with one element $1_{\mathbb{Q}} 1_C$.

In the context of $\mathbb{Q}[C]$, we use $x_{ij}$ to denote the single element sum $1_\mathbb{Q} x_{ij}$ (and similarly for $y_{ij}$, $u_i$, $v_i$, $a \in A$ and $b \in B$) and we use $0$ (or $0_{\mathbb{Q}[C]}$) and $1$ (or $1_{\mathbb{Q}[C]}$) to denote the additive and multiplicative identities.

The positive cone $P(\mathbb{Q}[C])$ is defined by $\sum_{i \in I} \alpha_i c_i \in P(\mathbb{Q}[C])$ if and only if $I = \emptyset$ or $\alpha_j >_\mathbb{Q} 0$ where $j$ is such that $c_j$ is the $\leq_C$-least element among the $c_i$ with $i \in I$. Since $I$ is finite there is such a $\leq_C$-least element. $\mathrm{RCA}_0$ suffices to prove that this gives a full order on $\mathbb{Q}[C]$.

Now that we have a fully ordered ring, we can use the machinery of the previous section to work with $\mathrm{Tri}_{\mathbb{Q}[C]}$. The goal is to embed $A * B$ into $\mathrm{Tri}_{\mathbb{Q}[C]}$ and then use our formal ordering of $\mathrm{Tri}_{\mathbb{Q}[C]}$ to order $A * B$. The embedding is given by uniformly associating to each element of $A * B$ a function in $\mathrm{Tri}_{\mathbb{Q}[C]}$. To do this we specify four matrices in $\mathrm{Tri}_{\mathbb{Q}[C]}$ denoted $X, Y, U$ and $V$.

$$X(i,j) = \begin{cases} 1 & i = j \\ 0 & i > j \\ x_{ij} & i < j \end{cases}$$

$$Y(i,j) = \begin{cases} 1 & i = j \\ 0 & i > j \\ y_{ij} & i < j \end{cases}$$

$$U(i,j) = \begin{cases} u_i & i = j \\ 0 & i \neq j \end{cases}$$

$$V(i,j) = \begin{cases} v_i & i = j \\ 0 & i \neq j \end{cases}$$

As matrices, these functions look like:

$$X = \begin{pmatrix} 1 & x_{12} & x_{13} & \cdots \\ 0 & 1 & x_{23} & \cdots \\ 0 & 0 & 1 & \cdots \\ \vdots & \vdots & \vdots & \ddots \end{pmatrix}$$

$$U = \begin{pmatrix} u_1 & 0 & 0 & \cdots \\ 0 & u_2 & 0 & \cdots \\ 0 & 0 & u_3 & \cdots \\ \vdots & \vdots & \vdots & \ddots \end{pmatrix}$$

$U$ is upper triangular and has positive elements on the diagonal since $u_i$ is positive in our order on $\langle u_i \rangle$. Also, since $1_\mathbb{Q} u_i \cdot 1_\mathbb{Q} u_i^{-1} = 1_{\mathbb{Q}[C]}$, $U$ has invertible elements along the diagonal. Thus, $U \in \mathrm{Tri}_{\mathbb{Q}[C]}$. Similarly, $X, Y, V \in \mathrm{Tri}_{\mathbb{Q}[C]}$.

These matrices are used to define the embedding in several steps. For each $a \in A$, define $\alpha(a) : \mathbb{N}^+ \times \mathbb{N}^+ \to \mathbb{Q}[C]$ by:

$$\alpha(a)(i,j) = \begin{cases} 1 & i = j \text{ and } i \text{ is odd} \\ a & i = j \text{ and } i \text{ is even} \\ 0 & i \neq j \end{cases}$$

As a matrix, this looks like:

$$\alpha(a) = \begin{pmatrix} 1 & 0 & 0 & \dots \\ 0 & a & 0 & \dots \\ 0 & 0 & 1 & \dots \\ \vdots & \vdots & \vdots & \ddots \end{pmatrix}$$

Regardless of whether $a$ is positive or negative in $A$, $1_\mathbb{Q} a$ is positive in $\mathbb{Q}[C]$ since $1_\mathbb{Q} > 0_\mathbb{Q}$. Also, $a$ is invertible in $\mathbb{Q}[C]$ because $1_\mathbb{Q} a \cdot 1_\mathbb{Q} a^{-1} = 1_{\mathbb{Q}[C]}$. Hence, $\alpha(a) \in \text{Tri}_{\mathbb{Q}[C]}$.

For each $b \in B$ define $\beta(b) \in \text{Tri}_{\mathbb{Q}[C]}$ similarly:

$$\beta(b)(i,j) = \begin{cases} 1 & i = j \text{ and } i \text{ is odd} \\ b & i = j \text{ and } i \text{ is even} \\ 0 & i \neq j \end{cases}$$

We define two more maps on $A$ and $B$. For $a \in A$ and $b \in B$, define

$$\alpha'(a) = X^{-1} \cdot \alpha(a) \cdot X \qquad \beta'(b) = Y^{-1} \cdot \beta(b) \cdot Y$$
$$\alpha''(a) = U^{-1} \cdot \alpha'(a) \cdot U \qquad \beta''(b) = V^{-1} \cdot \beta'(b) \cdot V.$$

Later, we will use results from the previous section to produce explicit formulas for the entries in these matrices.

Because RCA$_0$ proves that $\text{Tri}_{\mathbb{Q}[C]}$ is closed under inverses and products, $\alpha''(a)$ and $\beta''(b)$ are both in $\text{Tri}_{\mathbb{Q}[C]}$. Also, since we have explicit formulas for inverses and products in $\text{Tri}_{\mathbb{Q}[C]}$, $\alpha''(a)$ and $\beta''(b)$ can be given uniformly from $A$ and $B$. The embedding of $A * B$ into $\text{Tri}_{\mathbb{Q}[C]}$ is given by associating to each word $a_1 b_1 \cdots a_n b_n$ over $A$ and $B$ the product $\alpha''(a_1)\beta''(b_1) \cdots \alpha''(a_n)\beta''(b_n)$ in $\text{Tri}_{\mathbb{Q}[C]}$. The term embedding is being used loosely here since $\text{Tri}_{\mathbb{Q}[C]}$ is not a set. The correspondence is really a uniform construction of a function in $\text{Tri}_{\mathbb{Q}[C]}$ for each word over $A, B$. That said, we will continue to use the term embedding and will use $\gamma(w)$ to denote the element of $\text{Tri}_{\mathbb{Q}[C]}$ which corresponds to the word $w$.

We need to describe and check the properties of this embedding. It follows from the definitions that for $a \in A$, $\alpha(a)^{-1} = \alpha(a^{-1})$ and hence that $\gamma(a)^{-1} = \gamma(a^{-1})$. Similarly, $\gamma(b)^{-1} = \gamma(b^{-1})$ for $b \in B$, and $\gamma(w_1^{-1}) = \gamma(w_1)^{-1}$ and $\gamma(w_1 w_2) = \gamma(w_1)\gamma(w_2)$ for any words $w_1$ and $w_2$ over $A$ and $B$. Therefore, $\gamma$ is a group homomorphism. It is less trivial to check that $\gamma$ is one-to-one.

**PROPOSITION 30 (RCA$_0$).** *If $w_1 \neq w_2$ in $A * B$, then $\gamma(w_1) \neq \gamma(w_2)$ in $Tri_{\mathbb{Q}[C]}$.*

In order to prove this proposition, we need several lemmas. Throughout these lemmas $a \in A$, $b \in B$ and $w_1, w_2$ are arbitrary words in $A*B$. Our first goal is to derive formulas for $\alpha'(a)$ and by analogy $\beta'(b)$. Let $f = \alpha(a) \cdot X$ and $g = X^{-1}$, so $\alpha'(a) = g \cdot f$. More explicitly, $g$ is given by $g(i,i) = 1$, $g(i,j) = 0$ for $i > j$ and for $i < j$

$$g(i,j) = -x_{ij} + \sum_{i<k_1<j} x_{ik_1} x_{k_1 j} - \sum_{i<k_1<k_2<j} x_{ik_1} x_{k_1 k_2} x_{k_2 j} + \cdots + (-1)^{j-i}(x_{i(i+1)} \cdots x_{(j-1)j}).$$

As a matrix, this looks like:

$$g = \begin{pmatrix} 1 & -x_{12} & -x_{13} + x_{12}x_{23} & \cdots \\ 0 & 1 & -x_{23} & \cdots \\ 0 & 0 & 1 & \cdots \\ \vdots & \vdots & \vdots & \ddots \end{pmatrix}$$

$f$ can be given explicitly by:

$$f(i,j) = \begin{cases} 1 & i=j \wedge i \text{ is odd} \\ a & i=j \wedge i \text{ is even} \\ x_{ij} & i<j \wedge i \text{ is odd} \\ ax_{ij} & i<j \wedge i \text{ is even} \\ 0 & i>j \end{cases}$$

$$f = \begin{pmatrix} 1 & x_{12} & x_{13} & \cdots \\ 0 & a & ax_{23} & \cdots \\ 0 & 0 & 1 & \cdots \\ \vdots & \vdots & \vdots & \ddots \end{pmatrix}$$

**LEMMA 31 (RCA$_0$).**

$$\alpha'(a)(i,i) = \begin{cases} 1 & i \text{ is odd} \\ a & i \text{ is even} \end{cases}$$

**Proof.** This formula comes directly from the formulas for $f$ and $g$. □

The formulas for $\alpha'(a)(i,j)$ when $i < j$ are given below. We will prove Lemma 32 in Section 6. The proofs of Lemmas 33, 34 and 35 are similar and are omitted.

**LEMMA 32 (RCA$_0$).** *If $i < j$ and $i, j$ are both even, then*

$$\alpha'(a)(i,j) = (1-a) \sum_{\substack{n=i+1 \\ n \text{ odd}}}^{j-1} \left( -x_{in}x_{nj} + \sum_{i<k_1<n} (x_{ik_1}x_{k_1n}x_{nj}) - \right.$$

$$\left. - \sum_{i<k_1<k_2<n} (x_{ik_1}x_{k_1k_2}x_{k_2n}x_{nj}) + \cdots + (-1)^{n-i} x_{i(i+1)} \cdots x_{(n-1)n}x_{nj} \right)$$

**LEMMA 33 (RCA$_0$).** *If $i < j$, $i$ is even, and $j$ is odd then*

$$\alpha'(a)(i,j) = (1-a)(-x_{ij}) + (1-a) \sum_{\substack{n=i+1 \\ n \text{ even}}}^{j-1} \left( x_{in}x_{nj} - \sum_{i<k_1<n} (x_{ik_1}x_{k_1n}x_{nj}) + \right.$$

$$\left. + \sum_{i<k_1<k_2<n} x_{ik_1}x_{k_1k_2}x_{k_nn}x_{nj} - \cdots + (-1)^{n-i} x_{ii+1} \cdots x_{n-1n}x_{nj} \right)$$

**LEMMA 34 (RCA$_0$).** *If $i < j$ and $i, j$ are both odd, then*

$$\alpha'(a)(i,j) = (1-a) \sum_{\substack{n=i+1 \\ n \text{ even}}}^{j-1} \left( x_{in}x_{nj} - \sum_{i<k_1<n} (x_{ik_1}x_{k_1n}x_{nj}) + \right.$$

$$\left. + \sum_{i<k_1<k_2<n} (x_{ik_1}x_{k_1k_2}x_{k_2n}x_{nj}) + \cdots + (-1)^{n-i} x_{i(i+1)} \cdots x_{(n-1)n}x_{nj} \right)$$

**LEMMA 35 (RCA$_0$).** *If $i < j$, $i$ is odd, and $j$ is even then*

$$\alpha'(a)(i,j) = (1-a)(x_{ij}) + (1-a) \sum_{\substack{n=i+1 \\ n \text{ odd}}}^{j-1} \left( -x_{in}x_{nj} + \sum_{i<k_1<n} (x_{ik_1}x_{k_1n}x_{nj}) - \right.$$

$$\left. - \sum_{i<k_1<k_2<n} x_{ik_1}x_{k_1k_2}x_{k_nn}x_{nj} + \cdots + (-1)^{n-i} x_{i(i+1)} \cdots x_{(n-1)n}x_{nj} \right)$$

The same results hold for $\beta'(b)$ with $b$ substituted into the formulas for $a$ and $y_{ij}$ substituted for $x_{ij}$. From these formulas, it is clear that if $a = 1_A$ then $\alpha'(a) = I$ in $\mathrm{Tri}_{\mathbb{Q}[C]}$, and similarly for $b$. Also, if $a \neq 1_A$, then in particular, the diagonal elements of $\alpha'(a)$ are not all 1, so $\alpha'(a) \neq I$. A closer look at these formulas reveals the following lemma.

**LEMMA 36 (RCA$_0$).** *If $a \neq 1_A$ and $b \neq 1_B$ then for any $i, j$ with $i \leq j$, $\alpha'(a)(i,j) \neq 0$ and $\beta'(b)(i,j) \neq 0$.*

We are now ready to prove Proposition 30.

**Proof.** To show that $w_1 \neq w_2$ in $A * B$ implies that $\gamma(w_1) \neq \gamma(w_2)$ in $\mathrm{Tri}_{\mathbb{Q}[C]}$, it suffices to show that $\gamma(w) \neq I$ for an arbitrary nonidentity element $w$. Let $w = a_1 b_1 \cdots a_t b_t$ be an arbitrary nonidentity word in $A * B$ that is reduced, except that possibly $a_1 = 1_A$ or $b_t = 1_B$. It suffices to show for $i < j$ that $\gamma(w)(i,j) \neq 0$.

The multiplication formula in $\mathrm{Tri}_{\mathbb{Q}[C]}$ extends to the following formula for the product of $m$ functions $f_1, \ldots, f_m$ in $\mathrm{Tri}_{\mathbb{Q}[C]}$: for $i > j$, $f_1 \cdots f_m(i,j) = 0$ and for $i \leq j$

$$f_1 \cdots f_m(i,j) = \sum_{i \leq k_1 \leq \cdots \leq k_{m-1} \leq j} f_1(i, k_1) f_2(k_1, k_2) \cdots f_m(k_{m-1}, j).$$

We consider the case in which $a_1 \neq 1_A$ and $b_t \neq 1_B$. By the multiplication formula, if $c = \alpha''(a_1) \beta''(b_1) \cdots \alpha''(a_t) \beta''(b_t)$, then for $i \leq j$

$$c(i,j) = \sum_{i \leq k_i \leq \cdots \leq k_{2t-1} \leq j} \left( \alpha''(a_1)(i, k_1) \beta''(b_1)(k_1, k_2) \cdots \right.$$
$$\left. \cdots \alpha''(a_t)(k_{2t-2}, k_{2t-1}) \beta''(b_t)(k_{2t-1}, j) \right).$$

Applying the formulas for multiplication and inverses, we have

$$\alpha''(a)(i,j) = \frac{1}{u_i} \alpha'(a)(i,j) u_j$$

$$\beta''(b)(i,j) = \frac{1}{v_i} \beta'(b)(i,j) v_j.$$

As before, the notation $\frac{1}{u_i}$ stands for $u_i^{-1}$. Putting these formulas together gives us:

$$c(i,j) = \sum_{i \leq k_1 \leq \cdots \leq k_{2t-1} \leq j} \left( \frac{u_{k_1}}{u_i} \frac{v_{k_2}}{v_{k_1}} \frac{u_{k_3}}{u_{k_2}} \cdots \frac{v_j}{v_{k_{2t-1}}} \alpha'(a_1)(i, k_1) \cdot \right.$$
$$\left. \cdot \beta'(b_1)(k_1, k_2) \cdots \alpha'(a_t)(k_{2t-2}, k_{2t-1}) \beta'(b_t)(k_{2t-1}, j) \right).$$

Viewing $c(i,j)$ as a polynomial in the variables $u_r, v_r, 1/u_r$ and $1/v_r$ for $i \leq r \leq j$, it is clear that none of the terms in the polynomial cancel. Also, since $\alpha'(a_m)(r,s) \neq 0$, $\beta'(b_m)(r,s) \neq 0$ for any $i \leq r \leq s \leq j$, and since any group ring has no zero divisors, none of the terms drop out because they are zero. The remaining cases, $a_1 = 1_A, b_t \neq 1_B$ etc., are similar. Thus $c \neq I$. □

Recall that comparing arbitrary elements of $\mathrm{Tri}_{\mathbb{Q}[C]}$ involves comparing sequences with order type $\omega \cdot \omega$. The key to proving Theorem 16 in $\mathrm{RCA}_0$ is to show that if $w_1 \neq w_2 \in A * B$ then comparing $\gamma(w_1)$ and $\gamma(w_2)$ requires only comparing sequences of elements of $\mathbb{Q}[C]$ with order type $\omega$.

DEFINITION 37 ($\mathrm{RCA}_0$). If $r \in \mathbb{Q}[C]$ then define $r^{+n}$ to be the element of $\mathbb{Q}[C]$ that looks just like $r$ except the subscripts on $x_{ij}, y_{ij}, u_i$ and $v_i$ are all adjusted by $+n$. That is, $x_{ij} \mapsto x_{(i+n)(j+n)}$, $u_i \mapsto u_{i+n}$, etc.

PROPOSITION 38 ($\mathrm{RCA}_0$). If $f \in \mathrm{Tri}_{\mathbb{Q}[C]}$ is in the image of $\gamma$ then

$$f(1,j)^{+2n} = f(1+2n, j+2n)$$
$$f(2,j)^{+2n} = f(2+2n, j+2n).$$

DEFINITION 39 ($\mathrm{RCA}_0$). If the conditions in the conclusion of Proposition 38 hold for $f$, then we say $f$ possesses the *shift property*.

The proof of Proposition 38 is broken into several lemmas.

LEMMA 40 ($\mathrm{RCA}_0$). If $f, g \in \mathrm{Tri}_{\mathbb{Q}[C]}$ possess the shift property, then so does $f \cdot g$.

**Proof.** Consider $(f \cdot g)(1, j)$. If $j = 1$, then we have:

$$\begin{aligned}(f \cdot g)(1+2n, 1+2n) &= f(1+2n, 1+2n) g(1+2n, 1+2n) \\ &= f(1,1)^{+2n} g(1,1)^{+2n} \\ &= ((f \cdot g)(1,1))^{+2n}.\end{aligned}$$

If $j > 1$ then we have:

$$(f \cdot g)(1 + 2n, j + 2n) = \sum_{m=1+2n}^{j+2n} f(1+2n, m)g(m, j+2n)$$

$$= \sum_{m=1}^{j} f(1+2n, m+2n)g(m+2n, j+2n)$$

$$= \sum_{m=1}^{j} f(1, m)^{+2n} g(m, j)^{+2n}$$

$$= \sum_{m=1}^{j} (f(1, m)g(m, j))^{+2n} = ((f \cdot g)(1, j))^{+2n}.$$

The cases for $(f \cdot g)(2, j)$ are similar. □

LEMMA 41 (RCA$_0$). *If $a \in A$ then $\alpha'(a)$ and $\alpha'(a^{-1})$ have the shift property.*

**Proof.** This proof utilizes the formulas which we derived for $\alpha'(a)$. Along the principal diagonal, we have:

$$\alpha'(a)(i, i) = \begin{cases} 1 & i \text{ is odd} \\ a & i \text{ is even} \end{cases}$$

This satisfies the shift property for the cases $\alpha'(a)(1,1)$ and $\alpha'(a)(2,2)$. If $j > 1$ and odd, then using our formulas:

$$\alpha'(a)(1, j) = (1-a) \sum_{\substack{m=2 \\ m \text{ even}}}^{j-1} \left( x_{1m} x_{mj} - \sum_{1 < k_1 < m} (x_{1k_1} x_{k_1 m} x_{mj}) + \sum_{1 < k_1 < k_2 < m} (x_{1k_1} x_{k_1 k_2} x_{k_2 m} x_{mj}) + \cdots + (-1)^{m-1} x_{12} \cdots x_{m-1 m} x_{mj} \right).$$

When we write the formula for $\alpha'(a)(1 + 2n, j + 2n)$ instead of letting $m$ range from $2 + 2n$ to $j - 1 + 2n$, we let it range from $2$ to $j - 1$ and adjust the subscripts inside the sum.

$$\alpha'(a)(1+2n, j+2n) = (1-a) \sum_{\substack{m=2 \\ m \text{ even}}}^{j-1} \Big( x_{(1+2n)(m+2n)} x_{(m+2n)(j+2n)} -$$

$$- \sum_{1 < k_1 < m} \left( x_{(1+2n)(k_1+2n)} x_{(k_1+2n)(m+2n)} x_{(m+2n)(j+2n)} \right) + \cdots$$

$$+ \cdots (-1)^{m+2n-(1+2n)} \left( x_{(1+2n)(1+2n+1)} \cdots x_{(m+2n)(j+2n)} \right) \Big)$$

Once you observe that $m + 2n - (1 + 2n) = m - 1$, it is clear that these two sums can be obtained from one another by a shift in the indices of $+2n$. The other cases follow similarly using the formulas for $\alpha'(a)$ and $\alpha'(a^{-1})$. □

**LEMMA 42 (RCA$_0$).** *If $a \in A$ then $\alpha''(a)$ and $\alpha''(a^{-1})$ have the shift property.*

**Proof.** This follows from the fact that

$$\alpha''(a)(i,j) = \frac{u_j}{u_i} \alpha'(a)(i,j)$$

We have

$$\alpha''(a)(i+2n, j+2n) = \frac{u_{j+2n}}{u_{i+2n}} \alpha'(a)(i+2n, j+2n)$$

$$= \frac{u_{j+2n}}{u_{i+2n}} \alpha'(a)(i,j)^{+2n}$$

$$= \alpha''(a)(i,j)^{+2n}.$$

The case for $\alpha''(a^{-1})$ is similar. □

**LEMMA 43 (RCA$_0$).** *If $b \in B$ then $\beta'(b), \beta'(b^{-1}), \beta''(b)$ and $\beta''(b^{-1})$ have the shift property.*

**Proof.** The proof is the same as for $\alpha'(a)$ and $\alpha''(a)$. □

We can now prove Proposition 38.

**Proof.** By assumption $\gamma(w) = f$ for some $w \in A*B$. From the facts that $w$ is a word over $A$ and $B$, that $\gamma$ is a homomorphism, that $\gamma(a), \gamma(a^{-1}), \gamma(b)$ and $\gamma(b^{-1})$ have the shift property for all $a \in A$ and $b \in B$, and that the shift property is preserved under multiplication, it follows that $f$ has the shift property. □

It remains to show how to pull the order on $\text{Tri}_{\mathbb{Q}[C]}$ back to $A * B$. Suppose that $f \in \text{Tri}_{\mathbb{Q}[C]}$, $f \neq I$, and $f$ has the shift property. Since $f \neq I$, there is some pair $\langle i, j \rangle$ such that $f(i,j) \neq I(i,j)$. In order to tell if $f \in P(\text{Tri}_{\mathbb{Q}[C]})$ we need to look down the diagonals until we find the first such pair. However, because $f$ has the shift property, if $f$ and $I$ agree on the first two entries in any diagonal, they will agree on all entries in that diagonal. Comparing $f$ and $I$ is now easy. Thinking of them as matrices, we compare the entries in the following order:

$$\begin{pmatrix} 1 & 3 & 5 & 7 & \cdots \\ \cdot & 2 & 4 & 6 & \cdots \\ -&- & irrelevant & -&- \\ \vdots & \vdots & \vdots & \vdots & & \vdots \end{pmatrix}$$

We only need to search through a sequence of elements with order type $\omega$. If we know that $f \neq I$ then we can find the first place they differ in this sequence. We finally show how to define $P(A * B)$ from $P(\text{Tri}_{\mathbb{Q}[C]})$.

$$P(A * B) = \{1_{A*B}\} \cup \{x \in A * B \mid x \neq 1_{A*B} \wedge \gamma(x) \in P(\text{Tri}_{\mathbb{Q}[C]})\}$$

$\mathsf{RCA}_0$ proves the existence of this set because for any $x \neq 1_{A*B}$, we know that $\gamma(x) \neq I$ and $\gamma(x)$ has the shift property. Therefore, $\mathsf{RCA}_0$ proves there is a first place in the sequence of matrix elements indicated above in which the entry differs from the entry in the identity matrix. We compare this entry with the corresponding entry in the identity matrix to determine if $\gamma(x) \in P(\text{Tri}_{\mathbb{Q}[C]})$. It remains to show that this set is in fact the positive cone of a full order on $A * B$.

CLAIM 44. $P(A * B)$ is normal, pure and closed under multiplication.

Verifying these conditions is similar, so we restrict ourselves to checking that $P(A * B)$ is closed under multiplication. Assume $x, y \in P(A * B)$. Since $P(\text{Tri}_{\mathbb{Q}[C]})$ is closed under multiplication and $\gamma(x), \gamma(y) \in P(\text{Tri}_{\mathbb{Q}[C]})$, we have $\gamma(x)\gamma(y) = \gamma(xy) \in P(\text{Tri}_{\mathbb{Q}[C]})$. Also, assuming that at least one of $x, y$ is not $1_{A*B}$, then $x$ and $y$ cannot be inverses because $P(\text{Tri}_{\mathbb{Q}[C]})$ is pure. Thus, $\gamma(xy) \in P(\text{Tri}_{\mathbb{Q}[C]})$ implies that $xy \in P(A*B)$ and so $P(A*B)$ is closed under multiplication.

CLAIM 45. $P(A * B)$ is full.

Assume $\gamma(x) \notin P(\text{Tri}_{\mathbb{Q}[C]})$. We need to show that $\gamma(x)^{-1} = \gamma(x^{-1}) \in P(\text{Tri}_{\mathbb{Q}[C]})$. Notice that $\gamma(x) \neq I$. Before splitting into cases, note that $\gamma(x)(1,1) = 1$ for every $x \in A * B$, and if $x \in A*B$ is the word $a_1 b_1 \cdots a_k b_k$, then $\gamma(x)(2,2) \in \mathbb{Q}[C]$ is the single element sum $1_{\mathbb{Q}} ab$, where $ab \in C$ is the element whose projection onto $A$ is $a = a_1 \cdot_A a_2 \cdots_A a_k$ and whose projection onto $B$ is $b = b_1 \cdot_B b_2 \cdots_B b_k$. The proof splits into two cases.

CASE 46. $\gamma(x)(2,2) \neq 1$.

Since $\gamma(x)(1,1) = 1$ and $\gamma(x) \notin P(\text{Tri}_{\mathbb{Q}[C]})$, we know that $\gamma(x)(2,2) < 1$ in $\mathbb{Q}[C]$. Assume that $x$ is the word $a_1 b_1 \cdots a_k b_k$ and $\gamma(x)(2,2) = 1_\mathbb{Q} ab$ as above. The inequality $1_\mathbb{Q} ab < 1_\mathbb{Q} 1_C$ in $\mathbb{Q}[C]$ means that $1_\mathbb{Q} 1_C - 1_\mathbb{Q} ab \in P(\mathbb{Q}[C])$. Because membership in $P(\mathbb{Q}[C])$ is determined by looking at the coefficient for the $\leq_C$-least element in this sum, we must have $1_C <_C ab$ in the group $C$. However, $\gamma(x)^{-1}(2,2) = 1_\mathbb{Q}(ab)^{-1}$. From $1_C <_C ab$, it follows that $(ab)^{-1} <_C 1_C$ and hence that $1_\mathbb{Q} 1_C < 1_\mathbb{Q}(ab)^{-1}$. Therefore, $\gamma(x)^{-1}(2,2) > 1$. Since $\gamma(x)^{-1}(1,1) = 1$, we have $\gamma(x)^{-1} \in P(\text{Tri}_{\mathbb{Q}[C]})$ as required.

CASE 47. $\gamma(x)(1,1) = \gamma(x)(2,2) = 1$.

Because $\gamma(x)$ has the shift property, there is a least $j > 1$ such that either $\gamma(x)(1,j) \neq 0$ or $\gamma(x)(2,j) \neq 0$ and $\gamma(x)(1,j) = 0$. Assume that $\gamma(x)(1,j) \neq 0$. The other case is similar. Since $\gamma(x) \notin P(\text{Tri}_{\mathbb{Q}[C]})$ it must be that $\gamma(x)(1,j) < 0$. Using the fact that $\gamma(x)(n,n) = 1$ for all $n$, the formula for $\gamma(x)^{-1}(1,j)$ gives:

$$\gamma(x)^{-1}(i,j) = -\gamma(x)(1,j) + \sum_{1 < k_1 < j} \gamma(x)(1,k_1)\gamma(x)(k_1,j) - \cdots$$
$$\cdots + (-1)^{j-1}(\gamma(x)(1,2) \cdots \gamma(x)(j-1,j)).$$

All the terms drop out except for the first one because $\gamma(x)(1,k) = 0$ for any $1 < k < j$. Thus, $\gamma(x)^{-1}(1,j) = -\gamma(x)(1,j) > 0$. The check that $\gamma(x)^{-1}(k,k+s) = 0$ for the appropriate $k, s$ is similar.

We have completed the proof of Theorem 16.

# 6 Proofs for Tri$_{\mathbb{Q}[C]}$

In this section, we prove some of the technical formulas about $\text{Tri}_K$ and $\text{Tri}_{\mathbb{Q}[C]}$ that we omitted in Sections 4 and 5. Specifically, we prove Lemmas 28, 29 and 32. We have restated these lemmas below. Recall that $K$ is an f.o. ring. If $f \in \text{Tri}_K$, then $f(n,n)$ is invertible for all $n \in \mathbb{N}^+$ and hence we can write it in the denominator of fractions.

LEMMA 48 (RCA$_0$). *If $f \in \text{Tri}_K$, then $f$ has an inverse $g \in \text{Tri}_K$, in the* sense that $f \cdot g = g \cdot f = I$, given by:

$$g(i,j) = \begin{cases} 0 & j < i \\ f(i,j)^{-1} & i = j \\ \begin{aligned}-\frac{f(i,j)}{f(i,i)f(j,j)} + \sum_{i<k_1<j} \frac{f(i,k_1)f(k_1,j)}{f(i,i)f(k_1,k_1),f(j,j)} - \\ -\sum_{i<k_1<k_2<j} \frac{f(i,k_1)f(k_1,k_2)f(k_2,j)}{f(i,i)f(k_1,k_1)f(k_2,k_2)f(j,j)} + \cdots \\ \cdots + (-1)^{j-i} \frac{f(i,i+1)\cdots f(j-1,j)}{f(i,i)f(i+1,i+1)\cdots f(j,j)}\end{aligned} & i < j \end{cases}$$

**Proof.** We verify that $(f \cdot g)(i,j) = I(i,j)$ for all $i$ and $j$. If $i > j$ then we have already noted that for $f, g \in \text{Tri}_K$, $(f \cdot g)(i,j) = 0$.

For each $j$, the case for $i \leq j$ proceeds by induction on $j - i$. For the base case when $j - i = 0$, we have $i = j$. Since $g(i,i) = f(i,i)^{-1}$, we have $(f \cdot g)(i,i) = f(i,i) \cdot f(i,i)^{-1} = 1$ as required. For the induction case, assume that $j - i = l > 0$ and that the formula is correct for $g(j-k,j)$ for all $0 \leq k < l$. We need to show that $(f \cdot g)(j-l,j) = 0$. That is, we need to show that

$$f(j-l,j-l)g(j-l,j) + f(j-l,j-l+1)g(j-l+1,j) + \cdots$$
$$\cdots + f(j-l,j)g(j,j) = 0.$$

Solving this equation for $g(j-l,j)$, we need to show that

$$g(j-l,j) = \sum_{n=0}^{l-1} -\frac{f(j-l,j-n)}{f(j-l,j-l)} g(j-n,j).$$

To finish the proof, we need to show that this sum is equal to

$$-\frac{f(j-l,j)}{f(j-l,j-l)f(j,j)} + \sum_{k_1=1}^{l-1} \frac{f(j-l,j-k_1)f(j-k_1,j)}{f(j-l,j-l)f(j-k_1,j-k_1)f(j,j)} -$$
$$\sum_{k_1=1}^{l-2} \sum_{k_2=k_1+1}^{l-1} \frac{f(j-l,j-k_2)f(j-k_2,j-k_1)f(j-k_1,j)}{f(j-l,j-l)f(j-k_1,j-k_1)f(j-k_2,j-k_2)f(j,j)} + \cdots$$
$$\cdots + (-1)^l \frac{f(j-l,j-l+1)f(j-l+1,j-l+2)\cdots f(j-1,j)}{f(j-l,j-l)f(j-l+1,j-l+1)\cdots f(j,j)}. \quad (1)$$

To prove this equality, we split Equation (1) into a sum with summands of the form

$$-\frac{f(j-l,j-n)}{f(j-l,j-l)} \cdot X$$

and show that $X = g(j - n, j)$. When $n = 0$, $-f(j - l, j)/f(j - l, j - l)$ appears only in the first summand $-f(j - l, j)/(f(j - l, j - l)f(j, j))$ of Equation (1). In this case $X = f(j, j)^{-1} = g(j, j)$ as required.

When $n = 1$, the term $-f(j - l, j - 1)/f(j - l, j - l)$ appears only in the second sum of Equation (1) and only when $k_1 = 1$. Thus in this case, we have $X = -f(j - 1, j)/(f(j - 1, j - 1)f(j, j))$ which by the induction hypothesis is $g(j - 1, j)$.

In general, for $1 \leq n < l$, the term $-f(j-l, j-n)/f(j-l, j-l)$ does not appear in the first summand of Equation (1), but does appear in each of the other summands up to the $(n+1)^{\text{st}}$ one. If we examine how it appears in each of these summands, we see that the second term contributes

$$- \frac{f(j-n, j)}{f(j-n, j-n)f(j, j)}$$

to $X$, which is the first term in $g(j - n, j)$. The third term contributes something to $X$ whenever $k_2 = n$ and hence we get a total contribution of

$$\sum_{k_1=1}^{n-1} \frac{f(j-n, j-k_1)f(j-k_1, j)}{f(j-n, j-n)f(j-k_1, j-k_1)f(j, j)}.$$

which equals the second term in $g(j - n, j)$. This process continues until we reach the $(n+1)^{\text{st}}$ term, which contributes to $X$ only when $k_1 = 1, k_2 = 2, \ldots k_n = n$. This give us

$$(-1)^n \frac{f(j-n, j-n+1) \cdots f(j-1, j)}{f(j-n, j-n) \cdots f(j, j)}$$

which is the last term of $g(j - n, j)$ and shows that $X = g(j - n, j)$ as required. We have now shown that $f \cdot g = I$. From here, we have that $g \cdot f \cdot g = g$. By a simpler induction, it can be shown that if $h \cdot g = g$ then $h = I$. Hence $g \cdot f = I$ as well. □

LEMMA 49 ($RCA_0$).

1. If $f, g \in P(Tri_K)$ then $f \cdot g \in P(Tri_K)$.

2. If $f \in P(Tri_K)$ and $f \neq I$ then $f^{-1} \notin P(Tri_K)$.

3. If $f \in P(Tri_K)$ and $g \in Tri_K$ then $gfg^{-1} \in P(Tri_K)$.

**Proof.** We prove the first statement of this lemma. (The proofs of the second and third statement are similar case analyses of the entries in the matrices $f^{-1}$ and $gfg^{-1}$.) To prove the first statement of the lemma, assume $f, g \in P(\text{Tri}_K)$. For notational purposes, let $h = f \cdot g$. We need to show $h \in P(\text{Tri}_K)$. Without loss of generality, assume that $f, g, h \neq I$. There are two cases to consider.

CASE 50. $\exists i (f(i,i) \neq 1 \vee g(i,i) \neq 1)$

Let $i$ be the least such number. Then, $h(i,i) = f(i,i)g(i,i) > 1$ and for all $j < i$, $h(j,j) = 1$. Thus $h \in P(\text{Tri}_K)$.

CASE 51. $\forall i (f(i,i) = 1 \wedge g(i,i) = 1)$

Let the pair $\langle i,j \rangle$ be a witness for $f > I$. Without loss of generality, assume that $g(i,j) \geq 0$ and that $g(k, k+s) = 0$ for all $k$ and $s > 0$ such that $i + s < j$ or $i + s = j$ and $k < i$. That is, assume that the witness to $g > I$ comes later in the order on the diagonals than the witness for $f$. We are going to show that $h(i,j) > 0$, that $h(k, k+s) = 0$ for $k, s$ as above and that $h(n,n) = 1$ for all $n$.

Since $f(n,n) = g(n,n) = 1$, it is clear that $h(n,n) = 1$. To show that $h(i,j) > 0$, we examine

$$h(i,j) = \sum_{n=i}^{j} f(i,n) g(n,j).$$

By the assumptions made above on $f$ and $g$, we have that $f(i,i) = g(j,j) = 1$ and $f(i, i+1)$ through $f(i, j-1)$ are all 0. Thus, this sum reduces to $g(i,j) + f(i,j)$. Since $g(i,j) \geq 0$ and $f(i,j) > 0$, we have that $h(i,j) > 0$ are required.

Suppose $s > 0$, $i + s < j$ or $i + s = j$ and $k < i$. We have the following equalities:

$$h(k, k+s) = \sum_{n=k}^{k+s} f(k,n) g(n, k+s)$$

$$= \sum_{n=0}^{s} f(k, k+n) g(k+n, k+s)$$

$$= f(k,k)g(k, k+s) + f(k, k+s)g(k+s, k+s) +$$

$$+ \sum_{n=1}^{s-1} f(k, k+n) g(k+n, k+s).$$

The first term in the last equation is 0 because $g(k, k+s) = 0$. The second term is 0 because $f(k, k+s) = 0$. For the third term, since $i + s \leq j$ we have that $i + n < j$ for all $n$ in the sum. Thus $f(k, k+n) = 0$ and the third term is 0. This shows that $h(k, k+s) = 0$ and finishes the proof of the first statement of the lemma. □

Finally, we give a proof of Lemma 32. We refer the reader back to Section 5 for the notation in this lemma, in particular for the formulas for the functions $f = \alpha(a) \cdot X$ and $g = X^{-1}$.

**LEMMA 52 (RCA$_0$).** *If $i < j$ and $i, j$ are both even, then*

$$\alpha'(a)(i,j) = (1-a) \sum_{\substack{n=i+1 \\ n \text{ odd}}}^{j-1} \left( -x_{in}x_{nj} + \sum_{i<k_1<n} (x_{ik_1}x_{k_1n}x_{nj}) - \right.$$

$$\left. - \sum_{i<k_1<k_2<n} (x_{ik_1}x_{k_1k_2}x_{k_2n}x_{nj}) + \cdots + (-1)^{n-i}x_{i(i+1)}\cdots x_{(n-1)n}x_{nj} \right)$$

**Proof.** The proof consists of grinding through the calculations one step at a time, and breaking the sum up into pieces.

$$\alpha'(a)(i,j) = \sum_{n=i}^{j} g(i,n)f(n,j)$$

$$= \underbrace{g(i,i)f(i,j)}_{(I)} + \underbrace{g(i,j)f(j,j)}_{(II)} + \underbrace{\sum_{n=i+1}^{j-1} g(i,n)f(n,j)}_{(III)}$$

Since $i$ is even, (I) is $ax_{ij}$. Since $j$ is even, $f(j,j) = a$, and so (II) equals

$$a\left( -x_{ij} + \sum_{i<k_1<j} (x_{ik_1}x_{k_1j}) - \cdots + (-1)^{j-i}(x_{i(i+1)}\cdots x_{(j-1)j}) \right)$$

(III) breaks into two cases: when $n$ is even and when $n$ is odd.

$$\underbrace{\sum_{\substack{n=i+1 \\ n \text{ odd}}}^{j-1} \big(g(i,n)\cdot x_{nj}\big)}_{(IV)} + \underbrace{\sum_{\substack{n=i+1 \\ n \text{ even}}}^{j-1} \big(g(i,n)\cdot ax_{nj}\big)}_{(V)} =$$

$$\sum_{\substack{n=i+1 \\ n \text{ odd}}}^{j-1}\left(-x_{in}x_{nj} + \sum_{i<k_1<n}(x_{ik_1}x_{k_1n}x_{nj}) - \cdots + (-1)^{n-i}(x_{i(i+1)}\cdots x_{nj})\right) +$$

$$a\cdot \sum_{\substack{n=i+1 \\ n \text{ even}}}^{j-1}\left(-x_{in}x_{nj} + \sum_{i<k_1<n}(x_{ik_1}x_{k_1n}x_{nj}) - \cdots + (-1)^{n-i}(x_{i(i+1)}\cdots x_{nj})\right)$$

There are a couple of important observations. First, (I) cancels with the first term in (II). Second, all of the terms in (V) appear in and cancel with terms in (II). Third, since $j$ is even, it follows that $j-1$ is odd and so the last term in (II) does not cancel. Performing the cancelations, we are left with

$$a\cdot\left(\sum_{\substack{i<k_1<j \\ k_1 \text{ odd}}}(x_{ik_1}x_{k_1j}) - \sum_{\substack{i<k_1<k_2<j \\ k_2 \text{ odd}}}(x_{ik_1}x_{k_1k_2}x_{k_2j}) + \right.$$

$$\left. + \cdots + (-1)^{j-i}(x_{i(i+1)}\cdots x_{(j-1)j})\right) +$$

$$+ \sum_{\substack{n=i+1 \\ n \text{ odd}}}^{j-1}\left(-x_{in}x_{nj} + \sum_{i<k_1<n}(x_{ik_1}x_{k_1n}x_{nj}) - \cdots + (-1)^{n-i}(x_{i(i+1)}\cdots x_{nj})\right).$$

This equation yields the formula in the statement of the lemma once the following general rewriting principles are applied.

$$\sum_{\substack{i<k_1<j \\ k_1 \text{ odd}}} x_{ik_1}x_{k_1j} \;\Rightarrow\; -\sum_{\substack{n=i+1 \\ n \text{ odd}}}^{j-1} -x_{in}x_{nj}$$

$$-\sum_{\substack{i<k_1<k_2<j \\ k_2 \text{ odd}}} x_{ik_1}x_{k_1k_2}x_{k_2j} \;\Rightarrow\; -\sum_{\substack{n=i+1 \\ n \text{ odd}}}^{j-1}\left(\sum_{i<k_1<n} x_{ik_1}x_{k_1n}x_{nj}\right)$$

These principles continue for longer linear sequences of subscripted $k$'s. For example, a similar rewriting rule can be applied to the sum over $i < k_1 < k_2 < k_3 < j$ with $k_3$ odd. □

# BIBLIOGRAPHY

[1] M.A. Dabkowska, M.K. Dabkowski, V.S. Harizanov and A.A. Togha, "Spaces of orders and their Turing degree spectra", to appear in *Annals of Pure and Applied Logic*.

[2] R. Downey and S.A. Kurtz, "Recursion theory and ordered groups", *Annals of Pure and Applied Logic*, vol. 32, 1986, 137-151.

[3] R. Downey, D. Hirschfeldt, S. Lempp and R. Solomon, "Reverse mathematics and the Nielsen-Schreier Theorem", in S.S. Goncharov, ed., *Proceedings of the International Conference in Logic Honoring Ershov on his 60th Birthday and Maltsev on his 90th Birthday*, Novosibirsk, 2002, 59-71.

[4] L. Fuchs, *Partially ordered algebraic systems*, Pergamon Press, Oxford, 1963.

[5] A. Kokorin and V. Kopytov, *Fully ordered groups*, Halsted Press, New York, 1974.

[6] W. Magnus, A. Karrass and D. Solitar, *Combinatorial group theory*, Dover, 1976.

[7] G. Revesz, "Full orders on free groups", in S. Wolfenstein, ed., *Algebra and Order: Proceedings of the First International Symposium on Ordered Algebraic Structures Luminy-Marseille 1984*, 1986, 105-111.

[8] S.G. Simpson, *Subsystems of second order arithmetic*, Springer-Verlag, 1998.

[9] R. Solomon, "Reverse mathematics and ordered groups", *Notre Dame Journal of Formal Logic*, vol. 39, 1998, 157-189.

[10] R. Solomon, "Ordered groups: a case study in reverse mathematics", *Bulletin of Symbolic Logic*, vol. 5 (1) March 1999, 45-58.

# Towards $NP-P$ via Proof Complexity and Search

SAMUEL R. BUSS

ABSTRACT. This is a survey of work on proof complexity and proof search, as motivated by the $P$ versus $NP$ problem. We discuss propositional proof complexity, Cook's program, proof automatizability, proof search, algorithms for satisfiability, and the state of the art of our (in)ability to separate $P$ and $NP$.

## 1 Introduction

One of the principal open problems in computer science, and indeed in all of mathematics, is the $P$ versus $NP$ problem. This is really only one problem out of a host of related problems, including questions such as whether polynomial time equals polynomial space, whether particular concrete boolean problems require superlinear or even exponential size circuits, whether pseudo-random number generators exist, etc. It is traditional to focus on the $P$ versus $NP$ problem as being the central open problem, partly because of the naturalness of the problem, and partly for historical reasons.

Many of the methods that have been used to attack the $P$ versus $NP$ problem have been combinatorial, algebraic, or probabilistic; this includes, for instance switching lemmas, or polynomial approximations, and a host of results about randomization. In this paper, we survey instead various "logico-algorithmic" attempts to solve the $P$ versus $NP$ problem. This is motivated by the analogy with the Gödel incompleteness theorems, and by the hope that perhaps some kind of extension of self-reference arguments can separate $NP$ from $P$.

As far I know, the first person to express in print the idea that something akin to $P \neq NP$ might follow from a self-reference argument was G. Kreisel [69]. Kreisel wrote:

> "Suppose we ... have a proof system; ... the 'faith' is that in a natural way this will yield a feasible proof procedure for feasible theorems.

"Conjecture: Under reasonable conditions on feasibility, there is an analogue to Gödel's second incompleteness theorem, that is the article of faith above is unjustified."

Surprisingly enough, this comment by Kreisel, and the earlier comment below by Gödel, were made prior to the first formal definitions of the class $NP$ by Cook [33] and Levin [72]. Kreisel did not expand any further on his remarks above, and Gödel's letter was not circulated. Thus, neither of them influenced the development of computational complexity. For more comprehensive treatments of the prehistory of the $P$ versus $NP$ problem, see the articles by Cook [35] and Sipser [106].

What reasons do we have for believing $P \neq NP$? One is that $P = NP$ could make the practice of mathematics too easy. Mathematical research could be automated by formalizing mathematical questions completely, and then blindly searching for proofs of conjectured mathematical statements. If $P = NP$, this process could succeed whenever proofs are not too large.[1] This would be obviously be a major change in the practice of mathematics!

To the best of my knowledge, the observation about the profound effect that $P = NP$ would have on mathematics was first made by Gödel, who wrote in a 1956 letter to von Neumann that if proof search could be carried out feasibly, it would have "consequences of the greatest importance. ... The mental work of a mathematician concerning Yes-No questions could be completely replaced by a machine." (The translation is by Clote, see [32].)

In spite of the fact that most researchers believe $P \neq NP$, the evidence that $P \neq NP$ is tenuous. Indeed, it seems be based on three factors: firstly, conservativism about the possibility of fundamentally new algorithms; secondly, the analogy with the incompleteness theorems; and thirdly, the fact that no one has been able to prove that $P = NP$.

Accordingly, the title of the paper is a bit of a pun, meant to illustrate our doubts about our belief that $P \neq NP$: The titles uses "$NP$–$P$" this can also be interpreted as subtracting $P$ from $NP$. If they are equal, then we are heading towards the empty set!

As a caveat to the reader, the field of propositional proof complexity is quite large, and we have not attempted to be comprehensive in the citations to papers, as it would cause a substantial expansion in the length of this paper and its bibliography. At times we cite original results; at other times we cite only later, more definitive, results of a given area. We have tried to cite the papers that give the best entry to an area, but the reader who wishes to learn more about particular areas should be sure to look further into

---

[1] As a side remark: One needs more than just $P = NP$, rather one needs that natural $NP$ problems such as satisfiability have feasibly effective algorithms.

the literature. Other expository articles on propositional proof complexity include [11, 14, 22, 23, 24, 89, 103].

## 2 Propositional proof complexity

Gödel, in the above-quoted letter, was interested in the computational complexity of searching for formal proofs. Since that time, proof complexity and proof search have become increasingly important.

Suppose we are given a particular proof system $P$. The kinds of questions we might like to answer include the following.

- Given a formula $\varphi$, decide if it has a short $P$-proof.
- Given a formula $\varphi$, and a promise that it has a short $P$-proof, find a $P$-proof of $\varphi$.
- Characterize the formulas that have reasonable length (polynomial length) $P$-proofs.
- Compare the strength of $P$ with other proof systems.

Surprisingly, these questions are difficult even for propositional proof systems. Indeed, as is discussed below, under some special conditions, the existence of short propositional proofs is an $NP$-complete problem, and hence a feasible method to find optimal propositional proofs will also give a feasible algorithm to find (short) proofs in *any* proof system.

There are a variety of propositional proof systems that are commonly studied. Three of the most basic systems are

- Resolution — a proof system for proving disjunctive normal form (DNF) formulas.
- Frege proofs — "textbook"-style system using modus ponens as its only rule of inference.
- Extended Frege systems — Frege systems augmented with the ability to introduce new variables that abbreviate formulas.

For these, and other proof systems, it is important to have a good notion of the length or size of a proof. The most useful notion is generally as follows.

DEFINITION 1. The *length* of a proof is the number of symbols used in the proof.

The reader is likely to be familiar with resolution and with Frege systems. Extended Frege systems are less well-known: an alternate characterization of extended Frege systems is to define an extended Frege proof to be the same as a Frege proof, but with its length equal to the number of steps (or, formulas) in the proof rather than the number of symbols in the proof.

There are many other possible proof systems. An important generalization of the notion of a proof system is the *abstract proof systems* that were defined by Cook and Reckhow [36, 37].

DEFINITION 2. An *abstract proof system* is a polynomial time function $f$ whose range is equal to the set of tautologies. If $\varphi$ is a tautology, then an $f$-proof of $\varphi$ is any value $w$ such that $f(w) = \varphi$.

Any standard proof system $P$ can be viewed as an abstract proof system by defining $f(w)$ so that if $w$ is an encoding of a valid $P$-proof, then $f(w)$ is the last line of the proof. For $w$'s that do not code valid $P$-proofs, $f(w)$ can equal an arbitrary, fixed tautology. Assuming that $P$-proofs can be efficiently encoded, and can be recognized in polynomial time, the function $f$ is an abstract proof system that faithfully captures the notion of $P$-proofs.

Abstract proof systems can be quite strong; indeed, they capture any proof system with polynomial recognizable proofs, even somewhat nonstandard proof systems. For example, set theory can be viewed as a propositional proof system by letting any proof in $ZFC$ of a formula expressing "$\varphi$ is a tautology" serve as a proof of $\varphi$.

DEFINITION 3. A proof system is *super* if every tautology $\varphi$ has a proof which has length polynomially bounded by the length of $\varphi$.

THEOREM 4. (Cook-Reckhow [36]) *There exists a super proof system if and only if NP is closed under complementation; that is, if and only if $NP = coNP$.*

It is, of course, open whether any super proof system exists. Similarly, it is open whether there is any optimal proof system, that is to say, whether there is a "strongest" proof system that simulates any other given proof system with polynomial size proofs. Of course, if there is no optimal proof system, then $NP \neq coNP$.[2]

The above theorem has led to what is sometimes called "Cook's program" for proving $NP \neq coNP$.

**Cook's Program for separating $NP$ and $coNP$:** Prove superpolynomial lower bounds for proof lengths in stronger and stronger propositional proof systems, until they are established for all abstract proof systems.

Cook's program is an attractive and plausible approach; unfortunately, it has turned out to be quite hard to establish superpolynomial lower bounds

---

[2]For more on optimal proof systems, see [67, 99, 60, 19, 30, 49].

on common proof systems. Indeed, at the time that Cook and Reckhow introduced abstract proof systems, lower bounds were not known even for resolution. Subsequently, there has been a good deal of results, but the current state of the art is far away from being to establish superpolynomial lower bounds for common systems such as Frege systems.

Nonetheless, a number of propositional proof systems are known to require nearly exponential-size proofs. These include the following systems.

- Method of truth-tables.

- Tree-like resolution and regular resolution; Tseitin [108].

- Resolution; Haken [46], Urquhart [109], Chvátal-Szemerédi [31], Raz [90], Razborov [96, 97], and many others.

- Bounded depth Frege systems; Ajtai [1], Krajíček [61], Pitassi-Beame-Impaliazzo [84], Krajíček-Pudlák-Woods [68].

- Cutting planes systems; Pudlák [88].

- Nullstellensatz systems; Buss-Impagliazzo-Krajíček-Pudlák-Razborov-Sgall [27], Grigoriev [45].

- The polynomial calculus; Razborov [94], Impagliazzo-Pudlák-Sgall [56], Ben-Sasson-Impagliazzo [16], Buss-Grigoriev-Impagliazzo-Pitassi [25], Alekhnovich-Razborov [7].

- Bounded depth Frege systems with counting mod $m$ axioms (for fixed $m$); Ajtai [2], Beame-Impagliazzo-Krajíček-Pitassi-Pudlák [12], and Buss-Impagliazzo-Krajíček-Pudlák-Razborov-Sgall [27].

- Intuitionistic and modal logics; Hrubes [51, 50, 52].

- Ordered binary decision diagram (OBDD) proof systems; Atserias-Kolaitis-Vardi [9], Krajíček [65], Segerlind [104].

- Static and tree-like Lovász-Schrijver proofs; Dash [38], Itsykson-Kojevnikov [57], Beame-Pitassi-Segerlind [15], Lee-Shraibman [71], Chattopadhyay-Ada [29], Pitassi-Segerlind [85].

A reader unfamiliar with the above proof systems can consult the surveys [14, 103] for an overview. In short, *bounded depth Frege systems* are Frege systems in which the alternations of ∧'s and ∨'s is bounded; *cutting planes systems* are proof systems that reason about linear inequalities over the integers with 0/1-valued variables; *nullstellensatz systems* and the *polynomial calculus* are systems that reason with polynomials over finite fields;

and *counting axioms* express the impossibility of getting two different size remainders when partitioning a set into sets of size $m$.

Some of lower bound results listed above are inspired by analogies with two notable results from complexity theory. First, the lower bounds for bounded depth Frege system are lower bounds for proofs of the pigeonhole principle. These lower bounds are inspired by the Ajtai-Yao-Hastad switching lemma, but a more sophisticated switching lemma had to be developed, by [68, 84], to obtain the lower bounds for bounded depth Frege systems. Similarly, the lower bounds for cutting planes and Lovász-Schrijver systems, for intuitionistic and modal logics, and for OBDD systems are based on the clique-coloring tautologies; for these lower bounds, a Craig interpolation theorem was proved, and then a reduction was made to Razborov's theorem [91, 8] that monotone circuits for clique versus colorability require near-exponential size. Another popular set of tautologies for lower bounds is the Tseitin graph tautologies, which has been used for lower bounds on resolution and for a number of algebraic proof systems [108, 109, 45, 25, 57].

It is interesting to note that the lower bounds for the systems listed above are *all* based on counting. That is to say, the lower bounds are either for the propositional pigeonhole principle, or for tautologies based on clique-coloring principles, or for tautologies based on mod $p$ counting principals for some prime $p$. The clique-coloring principle expresses the fact that if there is a graph has a clique of size $n$, then an coloring requires at least $n$ colors; this is a variant of the pigeonhole principle of course.[3] Likewise, Tseitin principle is also based on the counting, in particular, on counting mod 2.

The main open problems at the "frontier" of Cook's program are perhaps the following:

**Open problems.**

1. Give exponential (or at least better than quasipolynomial) separations between depth $k$ Frege systems and depth $k+1$ Frege systems for low depth formulas, even for DNF formulas.

---

[3]The clique-coloring principle is a form of the pigeonhole principle since it follows from the fact that the vertices in a clique must all have distinct colors. However, a level of indirection is required to prove the clique-coloring principle from the pigeonhole principle. The cutting planes system can formalize a counting to a certain extent and do have simple polynomial-size proofs of the pigeonhole principle; however, the cutting planes system requires near-exponential size proofs for clique-coloring tautologies. The best lower bounds for OBDD proofs of the clique-coloring tautologies require yet another level of indirection: OBDD proofs have polynomial size proofs of the pigeonhole principle, but require exponential size for the clique-coloring tautologies when obfuscated by a permutation [65, 104].

**2.** Prove superpolynomial lower bounds for bounded depth Frege systems augmented with mod $p$ gates, for $p$ a prime.

The first problem is closely connected with the problem of proving non-conservativity results for fragments of bounded arithmetic. The second problem has been conjectured to be do-able, in light of the analogy with the exponential lower bounds for constant depth circuits that use $\mathrm{mod}\, p$ gates [92, 107]. Indeed, one might expect to get significant separation between the power of $\mathrm{mod}\, p$ gates and $\mathrm{mod}\, q$ gates for distinct primes $p$ and $q$. Overall, however, both problems have proved very resistant to progress. One partial step towards solving the first problem is due to Krajíček [61]: using a modified notion of depth called "$\Sigma$-depth", he showed that there are sets of clauses of $\Sigma$-depth $k$ which can be refuted with short $\Sigma$-depth $k+1$ sequent calculus proofs but require large refutations in the $\Sigma$-depth $k$ sequent calculus. In addition, Impagliazzo and Krajíček [53] using counting techniques of Paris and Wilkie [79] describe proofs of restricted pigeonhole principles in the fragments $T_1^k$ of bounded arithmetic: together with the lower bounds from [68, 84] this implies there is no polynomial simulation of depth $k+1$ Frege systems by depth $k$ Frege systems. This leaves open the question of whether there is a quasipolynomial separation between depth $d$ and $d+1$ Frege systems. The best result towards solving the second problem is due to Maciel and Pitassi [74] who prove a separation of tree-like Frege proofs in which cuts are restricted to bounded depth formulas based on a circuit complexity conjecture; however, their lower bounds apply only to proofs of unbounded depth formulas. For additional open problems in proof complexity, see Pudlák [89].

As mentioned above, the strongest present-day lower bounds for propositional proof length use mostly counting principles such as the pigeonhole principle and the the clique-coloring tautologies. On the other hand, Frege systems can formalize counting and prove the pigeonhole principle [20]. A similar (ostensibly weaker) system, is the $TC^0$-Frege proof system, which allows reasoning with bounded depth formulas that can contain special counting gates or threshold gates. Since both these systems can prove facts about counting, and have polynomial size proofs of the pigeonhole principles and the clique-coloring principles, it may well require a completely new approach to give superpolynomial lower bounds for $TC^0$-Frege or Frege proofs.

## 3 The hardness of proof search

Cook's program concerns primarily questions about proof length; it does not directly concern the very important questions of the complexity of proof search. In this section, we list some of the more important results about the difficulty of searching for proofs.

The first result is about the difficulty of finding a proof that is approximately as short as possible:

THEOREM 5. (Alekhnovich-Buss-Moran-Pitassi [4] and Dinur-Safra [41]) *For almost all common proof systems (resolution, Frege, nullstellensatz, sequent calculus, cut-free sequent calculus, etc.), it is impossible to approximate shortest proof length to within a factor of $2^{\log^{1-o(1)} n}$ in polynomial time, unless $P = NP$.*

Note that the theorem states more than it is difficult to search for a short proof; instead, it is already hard to determine whether such a proof exists (assuming $P \neq NP$). The proof of this theorem by [4] uses a reduction from the Minimum Monotone Circuit Satisfying Assignment problem [41].

It would be desirable to improve Theorem 5 by replacing the factor $2^{\log^{1-o(1)} n}$ with a factor $n^k$, for all fixed values $k > 0$.

The second result is about the difficulty of searching for Frege proofs.

DEFINITION 6. A proof system $P$ is *automatizable* if there is a procedure, which given a formula $\varphi$ with shortest $P$-proof of length $n$, finds some $P$-proof of $\varphi$ in time $poly(|\varphi| + n)$.

THEOREM 7. (Bonet-Pitassi-Raz [18]) *If Frege proofs (or, $TC^0$-Frege proofs) are automatizable, then there is a polynomial time algorithm for factoring Blum integers.*

Blum integers are products of two primes congruent to 3 mod 4. It is generally conjectured that factorization of Blum integers is not in $P$. By the theorem, this would imply further that $(TC^0\text{-})$Frege systems are not automatizable.

The proof of Theorem 7 uses automatizability to establish a feasible Craig interpolation property. The feasible Craig interpolation property can be used to give a polynomial time algorithm to break the Diffie-Hellman cryptographic protocol, which by results of Biham, Boneh, and Reingold would yield a polynomial time algorithm for factoring Blum integers. Similar methods were subsequently used by [17] to prove that bounded depth Frege systems are also not automatizable, under stronger assumptions on the hardness of factoring Blum integers.

Although Theorem 7 was originally stated for proof search (automatizability), it also can be recast as a theorem about the hardness of determining the *existence* of proofs.

DEFINITION 8. The *provability* problem for a proof system $P$ is the problem of, given a formula $\varphi$, deciding whether $P \vdash \varphi$. The *$k$-provability* problem for $P$ is the problem of, given a pair $\langle \varphi, 1^k \rangle$, determining whether $\varphi$ has a $P$-proof of length $\leq k$.

Note that $1^k$ just means that $k$ is specified in unary notation. The intent is that a polynomial time algorithm for the $k$-provability problem can run for $k^{O(1)}$ time. A minor modification of the methods of [18] yields the following theorem.

THEOREM 9. *If the $k$-provability problem for Frege systems (or, $TC^0$-Frege systems) is in P, then there is a polynomial time algorithm for factoring Blum integers.*

The third result we mention is about the hardness of proof search for resolution.

THEOREM 10. (Alekhnovich-Razborov [6]) *Neither resolution nor tree-like resolution is automatizable unless the weak parameterized hierarchy $W[P]$ is fixed-parameter tractable under randomized algorithms with one-sided error.*

For more on the the weak parameterized hierarchy, see [42]. The proof of Theorem 10 uses a reduction from the Minimum Monotone Circuit Satisfying Assignment problem. Like the earlier results, Theorem 10 does not really depend on automatizability per se; just on whether there is a polynomial time algorithm that solves the $k$-provability problem. Theorem 10 was extended to the polynomial calculus by [43]. It is still open whether (tree-like) resolution, bounded depth Frege, and the polynomial calculus are *quasi-automatizable*, where "quasi-automatizable" is defined like "automatizable", but with a quasipolynomial runtime bound.

For common propositional proof systems, such as resolution and (extended) Frege systems, the $k$-provability problem in is *NP*, since one can merely non-deterministically guess the proof. However, it is unlikely that their *provability* problems are in *NP*, since, by the completeness of these logics, the provable formulas are the tautologies. The set of tautologies is *coNP*-complete, and hence is generally conjectured to not be in *NP*.

There are some weaker systems for which the provable formulas are known to form an *NP*-complete set. These are described in the next theorem.

THEOREM 11. *The provability problems for the following systems are NP-complete.*

1. *The multiplicative fragment of linear logic.* (Lincoln-Mitchell-Scedrov-Shankar [73] and Kanovich [59].)

2. *The Lambek calculus, L.* (Pentus [83] and Savateev [100].)

3. *Reflected justification logic,* rLP. (Krupski [70] and Buss-Kuznets [28].)

## 4 Practical propositional proof search

The previous section described complexity results about the difficulty of finding propositional proofs. We now turn our attention to the opposite problem; namely, the question of how efficient can practical algorithms be for deciding validity. There is an enormous range of algorithms and software implementations aimed at deciding the validity of propositional formulas, far more than space will permit us to discuss. Thus, we discuss only a few core ideas from the field.

For the most part, the existing algorithms attack the *satisfiability* problem, instead of the validity problem. These two problems are essentially equivalent, since a formula $\varphi$ is valid if and only if $\neg\varphi$ is not satisfiable. However, in the case of a satisfiable formula, one is often interested in finding an example of a satisfying assignment rather than just obtaining a Yes/No answer about satisfiability. Thus, we define the satisfiability problem, SAT, in a slightly non-standard way:

DEFINITION 12. An instance of SAT is a set of clauses interpreted as a formula in conjunctive normal form. The *Satisfiability problem* (SAT) is the problem of, given a instance $\Gamma$ of SAT, finding either a satisfying assignment, or finding a refutation of $\Gamma$. For $k \geq 2$, the $k$-SAT problem is the SAT problem restricted to clauses of size $\leq k$.

The notion of what constitutes a refutation of $\Gamma$ is left unspecified; in effect it can be in any proof system. In particular, for an algorithm $M$ that, for satisfiable $\Gamma$, always finds a satisfying assignment, one can just take a failing run of $M$ as a refutation of $\Gamma$.

**Complexity upper bounds for** SAT. Let the set $\Gamma$ of clauses have $m$ clauses that involve $n$ distinct variables. The obvious brute force algorithm that exhaustively searches all truth assignments has runtime $O(m \cdot 2^n)$. One can do slightly better than this, and some of the best known asymptotic upper bounds are given in the next theorem. The first result of this type was due to Monien and Speckenmeyer [78]. The first part of the next theorem is a recent result due to Hertli [47], extending prior work of Paturi, Pudlák, and Zane [82]; Paturi, Pudlák, Saks and Zane [81]; and Schöning [101]; Iwama and Tamaki [58], Hertli, Moser and Scheder [48], and many others. The second part is due to Schuler [102]. The theorem is stated for probabilistic algorithms; there are deterministic algorithms known with similar, but slightly worse, bounds.

THEOREM 13.

1. [47] Let $k \geq 3$ and $\mu_k = \sum_{j=1}^{\infty} \left(j^2 + \frac{j}{(k-1)}\right)^{-1}$. There is a probabilistic algorithm for $k$-SAT with expected runtime $2^{cn+o(n)}$ with $c = 1 - \frac{\mu_k}{k-1}$.

For $k = 3$ and $k = 4$, the bound gives $c \approx 0.386$ and $c \approx 0.554$, respectively. As $k \to \infty$, $\mu_k \to \frac{\pi^2}{6}$.

2. [102] *There is a probabilistic algorithm for* SAT *with expected runtime* $poly(m,n)2^{(1-1/(1+\log m))n}$.

It is interesting to note that the first bound, for $k$-SAT depends only on the number of variables, not on the number of cluases.

It is natural to conjecture that the bounds in Theorem 13 are qualitatively optimal; that is to say, any algorithm for $k$-SAT or SAT must have run time at least $\Omega(2^{\epsilon n})$ for some constant $\epsilon$. (See [54, 55, 80].) For SAT, one might even conjecture that $\epsilon < 1$ is not achievable. Of course, the primary evidence for this conjecture is our inability to have discovered any better algorithms yet. Nonetheless, the conjecture is appealing: it says in effect that SAT is an extremely efficient way to encode nondeterminism, since it implies that brute-force search for satisfying assignments for SAT cannot be improved upon in any significant way.

**More practical algorithms for** SAT. There are large number of practical algorithms for solving SAT (and even annual contests comparing algorithms). These algorithms are aimed primarily at solving the kinds of satisfiability problems that arise in "real-world" applications, for instance from hardware or software verification. Most of the more successful algorithms are based on the DPLL algorithms; thusly named as an amalgamation of the initials of the authors from the two seminal papers that originated the techniques, Davis and Putnam [40] and Davis, Loveland, and Logemann [39]. The basic framework of the DPLL algorithm is rather simple, essentially just a brute-force search through all possible truth assignments, but backtracking as soon as any clause is falsified.

DPLL Algorithm:
$\quad\Gamma$ is a set of clauses, and $\sigma$ is a partial truth assignment.
DPLL_Search($\Gamma, \sigma$):
$\quad$ 1. If $\Gamma \upharpoonright \sigma$ is falsified, return false.
$\quad$ 2. If $\Gamma \upharpoonright \sigma$ is satisfied, exit. ($\sigma$ is a satisfying assignment.)
$\quad$ 3. Choose a literal $x$ that is not assigned by $\sigma$.
$\quad$ 4. Call DPLL_Search($\Gamma, \sigma \cup \{x \mapsto True\}$).
$\quad$ 5. Call DPLL_Search($\Gamma, \sigma \cup \{x \mapsto False\}$).
$\quad$ 6. Return false.

The notation $\Gamma \upharpoonright \sigma$ denotes the set $\Gamma$ as modified by the restriction $\sigma$. That is to say, every literal set false by $\sigma$ is erased and every clause that contains a literal set true is removed. If $\Gamma \upharpoonright \sigma$ is empty, then all clauses are satisfied, so $\Gamma$ is satisfied. If $\Gamma \upharpoonright \sigma$, contains an empty clause, then it is falsified.

There are many possible strategies for the literal selection step (Step 3); two of the basic ones are *unit propagation* and *pure literal selection*. For unit propagation, one chooses any literal $x$ that occurs in a singleton clause in $\Gamma \upharpoonright \sigma$. For pure literal selection, one chooses any literal $x$ such that $\overline{x}$ does not appear in $\Gamma \upharpoonright \sigma$. If $x$ is chosed by unit propagation or as a pure literal, then the subsequent Step 5 can be skipped. Both these strategies can be used with no essential loss in efficiency.

A major improvement to the DPLL algorithm is the method of *clause learning* introduced by Marques-Silva and Sakallah [77]. Clause learning is method of learning (or, inferring) new clauses when a contradiction is found in Step 1 of the DPLL algorithm. When $\Gamma \upharpoonright \sigma$ is falsified, a "reason" for the falsification is extracted from the information gathered by the run of the DPLL algorithm. This "reason" is a clause $C$ such that $\Gamma \vDash C$. The clause $C$ is then *learned*; that is to say, it is added to the set $\Gamma$. (For a more complete introduction to clause learning, consult [77] or [13].)

The intuition behind clause learning is that a learned clause $C$ can now be reused without needing to be re-derived. This avoids repeating the same subsearch. The presence of new learned clauses also allows more opportunities for unit propagation, which also speeds up searching.

Another important feature of clause learning algorithms is that they keep track of the "level" at which variables are set; namely, when a variable assignment $x \mapsto True/False$ is added to $\sigma$, the variable $x$ is labeled with the level (the depth of recursion) at which its value was set. When a contradiction is discovered in Step 1, the DPLL algorithm can use the information about levels to backtrack multiple levels at once. This process is called "fast backtracking" or "non-chronological backtracking", and it makes DPLL much less sensitive to the order in which variables are selected since it is often possible to backtrack past all the variable assignments that did not contribute to the contradiction.

Figure 1 illustrates the power of clause learning, even for a rather difficult family of tautologies. The pigeonhole principles $\text{PHP}_n^m$ are sets of clauses, whose unsatisfiability express the pigeonhole principle that there is no one-to-one mapping from a set of size $m$ to a set of size $n$ (for $m > n$). For the results reported in Figure 1, $\text{PHP}_n^m$ contains the following clauses:

(1) $\{p_{i,1}, p_{i,2}, \ldots, p_{i,n}\}$, for $i = 1, \ldots, m$, and

(2) $\{\overline{p}_{i,j}, \overline{p}_{i',j}\}$, for $1 \leq i < i' \leq m$ and $j = 1, \ldots, n$.

Figure 1 shows the results of a DPLL satisfiability solver software package (written by the author) both with and without clause learning. The table

| Formula | No Learning | | Clause Learning | |
| --- | --- | --- | --- | --- |
| | Steps | Time (s) | Steps | Time |
| $PHP_3^4$ | 5 | 0.0 | 5 | 0.0 |
| $PHP_6^7$ | 719 | 0.0 | 129 | 0.0 |
| $PHP_8^9$ | 40319 | 0.3 | 769 | 0.0 |
| $PHP_9^{10}$ | 362879 | 2.5 | 1793 | 0.5 |
| $PHP_{10}^{11}$ | 3628799 | 32.6 | 4097 | 2.7 |
| $PHP_{11}^{12}$ | 39916799 | 303.8 | 9217 | 14.9 |
| $PHP_{12}^{13}$ | 479001599 | 4038.1 | 20481 | 99.3 |

Figure 1. This table shows the number of decision variables that are set ("steps") and the approximate runtime (in seconds) for determining satisfiability of pigeonhole principle statements, with and without clause learning. The number of steps with no learning is always $n! - 1$.

shows the number of decision variables set during the run of the DPLL algorithm: this counts only the number of variables whose value is not already forced to a value. That is to say, variables that are set by unit propagation or by the pure literal rule are not counted. As the table shows, learning can reduce the size of the search space enormously. The run times are substantially faster for clause learning as well, even though there is extra overhead associated with learning clauses and maintaining learned clauses. The software implementation used for the table is not as efficient at working with clauses as it might be; important optimizations like two-literal watching and garbage collection have not been implemented. Thus, a more complete modern DPLL implementation would likely see much quicker runtimes for clause learning.

Although the table shows an impressive improvement from the use of clause learning for the pigeonhole principles, this is not really where clause learning is the most advantageous. Instead, clause learning has a much greater benefit for many "practical" or "real-world" applications, for instance for solving instances of satisfiability that come from applications such as software or hardware verification. In fact, DPLL algorithms with clause learning can routinely solve instances of satisfiability with hundreds of thousands of variables, at least for problems that arise in industrial applications.

Algorithms for satisfiability have been compared extensively in annual "SAT competitions" or "SAT Race competitions"; it is quite frequently the case that solvers based on DPLL algorithms with clause learning are the best. The best algorithms use a wide variety of techniques to improve the

DPLL search and clause learning. These include:

- Literal selection heuristics, see [76, 75, 44].

- Clause learning strategies for deciding what clause(s) to learn, such as the first-UIP strategy [77].

- Clause forgetting strategies (garbage collection).

- Restart strategies, in which the DPLL search is interrupted and started again with the learned clauses.

- Execution optimizations. These include the two literal watching method [75], and many other optimizations of data structures and algorithms.

**A logical characterization of clause learning.** From its inception, the DPLL algorithm has been closely linked with propositional resolution. Indeed, the basic DPLL algorithm (without clause learning) corresponds exactly to the traversal of a tree-like resolution refutation. It is a more difficult problem to give a logical characterization of DPLL algorithms with clause learning, but some substantial progress has been made by [13, 110, 26]. Beame, Kautz and Sabharwal [13] characterized the process of learning clauses in terms of input resolution; Van Gelder [110] models the search process for DPLL with clause learning in a system called "pool resolution", which corresponds to resolution proofs that admit a regular depth-first traversal; and Buss, Hoffman, and Johannsen [26] combine these with a "w-resolution" inference that allows a precise characterization of certain types of DPLL algorithms with clause learning.

DEFINITION 14. A *w-resolution inference* with respect to the variable $x$ is an inference of the form

$$\frac{C \qquad D}{(C \setminus \{x\}) \cup (D \setminus \{\overline{x}\})}$$

In place of dag-like proofs, one can use tree-like proofs with *lemmas*, where a lemma is a formula that was derived earlier in the proof. That is to say, in a tree-like refutation of a set $\Gamma$ using lemmas, the leaves of the tree are either formulas from $\Gamma$ or are lemmas. It is clear that tree-like proofs with lemmas are equivalent to dag-like proofs.

DEFINITION 15. Let $T$ be a tree-like refutation with lemmas.

- An *input proof* is a proof in which every inference has at least one hypothesis which is a leaf.

- An *input clause* in $T$ is any clause which is derived by an input sub-proof of $T$.
- The proof $T$ is a WRTI-proof provided it is a tree-like proof, using w-resolution inferences, in which (only) input clauses are used as lemmas. (WRTI is an acronym for "W-Resolution Tree-like with Input lemmas".)

As the next theorem shows, using input lemmas is as strong as using arbitrary lemmas, and thus WRTI is as strong as full resolution.

THEOREM 16. (Buss-Hoffmann-Johannsen [26].) *General dag-like resolution proofs can be simulated by WRTI proofs.*

DEFINITION 17. A resolution refutation is *regular* provided that no variable is used twice as a resolution variable on any single path through the proof. In particular, a WRTI proof is regular provided that no variable is used twice as the active variable of a w-resolution on any branch in the refutation tree.

The next theorem allows an almost exact characterization of DPLL search with clause learning. We restrict to input lemmas since the clause learning algorithms can learn only clauses that can be derived by input proofs [13]. The restriction to regular refutations corresponds to the fact that a DPLL search procedure never branches twice on the same variable; that is to say, once a variable's value is set, it cannot be changed (until after backtracking resets the variable). We say the correspondence is "almost" exact, since we have to allow the DPLL search algorithm to be "non-greedy", by which is meant that the DPLL search procedure may ignore contradictions and continue to branch on variables even after a contradiction has been achieved.

THEOREM 18. (Buss-Hoffmann-Johannsen [26].) *Regular WRTI proofs are polynomially equivalent to non-greedy DPLL search with clause learning.*

The reader should refer to [26] for the technical details of what kinds of non-greedy DPLL search with clause learning are permitted. However all the standard clause learning algorithms from [77, 13] are allowed, including first-UIP, all-UIP, first cut clauses, rel sat clauses, decision clauses, etc.

There are a number of important theoretical questions about DPLL search algorithms. First, it would be nice to improve Theorem 18 to give a logical characterization of *greedy* DPLL clause learning. This ideally would also accommodate the more restrictive forms of clause learning and backtracking of common DPLL implementations. Second, it is open whether regular WRTI is as strong as general resolution. Similarly, it is open whether pool resolution directly simulates general resolution. (However, [10] gives an indirect simulation of general resolution by pool resolution.) It is known that regular resolution does not simulate either general resolution or pool resolution [5, 110]. Possibly this can be extended to prove that pool resolution does not simulate general resolution. Third, it would be highly desirable to find substantially more efficient SAT algorithms; either more efficient DPLL algorithms or more efficient algorithms of the types discussed in Theorem 13.

## 5 Trial Approaches to $\mathit{NP}$–$\mathit{P}$ via Proof Complexity

This concluding section discusses some of the approaches to separating $P$ and $\mathit{NP}$ using proof complexity. Of course, these approaches have all failed so far, and it appears that some fundamentally new idea will be needed to prove $P \neq \mathit{NP}$.[4] Nonetheless it is interesting to review some of the extant ideas.

**Via diagonalization and incompleteness.** Returning to the analogy between the $\mathit{NP}$–$\mathit{P}$ problem and Gödel incompleteness theorem about the impossibility of proving self-consistency statements, it is natural to consider the following *partial* consistency statements.

DEFINITION 19. Fix a proof system $P$. The statement $\mathrm{Con}(P, n)$ is a tautology expressing the principle that there is no $P$-proof of a contradiction of length $\leq n$.

The size of the formula $\mathrm{Con}(P, n)$ is bounded by a polynomial of $n$ (at least, for standard proof systems $P$). Although one might hope that partial self-consistency statements might require super-polynomial proofs, this does not in fact happen:

THEOREM 20. (Cook [34], Buss [21].) *For $P$ either a Frege system or an extended Frege system, there are polynomial size $P$-proofs of* $\mathrm{Con}(P, n)$.

The idea of the proof is to express a partial truth definition for formulas of length $\leq n$, using a formula whose length is polynomial in $n$. This proof method is analogous to the constructions of Pudlák [86, 87] for partial self-consistency proofs in first-order theories of arithmetic and in set theory. (Conversely, there are lower bounds on the lengths of first order proofs of partial consistency statments which match the known upper bounds to within a polynomial, due to H. Friedman and Pudlák.)

**Hard tautologies from pseudo-random number generators.**
Krajíček [63] and independently Alekhnovich, Ben-Sasson, Razborov, and Widgerson [3] made the following conjecture about the hardness of tautologies based on pseudo-random number generators. For the purpose of the

---

[4]However, conversely, to prove $P = \mathit{NP}$ might "merely" require a new algorithm.

present discussion, it will suffice to treat a pseudo-random number generator as being any function $f$ which maps $\{0,1\}^n$ to $\{0,1\}^m$, where $m$ is a function of $n$ and $m > n$. More generally, a pseudo-random number generator is intended to be a function of this type for which is it difficult to distinguish strings in $Range(f)$ from strings not in $Range(f)$, even with multiple trials and with the aid of randomization.

Given a pseudo-random number generator $f$, define tautologies $\tau(f)_b$ by fixing a particular string $b \in \{0,1\}^m$ and forming a propositional formula

(1) $\quad \tau(f)_b := \text{``}f(\langle \vec{q} \rangle) \neq b\text{''},$

where $\vec{q}$ is the vector $q_1, \ldots, q_n$ of propositional variables that appear in $\tau(f)_b$. (In a moment, we shall use the notation $\tau(f)_b(\vec{q})$ when we need to specify the variables.) Assuming $f$ is computable by a polynomial size propositional formula, then $\tau(f)_b$ has size bounded by a polynomial of $n$. More generally, if $f$ is computable by a polynomial size circuit, then $\tau(f)_b$ can be expressed as a propositional formula with the aid of additional propositional variables that give intermediate values calculated by the circuit.

The suggestion [63, 3] is that the tautologies $\tau(f)_b$ might require superpolynomially (or even, exponentially) long propositional proofs.[5] Razborov [95] also conjectures that the Nisan-Widgerson functions give pseudo-random number generators such that the tautologies (1) require superpolynomially long Frege proofs. If this could be established for *all* proof systems, it would imply, of course, that $NP \neq coNP$.

Krajíček [64] introduces further tautologies based on iterating the idea behind the tautologies (1). Namely, let $k > 0$ and for $i = 1, \ldots, k$ let $\vec{q}^i$ be a vector of $n$ variables. Further let $B_i$ be a Boolean circuit that outputs a vector of $m$ variables, with the inputs to $B_i$ being $\vec{q}^1, \ldots, \vec{q}^{i-1}$. Let $b_i$ denote the value of $B_i(\vec{q}^1, \ldots, \vec{q}^{i-1})$. (So $b_0$ is just a string of fixed Boolean values, $b_0 \in \{0,1\}^m$.) Consider the following formulas:

(2) $\quad \tau(f)_{b_0}(\vec{q}^0) \vee \tau(f)_{b_1}(\vec{q}^1) \vee \cdots \vee \tau(f)_{b_k}(\vec{q}^k).$

These formulas will be tautologies in many situations, such as when $b_0$ is not in the range of $f$. The function $f$ is called *pseudo-surjective* provided that these formulas do not have proofs of size bounded by a polynomial of $n$. Krajíček conjectures that there do exist pseudo-surjective functions for Frege and extended Frege. As evidence for why this conjecture is reasonable, Krajíček shows that truth-table functions $f$ (see [93]) that are based on the explicit evaluation of Boolean circuits can form a kind of canonical pseudo-surjective functions (assuming pseudo-surjective functions exist); he further links their pseudo-surjectivity to the (un)provability of circuit lower bounds.

---

[5]The formulas $\tau(f)_b$ are tautologies if $b$ is not in the range of $f$. However, if $b$ is in the range of $f$, then $\tau(f)_b$ does not have any proof at all.

So far, superpolynomial lower bounds for the $\tau$-tautologies (1) have been given for resolution, some small extensions of resolution, and the polynomial calculus [64, 3, 95]. On the other hand, Krajíček [66] has pointed out that, for any propositional proof system that does not have polynomial size proofs of the pigeonhole principle tautologies $\text{PHP}_n^{n+1}$, it is possible to build pseudo-random number generators $f$ (in the weak sense as defined above), for which the tautologies (1) require superpolynomial size. This construction uses a direct reduction from (a failure of) the pigeonhole principle to construct onto functions mapping $\{0,1\}^n$ onto $\{0,1\}^{n+1}$. It is perhaps disheartening that this construction also covers all the proof systems for which more conventional pseudo-random functions $f$ had been used to get lower bounds on proof complexity.

Nonetheless, the pseudo-random number generator tautologies remain an interesting approach. The conjectures of [63, 3, 95] that the tautologies (1) and (2) require superpolynomial size proofs are very reasonable and seem likely to be true. In addition, the pseudo-random number generator tautologies are related to constructions used for natural proof methods.

**Natural proof methods.** The pseudo-random number generator method described above was inspired in large part by the "natural proof" method of Razborov and Rudich [98] which gives barriers to proofs of $P \neq NP$. The natural proof barriers are based on the assumption that certain kinds of computationally hard pseudo-random number generators exist. In related work, Razborov [93] used natural proofs to obtain results about the inability of certain theories of bounded arithmetic to prove superpolynomial lower bounds to circuit size. (We shall omit the technical details here; the reader can consult [93], [62], or [23] for more details.) Although Razborov's independence results for bounded arithmetic assumed the existence of strong pseudo-random generators, Krajíček [64] subsequently pointed out that the results could be obtained without this assumption by using the non-provability of the pigeonhole principle in these theories (and using an idea of Razborov [96] that if the pigeonhole principle fails, then large circuits can be compressed into small circuits).

It is interesting to note that natural proofs inject an element of self-reference into the $NP$–$P$ problem. Namely, the natural proofs construction of [98] shows that if a very strong form of $P \neq NP$ holds, then $P \neq NP$ is hard to prove in certain proof systems. Loosely speaking, one might interpret this as saying that the truth of $P \neq NP$ (if it is true) might be part of the reason it is hard to prove $P \neq NP$.[6] Of course, this is a further impetus for studying the $\tau$-tautologies discussed above.

---

[6]Admittedly, one could just as well interpret this as an indication that $P$ is equal to $NP$, but it is not an interpretation that we prefer to make.

## 6 Conclusions

It is frustrating that the proof complexity approach to complexity has not yet managed to surpass the "counting barrier" in any substantial way. Indeed, most lower bound results in proof complexity either can be categorized as a form of diagonalization (and thus are limited by the Baker-Gill-Solovay oracle constructions) or can be reduced to or subsumed by a failure of counting or of the pigeonhole principle. Of course, this is not a criticism of the proof complexity approach as compared to other approaches, since no other approaches have had much more success. Indeed, proof complexity remains a highly interesting and promising approach toward the NP–P problem and other complexity questions. In addition, as a method of attacking complexity questions, proof complexity has the advantage of being inherently different from circuit complexity, and thus provides a fundamentally different line of attack on the NP–P problem.

We conclude with a few rather general open problems. (1) Computational complexity has a rich structure that exploits randomization and cryptographic constructions in sophisticated ways. This includes results such as PCP theory, (de)randomization, expander graphs, communication complexity, etc. It would be highly desirable to bring more of these methods into proof complexity. (2) Algebraic complexity has been a successful tool for computational complexity: there should be more linkages possible between algebraic complexity and proof complexity. (3) One should develop randomized constructions that transcend the pigeonhole principle. In particular, proof complexity needs to find a way to break past the "counting barrier". Conversely, perhaps it can be shown that breaking past the "counting barrier" is as hard as separating computational classes.

**Acknowledgement.** We thank J. Krajíček and A. Razborov for comments on an earlier draft of this paper. We also thank the two anonymous referees for their careful reading and useful comments.

Supported in part by NSF grant DMS-0700533.

## BIBLIOGRAPHY

[1] M. AJTAI, *The complexity of the pigeonhole principle*, in Proceedings of the 29-th Annual IEEE Symposium on Foundations of Computer Science, 1988, pp. 346–355.

[2] ———, *The independence of the modulo p counting principles*, in Proceedings of the 26th Annual ACM Symposiom on Theory of Computing, Association for Computing Machinery, 1994, pp. 402–411.

[3] M. ALEKHNOVICH, E. BEN-SASSON, A. RAZBOROV, AND A. WIDGERSON, *Pseudorandom generators in propositional proof complexity*, in Proc. 41st IEEE Conf. on Foundations of Computer Science (FOCS), 2000, pp. 43–53.

[4] M. ALEKHNOVICH, S. BUSS, S. MORAN, AND T. PITASSI, *Minimum propositional proof length is NP-hard to linearly approximate*, Journal of Symbolic Logic, 66 (2001), pp. 171–191.

[5] M. ALEKHNOVICH, J. JOHANNSEN, T. PITASSI, AND A. URQUHART, *An exponential separation between regular and general resolution*, in Proc. 34th Annual ACM Symposium on Theory of Computing, 2002, pp. 448–456.

[6] M. ALEKHNOVICH AND A. A. RAZBOROV, *Resolution is not automatizable unless $W[P]$ is tractable*, in Proc. 42nd IEEE Conf. on Foundations of Computer Science (FOCS), 2001, pp. 210–219.

[7] ———, *Lower bounds for polynomial calculus: Nonbinomial case*, Proceedings of the Steklov Institute of Mathematics, 242 (2003), pp. 18–35.

[8] B. ALON AND R. BOPPANA, *The monotone circuit complexity of Boolean functions*, Combinatorica, 7 (1987), pp. 1–22.

[9] A. ATSERIAS, P. G. KOLAITIS, AND M. Y. VARDI, *Constraint propogation as a proof system*, in Proc. Tenth International Conf. on Principles and Practice of Constraint Programming, 1974, pp. 77–91.

[10] F. BACCHUS, P. HERTEL, T. PITASSI, AND A. VAN GELDER, *Clause learning can effectively p-simulate general propositional resolution*, in Proc. 23rd AAAI Conf. on Artificial Intelligence (AAAI 2008), AAAI Press, 2008, pp. 283–290.

[11] P. BEAME, *Proof complexity*, in Computational Complexity Theory, IAS/Park City Mathematical Series, Vol. 10, American Mathematical Society, 2004, pp. 199–246. Lecture notes scribed by Ashish Sabharwal.

[12] P. BEAME, R. IMPAGLIAZZO, J. KRAJÍČEK, T. PITASSI, AND P. PUDLÁK, *Lower bounds on Hilbert's Nullstellensatz and propositional proofs*, Proceedings of the London Mathematical Society, 73 (1996), pp. 1–26.

[13] P. BEAME, H. A. KAUTZ, AND A. SABHARWAL, *Towards understanding and harnessing the potential of clause learning*, J. Artificial Intelligence Research, 22 (2004), pp. 319–351.

[14] P. BEAME AND T. PITASSI, *Propositional proof complexity: Past, present and future*, in Current Trends in Theoretical Computer Science Entering the 21st Century, G. Paun, G. Rozenberg, and A. Salomaa, eds., World Scientific, 2001, pp. 42–70. Earlier version appeared in *C*omputational Complexity Column, Bulletin of the EATCS, 2000.

[15] P. BEAME, T. PITASSI, AND N. SEGERLIND, *Lower bounds for Lovász-Schrijver systems and beyond follow from multiparty communication complexity*, SIAM Journal on Computing, 37 (2007), pp. 845–869.

[16] E. BEN-SASSON AND R. IMPAGLIAZZO, *Random CNF's are hard for the polynomial calculus*, in Proceedings 40th IEEE Conference on Foundations of Computer Science (FOCS), 1999, pp. 415–421.

[17] M. BONET, R. GAVALDA, C. DOMINGO, A. MACIEL, AND T. PITASSI, *Non-automatizability for bounded-depth Frege systems*, Computational Complexity, 13 (2004), pp. 47–68.

[18] M. L. BONET, T. PITASSI, AND R. RAZ, *On interpolation and automatization for Frege systems*, SIAM Journal on Computing, 29 (2000), pp. 1939–1967.

[19] H. BUHRMAN, S. FENNER, L. FORTNOW, AND D. CAN MELKEBEEK, *Optimal proof systems and sparse sets*, in Proc. 17th Symp. on Theoretical Aspects of Computer Science (STACS), Lecture Notes in Computer Science #1770, Springer Verlag, 2000, pp. 407–418.

[20] S. R. BUSS, *Polynomial size proofs of the propositional pigeonhole principle*, Journal of Symbolic Logic, 52 (1987), pp. 916–927.

[21] ———, *Propositional consistency proofs*, Annals of Pure and Applied Logic, 52 (1991), pp. 3–29.

[22] ———, *Lectures in proof theory*, Tech. Report SOCS 96.1, School of Computer Science, McGill University, 1996. Notes from a series of lectures given at the at the McGill University Bellair's Research Institute, Holetown, Barbados, March 1995. Scribe notes written by E. Allender, M.L. Bonet, P. Clote, A. Maciel, P. McKenzie, T. Pitassi, R. Raz, K. Regan, J. Torán and C. Zamora.

[23] ———, *Bounded arithmetic and propositional proof complexity*, in Logic of Computation, H. Schwichtenberg, ed., Springer-Verlag, Berlin, 1997, pp. 67–121.

[24] ———, *Propositional proof complexity: An introduction*, in Computational Logic, U. Berger and H. Schwichtenberg, eds., Springer-Verlag, Berlin, 1999, pp. 127–178.
[25] S. R. BUSS, D. GRIGORIEV, R. IMPAGLIAZZO, AND T. PITASSI, *Linear gaps between degrees for the polynomial calculus modulo distinct primes*, Journal of Computer and System Sciences, 62 (2001), pp. 267–289.
[26] S. R. BUSS, J. HOFFMANN, AND J. JOHANNSEN, *Resolution trees with lemmas: Resolution refinements that characterize DLL-algorithms with clause learning*, Logical Methods of Computer Science, 4, 4:13 (2008), pp. 1–18.
[27] S. R. BUSS, R. IMPAGLIAZZO, J. KRAJÍČEK, P. PUDLÁK, A. A. RAZBOROV, AND J. SGALL, *Proof complexity in algebraic systems and bounded depth Frege systems with modular counting*, Computational Complexity, 6 (1996/1997), pp. 256–298.
[28] S. R. BUSS AND R. KUZNETS, *The NP-completeness of reflected fragments of justification logics*, in Proc. Logical Foundations of Computer Sciences (LFCS'09), Lecture Notes in Computer Science #5407, 2009, pp. 122–136.
[29] A. CHATTOPADHYAY AND A. ADA, *Multiparty communication complexity of disjointness*, Tech. Report TR08-002, Electronic Colloquium in Computational Complexity (ECCC), 2008. 15pp.
[30] Y. CHEN AND J. FLUM, *On p-optimal proof systems and logics for PTIME*, in Proc. 37th International Colloquium on Automata, Languages, and Programming (ICALP'10), Lecture Notes in Computer Science 6199, Springer, 2010, pp. 312–313.
[31] V. CHVÁTAL AND E. SZEMERÉDI, *Many hard examples for resolution*, Journal of the ACM, 35 (1988), pp. 759–768.
[32] P. CLOTE AND J. KRAJÍČEK, *Arithmetic, Proof Theory and Computational Complexity*, Oxford University Press, 1993.
[33] S. A. COOK, *The complexity of theorem proving techniques*, in Proceedings of the 3-rd Annual ACM Symposium on Theory of Computing, 1971, pp. 151–158.
[34] ———, *Feasibly constructive proofs and the propositional calculus*, in Proceedings of the Seventh Annual ACM Symposium on Theory of Computing, 1975, pp. 83–97.
[35] ———, *An overview of computational complexity*, Journal of the ACM, 26 (1983), pp. 401–408.
[36] S. A. COOK AND R. A. RECKHOW, *On the lengths of proofs in the propositional calculus, preliminary version*, in Proceedings of the Sixth Annual ACM Symposium on the Theory of Computing, 1974, pp. 135–148.
[37] ———, *The relative efficiency of propositional proof systems*, Journal of Symbolic Logic, 44 (1979), pp. 36–50.
[38] S. DASH, *Exponential lower bounds on the lengths of some classes of branch-and-cut proofs*, Mathematics of Operations Research, 30 (2005), pp. 678–700.
[39] M. DAVIS, G. LOGEMANN, AND D. LOVELAND, *A machine program for theorem proving*, Communications of the ACM, 5 (1962), pp. 394–397.
[40] M. DAVIS AND H. PUTNAM, *A computing procedure for quantification theory*, Journal of the Association for Computing Machinery, 7 (1960), pp. 201–215.
[41] I. DINUR AND S. SAFRA, *On the hardness of approximating label cover*, Information Processing Letters, 89 (2004), pp. 247–254.
[42] R. G. DOWNEY AND M. R. FELLOWS, *Parameterized Complexity*, Springer, 1999.
[43] N. GALESI AND M. LAURIA, *On the automatizability of the polynomial calculus*, Theory of Computing Systems, 47 (2010), pp. 491–506.
[44] E. GOLDBERG AND Y. NOVIKOV, *BerkMin: A fast and robust Sat-solver*, Discrete Applied Mathematics, 155 (2007), pp. 1549–1561.
[45] D. GRIGORIEV, *Nullstellensatz lower bounds for Tseitin tautologies*, in Proceedings of the 39th Annual IEEE Symposium on Foundations of Computer Science, IEEE Computer Society Press, 1998, pp. 648–652.
[46] A. HAKEN, *The intractability of resolution*, Theoretical Computer Science, 39 (1985), pp. 297–308.
[47] T. HERTLI, *3-SAT faster and simpler — Unique-SAT bounds for PPSZ hold in general*. Preprint, arXiv:1103.2165v2 [cs.CC]., May 2011.

[48] T. HERTLI, R. A. MOSER, AND D. SCHEDER, *Improving PPSZ for 3-SAT using critical variables*, in Proc. 28th Symp. on Theoretical Aspects of Computer Science (STACS), 2011, pp. 237–248.

[49] E. A. HIRSCH, D. ITSYKSON, I. MONAKHOV, AND A. SMAL, *On optimal heuristic randomized semidecision procedures, with applications to proof complexity and cryptography*, Tech. Report TR10-193, Electronic Colloquium in Computational Complexity (ECCC), 2010.

[50] P. HRUBEŠ, *A lower bound for intuitionistic logic*, Annals of Pure and Applied Logic, 146 (2007), pp. 72–90.

[51] ———, *A lower bound for modal logics*, Journal of Symbolic Logic, 72 (2007), pp. 941–958.

[52] ———, *On lengths of proofs in non-classical logics*, Annals of Pure and Applied Logic, 157 (2009), pp. 194–205.

[53] R. IMPAGLIAZZO AND J. KRAJÍČEK, *A note on conservativity relations among bounded arithmetic theories*, Mathematical Logic Quarterly, 48 (2002), pp. 375–377.

[54] R. IMPAGLIAZZO AND R. PATURI, *On the complexity of k-SAT*, Journal of Computer and System Sciences, 62 (2001), pp. 367–375.

[55] R. IMPAGLIAZZO, R. PATURI, AND F. ZANE, *Which problems have strongly exponential complexity*, Journal of Computer and System Sciences, 63 (2001), pp. 512–530.

[56] R. IMPAGLIAZZO, P. PUDLÁK, AND J. SGALL, *Lower bounds for the polynomial calculus and the Gröbner basis algorithm*, Computational Complexity, 8 (1999), pp. 127–144.

[57] D. M. ITSYKSON AND A. A. KOJEVNIKOV, *Lower bounds of static Lovász-Schrijver calculus proofs for Tseitin tautologies*, Journal of Mathematical Sciences, 145 (2007). Translated from *Zapiski Nauchnykh Seminarov POMI*, Vol 340, 2006, pp. 10-32.

[58] K. IWAMA AND S. TAMAKI, *Improved upper bounds for 3-SAT*, in Proc. 15th Annual ACM-SIAM Symposium on Discrete Algorithms (SODA), 2004, pp. 328–329.

[59] M. I. KANOVICH, *The complexity of Horn fragments of linear logic*, Pure and Applied Logic, 69 (1994), pp. 195–241.

[60] J. KÖBLER, J. MESSNER, AND J. TORAN, *Optimal proof systems imply complete sets for promise classes*, Information and Computation, 184 (2003), pp. 71–92.

[61] J. KRAJÍČEK, *Lower bounds to the size of constant-depth propositional proofs*, Journal of Symbolic Logic, 59 (1994), pp. 73–86.

[62] ———, *Interpolation theorems, lower bounds for proof systems, and independence results for bounded arithmetic*, Journal of Symbolic Logic, 62 (1997), pp. 457–486.

[63] ———, *On the weak pigeonhole principle*, Fundamenta Mathematica, 170 (2001), pp. 123–140.

[64] ———, *Dual weak pigeonhole principle, pseudo-surjective functions, and provability of circuit lower bounds*, Journal of Symbolic Logic, 69 (2004), pp. 265–286.

[65] ———, *An exponential lower bound for a constraint propogation proof system based on ordered binary decision diagrams*, Journal of Symbolic Logic, 73 (2008), pp. 227–237.

[66] ———, *A proof complexity generator*, in Proc. 13th International Congress of Logic, Methodology and Philosophy of Science, C. Glymour, W. Wanng, and D. Weststahl, eds., London, 2009, King's College Publications, pp. 185–190.

[67] J. KRAJÍČEK AND P. PUDLÁK, *Propositional proof systems, the consistency of first-order theories and the complexity of computations*, Journal of Symbolic Logic, 54 (1989), pp. 1063–1079.

[68] J. KRAJÍČEK, P. PUDLÁK, AND A. WOODS, *Exponential lower bound to the size of bounded depth Frege proofs of the pigeonhole principle*, Random Structures and Algorithms, 7 (1995), pp. 15–39.

[69] G. KREISEL, *Hilbert's programme and the search for automatic proof procedures*, in Symposium on Automatic Demonstration, Lecture Notes in Mathematics 125, Berlin, 1968, Springer Verlag, pp. 128–146.

[70] N. V. KRUPSKI, *On the complexity of the reflected logic of proofs*, Theoretical Computer Science, 357 (2006), pp. 136–142.

[71] T. LEE AND A. SHRAIBMAN, *Disjointness is hard in the multiparty number-on-the-forehead model*, in Proc. 23rd IEEE Conference on Computational Complexity (CCC), 2008, pp. 81–91.

[72] L. LEVIN, *Universal search problems*, Problems of Information Transmission, 9 (1973), pp. 265–266. In Russian. English translation by B.A. Trakhtenroty, in *A survey of Russian approaches to perebor (brute-force searches) algorithms*, Annals of the History of Computing 6 (1984) 384-400.

[73] P. LINCOLN, J. MITCHELL, A. SCEDROV, AND N. SHANKAR, *Decision problems for propositional linear logic*, Annals of Pure and Applied Logic, 60 (1993), pp. 151–177.

[74] A. MACIEL AND T. PITASSI, *A conditional lower bound for a system of constant-depth proofs with modular connectives*, in Proc. Twenty-First IEEE Symp. on Logic in Computer Science (LICS'06), IEEE Computer Society Press, 2006, pp. 189–198.

[75] S. MALIK, Y. ZHAO, C. F. MADIGAN, L. SHANG, AND M. W. MOSKEWICZ, *Chaff: Engineering an efficient SAT solver*, in 38th Conference on Design Automation (DAC'01), 2001, pp. 530–535.

[76] J. P. MARQUES-SILVA, *The impact of branching hueristics in propositional satisfiability algorithms*, in Proc. 9th Portuguese Conference on Artificial Intelligence (EPIA), Springer Verlag, 1999, pp. 62–74.

[77] J. P. MARQUES-SILVA AND K. A. SAKALLAH, *GRASP — A new search algorithm for satisfiability*, IEEE Transactions on Computers, 48 (1999), pp. 506–521.

[78] B. MONIEN AND E. SPECKENMEYER, *Solving satisfiability in less than $2^n$ steps*, Discrete Applied Mathematics, 10 (1985), pp. 287–295.

[79] J. B. PARIS AND A. J. WILKIE, *Counting problems in bounded arithmetic*, in Methods in Mathematical Logic, Lecture Notes in Mathematics #1130, Springer-Verlag, 1985, pp. 317–340.

[80] M. PĂTRAȘCU AND R. WILLIAMS, *On the possibility of faster SAT alorithms*, in Proc. 21st ACM/SIAM Symp. on Discrete Algorithms (SODA), 2010, pp. 1065–1075.

[81] R. PATURI, P. PUDLÁK, M. E. SAKS, AND F. ZANE, *An improved exponential-time algorithm for k-SAT*, Journal of the Association for Computing Machinery, 52 (2005), pp. 337–364.

[82] R. PATURI, P. PUDLÁK, AND F. ZANE, *Satisfiability coding lemma*, Chicago Journal of Theoretical Computer Science, (1999). Article 11.

[83] M. PENTUS, *Lambek calculus is NP-complete*, Theoretical Computer Science, 357 (2006), pp. 186–201.

[84] T. PITASSI, P. BEAME, AND R. IMPAGLIAZZO, *Exponential lower bounds for the pigeonhole principle*, Computational Complexity, 3 (1993), pp. 97–140.

[85] T. PITASSI AND N. SEGERLIND, *Exponential lower bounds and integrality gaps for tree-like Lovász-Schrijver procedures*, in Proc. 20th Annual ACM-SIAM Symposium on Discrete Algorithms (SODA), 2009, pp. 355–364.

[86] P. PUDLÁK, *On the lengths of proofs of finitistic consistency statements in first order theories*, in Logic Colloquium '84, North-Holland, 1986, pp. 165–196.

[87] ———, *Improved bounds to the lengths of proofs of finitistic consistency statements*, in Logic and Combinatorics, vol. 65 of Contemporary Mathematics, American Mathematical Society, 1987, pp. 309–331.

[88] ———, *Lower bounds for resolution and cutting planes proofs and monotone computations*, Journal of Symbolic Logic, 62 (1997), pp. 981–998.

[89] ———, *Twelve problems in proof complexity*, in Computer Science — Theory and Applications, Lecture Notes in Computer Science #5010, Berlin, Heidelberg, 2008, Springer, pp. 13–27.

[90] R. RAZ, *Resolution lower bounds for the weak pigeonhole principle*, in Proc. 34th ACM Symp. on Theory of Computing (STOC), 2002, pp. 553–562.

[91] A. A. RAZBOROV, *Lower bounds on the monotone complexity of some Boolean functions*, Soviet Mathematics Doklady, 31 (1985), pp. 354–357.

[92] ———, *Lower bounds on the size of bounded depth networks over a complete basis with logical addition*, Matematischi Zametki, 41 (1987), pp. 598–607. English translation in *Mathematical Notes of the Academy of Sciences of the USSR* 41(1987) 333-338.

[93] ———, *Unprovability of lower bounds on the circuit size in certain fragments of bounded arithmetic*, Izvestiya of the RAN, 59 (1995), pp. 201–224.

[94] ———, *Lower bounds for the polynomial calculus*, Computation Complexity, 7 (1998), pp. 291–324.

[95] ———, *Pseudorandom generators hard for k-DNF resolution and polynomial calculus resolution.* Manuscript, available at author's web site, 67 pp., 2003.

[96] ———, *Resolution lower bounds for the weak pigeonhole principle*, Theoretical Computer Science, 303 (2003), pp. 233–243.

[97] ———, *Resolution lower bounds for perfect matching principles*, Journal of Computer and System Sciences, 69 (2004), pp. 3–27.

[98] A. A. Razborov and S. Rudich, *Natural proofs*, Journal of Computer and System Sciences, 55 (1997), pp. 24–35.

[99] Z. Sadowski, *Optimal proof systems, optimal acceptors and recursive presentability*, Fundamenta Informaticae, 79 (2007), pp. 169–185.

[100] Y. Savateev, *Product-free Lambek calculus is NP-complete*, in Proc. 2009 Intl. Symp. on Logical Foundations of Computer Science (LFCS'09), Lecture Notes in Computer Science #5407, Springer Verlag, 2009, pp. 380–394.

[101] U. Schöning, *A probabilistic algorithm for k-SAT based on limited local search and restart*, Algorithmica, 32 (2008), pp. 615–623.

[102] R. Schuler, *An algorithm for the satisfiability problem of formulas in disjunctive normal form*, Journal of Algorithms, 54 (2005), pp. 40–44.

[103] N. Segerlind, *The complexity of propositional proofs*, Bulletin of Symbolic Logic, 13 (2007), pp. 417–481.

[104] ———, *On the relative efficiency of resolution-like proofs and ordered binary decision diagram proofs*, in Proc. 23rd Annual IEEE Conference on Computational Complexity (CCC'08), 2008, pp. 100–111.

[105] J. Siekmann and G. Wrightson, *Automation of Reasoning*, vol. 1&2, Springer-Verlag, Berlin, 1983.

[106] M. Sipser, *The history and status of the P versus NP question*, in Proceedings of the Twenty-fourth Annual ACM Symposium on Theory of Computing, ACM, 1992, pp. 603–618.

[107] R. Smolensky, *Algebraic methods in the theory of lower bounds for Boolean circuit complexity*, in Proceedings of the Nineteenth Annual ACM Symposium on the Theory of Computing, ACM Press, 1987, pp. 77–82.

[108] G. S. Tsejtin, *On the complexity of derivation in propositional logic*, Studies in Constructive Mathematics and Mathematical Logic, 2 (1968), pp. 115–125. Reprinted in: [105, vol 2], pp. 466-483.

[109] A. Urquhart, *Hard examples for resolution*, J. Assoc. Comput. Mach., 34 (1987), pp. 209–219.

[110] A. Van Gelder, *Pool resolution and its relation to regular resolution and DPLL with clause learning*, in Logic for Programming, Artificial Intelligence, and Reasoning (LPAR), Lecture Notes in Computer Science Intelligence 3835, Springer-Verlag, 2005, pp. 580–594.

# Equivalence Relations in Set Theory, Computation Theory, Model Theory, and Complexity Theory

Sy-David Friedman

One of Harvey's most influential articles is his joint work with Lee Stanley [8] in which he introduces a notion of *Borel reducibility* between isomorphism relations on the countable models of a theory in infinitary logic. Through the work of many researchers, this theory later blossomed into a rich field devoted to the more general study of Borel reducibility between Borel and analytic equivalence relations (and quasi-orders). For a look at some of this work see [11, 12, 17, 19, 23, 26, 27, 30].

The aim of the present article is to illustrate how a similar idea has recently been used to good effect in four new contexts: *effective* descriptive set theory, computation theory, model theory and complexity theory. This work has deepened research in these fields, produced a number of unexpected results and raised a host of interesting new open problems.

The author wishes to thank the John Templeton Foundation for its support through the *CRM Infinity Project*, Project ID# 13152, and the Austrian Science Fund (FWF) for its support through research project P 20835-N13.

## 1 Effective Descriptive Set Theory

We begin with a brief description of the classical, non-effective setting, before turning to the more recent work [6] in the effective context. The principal objects of study in the classical theory are analytic ($\Sigma_1^1$ with parameters) equivalence relations on Polish spaces (think of the reals). Such equivalence relations are compared using *Borel reducibility* in the following way:
$E_0$ *is Borel reducible to* $E_1$ iff there is a Borel function $f : X_0 \to X_1$ such that
$$xE_0y \text{ iff } f(x)E_1f(y).$$

$E_0$ and $E_1$ are *Borel bireducible* if each Borel reduces to the other. Then $\mathcal{B}$ denotes the resulting set of degrees, ordered under Borel reducibility. When discussing Borel reducibility we sometimes identify an equivalence relation with its degree. Work of Silver [37] and of Harrington-Kechris-Louveau [16] identifies an interesting initial segment of $\mathcal{B}$:

THEOREM 1. $\mathcal{B}$ has the initial segment

$$1 < 2 < \cdots < \omega < id < E_0,$$

where:
$n$ = Borel equivalence relations with exactly $n$ classes;
$\omega$ = Borel equivalence relations with exactly $\aleph_0$ classes;
id is $(^\omega\omega, =)$ (equality on reals);
$E_0$ is the equivalence relation $xE_0y$ iff $x(n) = y(n)$ for all but finitely many $n$. In fact, any Borel equivalence relation is Borel equivalent to one of the above or lies strictly above $E_0$ under Borel reducibility.

The question for the effective theory is: What happens if we replace "Borel" by "effectively Borel"? In what follows we simply write "Hyp" for "effectively Borel" (= lightface $\Delta_1^1$). We define:
If $E$ and $F$ are Hyp equivalence relations on the reals, then $E$ is *Hyp reducible to $F$*, written $E \leq_H F$, iff For some Hyp function $f$, $xEy$ iff $f(x)Ff(y)$
$\leq_H$ is reflexive and transitive. We write $E \equiv_H F$ for $E \leq_H F$ and $F \leq_H E$.

So the new object of study is $\mathcal{H}$, the degrees of Hyp equivalence relations on the reals under Hyp reducibility.

There are some surprises! Again we have degrees

$$1 < 2 < \cdots < \omega < id < E_0,$$

defined as follows:
$n$ is represented by $xE^ny$ iff $x(0) = y(0) < n-1$ or $x(0), y(0) \geq n-1$;
$\omega$ is represented by $xE^\omega y$ iff $x(0) = y(0)$;
id, $E_0$ are as before: $x$id$y$ iff $x = y$, $xE_0y$ iff $x(n) = y(n)$ for all but finitely many $n$.

PROPOSITION 2. *There are Hyp equivalence relations strictly between 1 and 2!*

Here is why: Let $E$ be a Hyp equivalence relation. Recall that the $\mathcal{H}$-degree $n$ is represented by the equivalence relation $E^n$ where:
$xE^ny$ iff $x(0) = y(0) < n-1$ or $x(0), y(0) \geq n-1$.
*Fact 1.* $E^n$ is Hyp reducible to $E$ iff at least $n$ distinct $E$-equivalence classes contain Hyp reals.
*Proof.* Suppose that $E^n$ Hyp reduces to $E$ via the Hyp function $f$. Each of the $n$ equivalence classes of $E^n$ contains a Hyp real; let $x_0, \ldots, x_{n-1}$ be

Hyp, pairwise $E^n$-inequivalent reals. Then the reals $f(x_i)$, $i < n$, are Hyp, pairwise $E$-inequivalent reals. Conversely, if $y_0, \ldots, y_{n-1}$ are Hyp, pairwise $E$-inequivalent reals then send the $E^n$-equivalence class of $x_i$ to the real $y_i$; this is a Hyp reduction of $E^n$ to $E$. □

*Fact 2.* $E$ is Hyp reducible to $E^2$ iff $E$ has at most 2 equivalence classes.

*Proof.* If $E$ is Hyp reducible to $E^2$, then $E$ has at most 2 equivalence classes because $E^2$ has only 2 equivalence classes. Conversely, suppose that the equivalence classes of $E$ are $A_0$ and $A_1$. We may assume that $A_0$ has a Hyp element $x$. Then $A_0$ is Hyp as it consists of those reals $E$-equivalent to $x$ and $A_1$ is Hyp as it consists of those reals not $E$-equivalent to $x$. Now we can reduce $E$ to $E^2$ by choosing $E^2$-inequivalent Hyp reals $y_0, y_1$ and sending the elements of $A_0$ to $y_0$ and the elements of $A_1$ to $y_1$. □

So to get a Hyp equivalence relation between 1 and 2 we need only find one with two equivalence classes but with all Hyp reals in just one class. The existence of such an equivalence relation follows from a classical fact from Hyp theory (see [35], page 52, Theorem 1.1):

*Fact 3.* There are nonempty Hyp sets of reals which contain no Hyp element.

*Proof.* Let $A$ be the set of non-Hyp reals. Then $A$ is $\Sigma_1^1$ and therefore the projection of a $\Pi_1^0$ subset $P$ of Reals $\times$ Reals. $P$ is nonempty. A Hyp real $h = (h_0, h_1)$ in $P$ would give a Hyp real $h_0$ in $A$, contradiction. □

Now we ask a harder question: Are there incomparable degrees between 1 and 2? To answer this we prove:

THEOREM 3. *([6]) There exist Hyp sets of reals $A, B$ such that for no Hyp function $F$ do we have $F[A] \subseteq B$ or $F[B] \subseteq A$.*

Given this Theorem, define $E_A$ to be the equivalence relation with equivalence classes $A$ and $\sim A$ (the complement of $A$); define $E_B$ similarly. Note that the sets $A, B$ contain no Hyp reals, else there would be a constant Hyp function $F$ mapping one of them into the other. So a Hyp reduction of $E_A$ to $E_B$ would have to send the elements of $\sim A$ (which contains Hyp reals) to elements of $\sim B$, and therefore the elements of $A$ to elements of $B$, contradicting the Theorem. Similarly there is no Hyp reduction of $E_B$ to $E_A$.

*Proof Sketch of Theorem 3.* First we quote a result of Harrington [15] (also see [33], Theorem XIII.3.5). For reals $a, b$ and a recursive ordinal $\alpha$ we say that $a$ is $\alpha$-*below* $b$ iff $a$ is recursive in the $\alpha$-jump of $b$.

*Fact.* For any recursive ordinal $\alpha$ there are $\Pi_1^0$ singletons $a, b$ such that $a$ is not $\alpha$-below $b$ and $b$ is not $\alpha$-below $a$.

Now using Barwise Compactness, find a nonstandard $\omega$-model $M$ of $ZF^-$ with standard ordinal $\omega_1^{CK}$ in which are there are $\Pi_1^0$ singletons $a, b$ such that for all recursive $\alpha$, $a$ is not $\alpha$-below $b$ and $b$ is not $\alpha$-below $a$ (i.e., $a$

and $b$ are Hyp incomparable.) Let $a, b$ be the unique solutions in $M$ to the $\Pi_1^0$ formulas $\varphi_0, \varphi_1$, respectively. The desired sets $A, B$ are $\{x \mid \varphi_0(x)\}$ and $\{x \mid \varphi_1(x)\}$. If $F$ were a Hyp function mapping $A$ into $B$, then it would send the element $a$ of $A$ to an element $F(a)$ of $B \cap M$; but then $F(a)$ must equal $b$ and therefore $b$ is Hyp in $a$, contradicting the choice of $a, b$. □

Now fix $A, B$ as in the Theorem. Using them we can get incomparable Hyp equivalence relations between $n$ and $n+1$ for any finite $n$, by considering $E_A, E_B$ where the equivalence classes of $E_A$ are $A$ together with a split of $\sim A$ (the complement of $A$) into $n$ classes, each of which contains a Hyp real (similarly for $E_B$).

We now consider Hyp equivalence relations with infinitely many equivalence classes. Recall the Silver and Harrington-Kechris-Louveau dichotomies:

THEOREM 4. *(a) (Silver) A Borel equivalence relation is either Borel reducible to $\omega$ or Borel reduces id.*
*(b) (Harrington-Kechris-Louveau) A Borel equivalence relation is either Borel reducible to id or Borel reduces $E_0$.*

How effective are these results? Harrington's proof of (a) and the original proof of (b) show:

THEOREM 5. *(a) A Hyp equivalence relation is either Hyp reducible to $\omega$ or Borel reduces id.*
*(b) A Hyp equivalence relation is either Hyp reducible to id or Borel reduces $E_0$.*

The sets $A, B$ of Theorem 3 can be used to show that the Silver and Harrington-Kechris-Louveau dichotomies are *not* fully effective:

THEOREM 6. *([6]) (a) There are incomparable Hyp equivalence relations between $\omega$ and id.*
*(b) There are incomparable Hyp equivalence relations between id and $E_0$.*

*Proof Sketch.* (a) Consider the relations
$E_A(x, y)$ iff $(x \in A$ and $x = y)$ or $(x, y \notin A$ and $x(0) = y(0))$
$E_B$: The same, with $A$ replaced by $B$
Now $E^\omega$ Hyp reduces to $E_A$ by $n \mapsto (n, 0, 0, ...)$. Also $E_A$ Hyp reduces to id via the map $G(x) = x$ for $x \in A$, $G(x) = (x(0), 0, 0, ...)$ for $x \notin A$ (same for $B$)

There is no Hyp reduction of $E_A$ to $E_B$: If $F$ were such a reduction, then let $C$ be $F^{-1}[\sim B]$. As $\sim B$ is Hyp, $C$ is also Hyp and therefore $A \cap C$ is also Hyp. But $A \cap C$ must be countable as $F$ is a reduction. So if $A \cap C$ were nonempty it would have a Hyp element, contradicting the fact that $A$ has no Hyp element. Therefore $F$ maps $A$ into $B$, which is impossible by the choice of $A, B$. By symmetry, there is no Hyp reduction of $E_B$ to $E_A$.

(b) Now we define $E_A$ on $\mathbb{R} \times \mathbb{R}$ by: $(x,y)E_A(x',y')$ iff $x = x'$ and either $x \notin A$ or ($x \in A$ and $yE_0y'$). $E_B$ is the same, with $A$ replaced by $B$.
We need two Facts (see [18], Lemma 2.49 and [24], Theorem 2.2.5 (a), respectively):
1. If $h : \mathbb{R} \to \mathbb{R}$ is Baire measurable and constant on $E_0$ classes, then $h$ is constant on a comeager set.
2. If $B \subseteq \mathbb{R}^2$ is Hyp, then so is $\{x \mid \{y \mid (x,y) \in B\}$ is comeager$\}$.
Now suppose that $F$ were a Hyp reduction of $E_A$ to $E_B$. Let $\pi(x,y) = x$ for all $x$ and define $h : \mathbb{R} \to \mathbb{R}$ by: $h(x) = z$ iff $\{y \mid \pi(F(x,y)) = z\}$ is comeager.
Using 1 and 2, $h$ is a total Hyp function. We claim that $h[A] \subseteq B$, contradicting the choice of $A, B$: Assume $x \in A$. Then for comeager-many $y$, $\pi(F(x,y)) = h(x)$. So if $h(x) \notin B$ then $F$ maps more than one $E_A$ class into a single $E_B$ class, contradiction. By symmetry there is no Hyp reduction of $E_B$ to $E_A$. □

The overall picture of the degrees of Hyp sets of reals under Hyp reducibility is the following: Call a degree *canonical* if it is one of $1 < 2 < \cdots < \omega < \text{id} < E_0$. For any two canonical degrees $a < b$ there is a rich collection of degrees which are above $a$, below $b$ and incomparable with all canonical degrees in between.

However at least one nice thing happens: If a degree is above $n$ for each finite $n$, then it is also above $\omega$.

Because this field is so new (like the others introduced in this paper), there remain many open questions. Here are several:
1. If a Hyp equivalence relation is Borel reducible to $E_0$, then must it also be Hyp reducible to $E_0$? (This is true for finite $n$, $\omega$, id.)
2. Are there any nodes other than 1? I.e., is there a Hyp equivalence relation with more than one equivalence class which is comparable with all Hyp equivalence relations under Hyp reducibility?
3. Is there a minimal degree? Are there incomparables above each degree?
There is also a jump operation, which is in need of further study.

## 2 Computation Theory

We now turn to equivalence relations not on the reals but on the natural numbers, where computation theory play a central role. As seen in the last section, Hyp-reducibility for Hyp equivalence relations on the real numbers has a rich structure; however the analogous theory in the context of the natural numbers is trivial:

PROPOSITION 7. ([4], Section 2.2, Fact 2.10.2) *Any Hyp equivalence relation on the natural numbers is Hyp reducible to the equality relation on $\omega$.*

Therefore the central objects of interest in our study of equivalence relations on the natural nubmers are not the Hyp equivalence relations but instead the $\Sigma_1^1$ equivalence relations. Indeed, in the classical theory of Borel reducibility one considers not only the Borel equivalence relations but more generally analytic ($\Sigma_1^1$ with parameters) equivalence relations which are not Borel; indeed these appeared already in [8]:

Let $T$ be any theory in first-order logic (or any sentence of the infinitary logic $\mathcal{L}_{\omega_1\omega}$). Then the isomorphism relation on the countable models of $T$ is an analytic equivalence relation which need not be Borel.

There are analytic equivalence relations which are not Borel reducible to such an isomorphism relation; an example is $E_1$, the equivalence relation on $\mathbb{R}^\omega$ defined by:

$$\vec{x} E_1 \vec{y} \text{ iff } \vec{x}(n) = \vec{y}(n) \text{ for almost all } n.$$

Note that $E_1$ is even Hyp.

A motivating question for our study is the following:

*Question.* Is every $\Sigma_1^1$ equivalence relation on the natural numbers reducible to isomorphism on a Hyp class of *computable* structures?

Of course we can identify a computable structure with a natural number which serves as an index for it. The reducibility we use is: $E_0 \leq_H E_1$ iff there is a Hyp function $f : \mathcal{N} \to \mathcal{N}$ such that $m E_0 n$ iff $f(m) E_1 f(n)$. (We say that $E_0$ is *Hyp reducible* to $E_1$.)

**THEOREM 8.** *([5]) Every $\Sigma_1^1$ equivalence relation on $\mathcal{N}$ is Hyp reducible to isomorphism on computable trees.*

This answers the above Question positively.

*Proof Sketch:* Let $E$ be a $\Sigma_1^1$ equivalence relation on $\mathcal{N}$ and choose a computable $f : \mathcal{N}^2 \to$ Computable Trees such that $\sim m E n$ iff $f(m,n)$ is well-founded.

Now associate to pairs $m, n$ computable trees $T(m, n)$ so that:
$T(m, n)$ is isomorphic to $T(n, m)$;
$m E n$ implies that $T(m, n)$ is isomorphic to the "canonical" non-well-founded computable tree;
$\sim m E n$ implies that $T(m, n)$ is isomorphic to the "canonical" computable tree of rank $\alpha$, where $\alpha$ is least so that $f(m', n')$ has rank at most $\alpha$ for all $m' \in [m]_E$, $n' \in [n]_E$.

Now to each $n$ associate the tree $T_n$ gotten by gluing together the $T(n, i)$, $i \in \omega$. If $m E n$, then $T_m$ is isomorphic to $T_n$ as they are obtained by gluing together isomorphic trees. And if $\sim m E n$ then $T_m, T_n$ are not isomorphic as they are obtained by gluing together trees which on some component are non-isomorphic. $\square$

It can be shown that the isomorphism relation on computable trees (and therefore any $\Sigma_1^1$ equivalence relation on $\mathcal{N}$) Hyp-reduces to the isomorphism relation on each of the following Hyp classes:

1. Computable graphs
2. Computable torsion-free Abelian groups
3. Computable Abelian $p$-groups for a fixed prime $p$
4. Computable Boolean Algebras
5. Computable linear orders
6. Computable fields

These results came as a surprise, because in the classical setting, the analogue of 2 is an open problem and the analogue of 3 is false!

Fokina and I show in [4] that the global structure of $\Sigma_1^1$ equivalence relations on $\mathcal{N}$ under Hyp reducibility is very rich: it embeds the partial order of $\Sigma_1^1$ sets under Hyp many-one reducibility. But it is not known if there is a single isomorphism relation on computable structures which is neither Hyp nor complete under Hyp-reducibility! However we do have:

THEOREM 9. ([4]) *Every $\Sigma_1^1$ equivalence relation is Hyp bireducible to a bi-embeddability relation on computable structures.*

The proof is based on the analagous result in the non-effective setting:

THEOREM 10. ([11]) *Every analytic equivalence relation on the reals is Borel bireducible to a bi-embeddability relation on countable structures.*

I should also mention that there has been considerable prior work on *computably enumerable* equivalence relations, of which provable equivalence is a natural example. For those interesting results we refer to [13] and the references therein.

## 3 Model Theory

It is natural to expect that insights into the model-theoretic properties of a first-order theory could be derived from the descriptive set-theoretic behaviour of the isomorphism relation on its countable models under Borel reducibility. This idea was pursued by Laskowski [29], Marker [31] and in depth by Koerwien [28]. But the conclusion was rather negative: theories can be complicated model-theoretically and simple descriptive set-theoretically (an example is dense linear orderings), or vice-versa (an example is described in [28]).

A solution to this difficulty emerged through the study of isomorphism on a theory's *uncountable* models. The work of [10] (see Chapter V, Theorem 64) shows, for example, that a theory is classifiable and shallow in Shelah's model-theoretic sense exactly if the isomorphism relation on its models of size $\kappa$ (for an appropriate choice of regular uncountable cardinal $\kappa$) is "Borel" in a generalised sense.

Naturally, a prerequisite for this study is the development of a suitable descriptive set theory of the uncountable, which has turned out to be a fascinating area of independent interest. Armed with such a theory it becomes possible to bring in the methods of model-theoretic stability theory to uncover deep connections between the model theory and descriptive set theory of first-order theories.

I begin with the uncountable descriptive set theory. It is favourable to choose $\kappa$ to be uncountable and such that $\kappa^{<\kappa} = \kappa$. The *Generalised Baire Space* $\kappa^\kappa$ is the space of all functions $f : \kappa \to \kappa$ topologised with basic open sets of the form $N_s = \{f \mid s \subseteq f\}$, $s$ an element of $\kappa^{<\kappa}$. In this context the *Borel* sets are obtained by closing the open sets under the operations of complementation and unions of size at most $\kappa$. The $\Sigma_1^1$ sets are the projections of Borel sets, the $\Pi_1^1$ sets are the complements of the $\Sigma_1^1$ sets and the $\Delta_1^1$ sets are those which are both $\Sigma_1^1$ and $\Pi_1^1$. Borel sets are $\Delta_1^1$ but the converse is false. As usual, a set is *nowhere dense* if its closure contains no nonempty open set; a set is *meager* if it is the union of $\kappa$-many nowhere dense sets. The Baire Category Theorem holds in the sense that the intersection of $\kappa$-many open dense sets is dense. A set has the *Baire Property (BP)* if its symmetric difference with some open set is meager. Borel sets have the BP. A *perfect set* is the range of a continuous injection from $2^\kappa$ (the Generalised Cantor Space) into $\kappa^\kappa$. A set has the *Perfect Set Property (PSP)* iff it either has size at most $\kappa$ or contains a perfect subset.

THEOREM 11. *(see [10]) (a) It is consistent that all $\Delta_1^1$ sets have the BP.*
*(b) For any stationary subset $S$ of $\kappa$, the filter $CUB(S)$, the closed unbounded filter restricted to $S$, is a $\Sigma_1^1$ set without the BP.*
*(c) In $L$, $CUB(S)$ for stationary $S$ is not $\Delta_1^1$, but there are nevertheless $\Delta_1^1$ sets without the BP and without the PSP.*
*(d) It is consistent relative to an inaccessible cardinal that all $\Sigma_1^1$ sets have the PSP (and the use of an inaccessible is necessary).*

*Remark.* Part (a) was proved independently by Lücke-Schlicht, in the case $S = \kappa$ part (b) is due to Halko-Shelah and part (d) was proved independently by Schlicht.

I turn now to Borel reducibility. Suppose that $X_0, X_1$ are Borel subsets of $\kappa^\kappa$. Then $f : X_0 \to X_1$ is a *Borel function* iff $f^{-1}[Y]$ is Borel whenever $Y$ is Borel. This implies that the graph of $f$ is Borel, as $(x,y)$ belongs to the graph of $f$ iff for all $s \in \kappa^{<\kappa}$, either $y$ does not belong to $N_s$ or $x$ belongs to $f^{-1}[N_s]$.

If $E_0, E_1$ are equivalence relations on Borel sets $X_0, X_1$ respectively, then

we say that $E_0$ is *Borel reducible* to $E_1$, written $E_0 \leq_B E_1$, iff for some Borel $f : X_0 \to X_1$:
$$x_0 E_0 y_0 \text{ iff } f(x_0) E_1 f(x_1).$$

Now recall the following picture from the classical case:
$$1 <_B 2 <_B \cdots <_B \omega <_B \text{id} <_B E_0$$

forms an initial segment of the Borel equivalence relations under $\leq_B$ where $n$ denotes an equivalence relation with $n$ classes for $n \leq \omega$, id denotes equality on $\omega^\omega$ and $E_0$ denotes equality modulo finite on $\omega^\omega$.

At $\kappa$ we easily get the initial segment
$$1 <_B 2 <_B \cdots <_B \omega <_B \omega_1 <_B \cdots <_B \kappa$$

where for each nonzero cardinal $\lambda \leq \kappa$ we identify $\lambda$ with the $\equiv_B$ class of Borel equivalence relations with exactly $\lambda$-many classes. What happens above these equivalence relations? We might hope for:

*Silver Dichotomy* The equivalence relation id (equality on $\kappa^\kappa$) is the strong successor of $\kappa$ under $\leq_B$, i.e., if a Borel equivalence relation $E$ has more than $\kappa$ classes then id is Borel reducible to $E$.

THEOREM 12. *(a) The Silver Dichotomy implies the PSP for Borel sets. Therefore it fails in L and its consistency requires at least an inaccessible cardinal.*
*(b) The Silver Dichotomy is false with Borel replaced by $\Delta_1^1$.*

Is the Silver Dichotomy consistent? This question remains open.

We can also consider what happens above id. In the case $\kappa = \omega$ we have:
*Classical Glimm-Effros Dichotomy* $E_0 =$ (equality mod finite) is the strong successor of id, i.e., if a Borel equivalence relation $E$ is not Borel reducible to id (i.e., $E$ is not *smooth*) then $E_0$ Borel-reduces to $E$.

At $\kappa$, what shall we take $E_0$ to be? For infinite regular $\lambda \leq \kappa$, define $E_0^{<\lambda} =$ equality for subsets of $\kappa$ modulo sets of size $< \lambda$.

PROPOSITION 13. *For $\lambda < \kappa$, $E_0^{<\lambda}$ is Borel bireducible with id.*

So we can forget about $E_0^{<\lambda}$ for $\lambda < \kappa$ and set $E_0 = E_0^{<\kappa}$, equality modulo bounded sets.

As in the classical case we have:

PROPOSITION 14. *$E_0 = E_0^{<\kappa}$ is not Borel reducible to id.*

There are other versions of $E_0$: For regular $\lambda < \kappa$ define $E_\lambda^\kappa =$ equality modulo the ideal of $\lambda$-nonstationary sets. These equivalence relations are key for connecting model-theoretic stability with uncountable descriptive set theory.

How do the relations $E_\lambda^\kappa$ compare to each other under Borel reducibility for different $\lambda$? For simplicity, consider the special case $\kappa = \omega_2$.

**THEOREM 15.** *([10]) (a) It is consistent that $E_\omega^{\omega_2}$ and $E_{\omega_1}^{\omega_2}$ are incomparable under Borel reducibility. (b) Relative to a weak compact it is consistent that $E_\omega^{\omega_2}$ is Borel reducible to $E_{\omega_1}^{\omega_2}$.*

It is not known if it is consistent for $E_{\omega_1}^{\omega_2}$ to be Borel reducible to $E_\omega^{\omega_2}$. What is the relationship between $E_0$ and $E_\lambda^\kappa$?

**THEOREM 16.** *(a) The relations $E_\lambda^\kappa$ do not Borel reduce to $E_0$, as $E_0$ is Borel and the $E_\lambda^\kappa$ are not.*
*(b) If $\kappa = \mu^+$ for some cardinal $\mu$, then $E_0$ reduces to $E_\lambda^\kappa$, unless $\lambda$ is the cofinality of $\mu$.*
*(c) In $L$, the condition in (b) that $\lambda$ not be the cofinality of $\mu$ can be dropped.*

The structure of the $\Delta_1^1$ equivalence relations under Borel reducibility is (consistently) very rich:

**THEOREM 17.** *Consistently, there is an injective, order-preserving embedding from $(\mathcal{P}(\kappa), \subseteq)$ into the partial order of $\Delta_1^1$ equivalence relations under Borel reducibility.*

The above summarises the current state of knowledge regarding uncountable descriptive set theory. As has been mentioned, there remain many open questions, some of which we list at the end of this section.

Now we return to the connection between uncountable descriptive set theory and model theory. Let $T$ be a countable, complete and first-order theory. Then $T$ is *classifiable* iff there is a "structure theory" for its models. (Example: Algebraically closed fields (transcendence degree).) $T$ is *unclassifiable* otherwise. (Example: Dense linear orderings.)
*Shelah's Characterisation (Main Gap):* $T$ is classifiable iff $T$ is superstable without the OTOP and without the DOP.

A classifiable $T$ is *deep* iff it has the maximum number of models in all uncountable powers. (Example: Acyclic undirected graphs, every node has infinitely many neighbours.) $T$ is *shallow* otherwise. (Remark: Actually, Shelah defined "deep" differently, in terms of rank. The fact that his definition is equivalent to the previous is one of the most profound results of his classification theory.)

Now for simplicity assume $\kappa = \lambda^+$ where $\lambda$ is uncountable and regular and the GCH holds at $\lambda$. $\mathrm{Isom}_T^\kappa$ is the isomorphism relation on the models of $T$ of size $\kappa$.

**THEOREM 18.** *([10])*
*(a) $T$ is classifiable and shallow iff $\mathrm{Isom}_T^\kappa$ is Borel.*
*(b) $T$ is classifiable iff for all regular $\mu < \kappa$, $E_{S_\mu^\kappa}$ is not Borel reducible to*

$\mathrm{Isom}_T^\kappa$.

(c) In L, T is classifiable iff $\mathrm{Isom}_T^\kappa$ is $\Delta_1^1$.

The proof uses Ehrenfeucht-Fraissé games. The Game $\mathrm{EF}_t^\kappa(\mathcal{A}, \mathcal{B})$ is defined as follows, where $\mathcal{A}$, $\mathcal{B}$ are structures of size $\kappa$ and $t$ is a tree. Player $I$ chooses size $< \kappa$ subsets of $A \cup B$ and nodes along an initial segment of a branch through $t$; player $II$ builds a partial isomorphism between $\mathcal{A}$ and $\mathcal{B}$ which includes the sets that player $I$ has chosen. Player $II$ wins iff he survives until a cofinal branch is reached.

The tree $t$ *captures* $\mathrm{Isom}_T^\kappa$ iff for all size $\kappa$ models $\mathcal{A}$, $\mathcal{B}$ of $T$, $\mathcal{A} \simeq \mathcal{B}$ iff Player $II$ has a winning strategy in $\mathrm{EF}_t^\kappa(\mathcal{A}, \mathcal{B})$.

Now there are 4 cases:

*Case 1: T is classifiable and shallow.*

Then Shelah's work [36] shows that some well-founded tree captures $\mathrm{Isom}_T^\kappa$. We use this to show that $\mathrm{Isom}_T^\kappa$ is Borel.

*Case 2: T it classifiable and deep.*

Then Shelah's work shows that no fixed well-founded tree captures $\mathrm{Isom}_T^\kappa$. We use this to show that $\mathrm{Isom}_T^\kappa$ is not Borel.

Shelah's work also shows that $L_{\infty\kappa}$ equivalent models of $T$ of size $\kappa$ are isomorphic. This means that the tree $t = \omega$ (with a single infinite branch) captures $\mathrm{Isom}_T^\kappa$. As the games $\mathrm{EF}_\omega^\kappa(\mathcal{A}, \mathcal{B})$ are determined, this shows that $\mathrm{Isom}_T^\kappa$ is $\Delta_1^1$.

We must also show: $E_{S_\mu^\kappa}$ (equality modulo the $\mu$-nonstationary ideal) is not Borel reducible to $\mathrm{Isom}_T^\kappa$ for any regular $\mu < \kappa$. This is because (in this case) $\mathrm{Isom}_T^\kappa$ is absolutely $\Delta_1^1$, whereas $\mu$-stationarity is not.

Now we look at the unclassifiable cases. Recall: Classifiable means superstable without DOP and without OTOP.

*Case 3: T is unstable, superstable with DOP or superstable with OTOP.*

Work of Hyttinen-Shelah [20] and Hyttinen-Tuuri [21] shows that in this case no tree of size $\kappa$ without branches of length $\kappa$ captures $\mathrm{Isom}_T^\kappa$. This can be used to show $\mathrm{Isom}_T^\kappa$ is not $\Delta_1^1$.

But $E_{S_\lambda^\kappa} \leq_B \mathrm{Isom}_T^\kappa$ is harder. Following Shelah, there is a Borel map $S \mapsto \mathcal{A}(S)$ from subsets of $\kappa$ to Ehrenfeucht-Mostowski models of $T$ built on linear orders so that $\mathcal{A}(S_0) \simeq \mathcal{A}(S_1)$ iff $S_0 = S_1$ modulo the $\lambda$-nonstationary ideal.

*Case 4: T is stable but not superstable.*

This is the hardest case and requires some new model theory. In our joint paper [10], Hyttinen replaces Ehrenfeucht-Mostowski models built on linear orders with primary models built on trees of height $\omega + 1$ to show $E_{S_\omega^\kappa} \leq_B \mathrm{Isom}_T^\kappa$. (We don't know if $E_{S_\lambda^\kappa} \leq_B \mathrm{Isom}_T^\kappa$ or if $\mathrm{Isom}_T^\kappa$ could be $\Delta_1^1$ in this case.)

Now we have all we need to prove the Theorem mentioned earlier:

(a) $T$ is classifiable and shallow iff $\text{Isom}_T^\kappa$ is Borel.

We mentioned that if $T$ is classifiable and shallow then $\text{Isom}_T^\kappa$ is Borel and if it is classifiable and deep it is not. If $T$ is not classifiable, then some $E_{S_\mu^\kappa}$ Borel reduces to $\text{Isom}_T^\kappa$, so the latter cannot be Borel.

(b) $T$ is classifiable iff for all regular $\mu < \kappa$, $E_{S_\mu^\kappa}$ is not Borel reducible to $\text{Isom}_T^\kappa$.

We mentioned that if $T$ is not classifiable then $E_{S_\mu^\kappa}$ is Borel reducible to $\text{Isom}_T^\kappa$ where $\mu$ is either $\lambda$ or $\omega$. We also mentioned that if $T$ is classifiable and deep then no $E_{S_\mu^\kappa}$ is Borel reducible to $\text{Isom}_T^\kappa$, by an absoluteness argument. When $T$ is classifiable and shallow there is no such reduction as $\text{Isom}_T^\kappa$ is Borel.

(c) In $L$, $T$ is classifiable iff $\text{Isom}_T^\kappa$ is $\Delta_1^1$.

We mentioned that if $T$ is classifiable then $\text{Isom}_T^\kappa$ is $\Delta_1^1$, in ZFC. If $T$ is not classifiable, then $E_{S_\mu^\kappa}$ Borel reduces to $\text{Isom}_T^\kappa$ for some $\mu$, and in $L$, $E_{S_\mu^\kappa}$ is not $\Delta_1^1$.

This summarises the work in [10]. Some surprisingly basic and very interesting open questions remain in this new area. Below are some of them. Assume $\kappa^{<\kappa} = \kappa$, as before.

1. Under what conditions on an uncountable $\kappa$ does Vaught's Conjecture hold in the following form: If an isomorphism relation on the models of size $\kappa$ has more than $\kappa$ classes, then id is Borel reducible to it?

2. Is the Silver Dichotomy for uncountable $\kappa$ consistent?

3. Is it consistent for there to be Borel equivalence relations which are incomparable under Borel reducibility for an uncountable $\kappa$?

4. Is it consistent that $S_{\omega_1}^{\omega_2}$ Borel reduces to $S_\omega^{\omega_2}$?

5. We proved that the isomorphism relation of a theory $T$ is Borel if and only if $T$ is classifiable and shallow. Is there a connection between the depth of a shallow theory and the Borel degree of its isomorphism relation? Is one monotone in the other?

6. Can it be proved in ZFC that if $T$ is stable unsuperstable then isomorphism for the size $\kappa$ models of $T$ ($\kappa$ uncountable) is not $\Delta_1^1$?

7. If $\kappa = \lambda^+$, $\lambda$ regular and uncountable, then does equality modulo the $\lambda$-nonstationary ideal Borel reduce to isomorphism for the size $\kappa$ models of $T$ for all stable unsuperstable $T$?

8. Let DLO be the theory of dense linear orderings without end points and RG the theory of random graphs. Does the isomorphism relation of RG Borel reduce to that of DLO for an uncountable $\kappa$?

## 4 Complexity Theory

We consider NP equivalence relations on finite strings. One motivation for this topic is the following: Borel reducibility allows us to compare isomorphism relations on Borel classes of countable structures. Is there an analogous reducibility for "nice" classes of *finite* structures?

The resulting theory of "strong isomorphism reductions" is introduced in [9] and studied systematically in [2]. We consider polynomial-time definable classes $C$ of structures for a finite vocabulary $\tau$, where the structures in $C$ have universe $\{1,\ldots,n\}$ for some finite $n > 0$ and where $C$ is *invariant*, i.e., closed under isomorphism. To avoid trivialities we also assume that $C$ contains arbitrarily large structures. Some examples of such classes are:
1. The classes SET, BOOLE, FIELD, GROUP, ABELIAN and CYCLIC of sets (structures of empty vocabulary), Boolean algebras, fields, groups, abelian groups, and cyclic groups, respectively.
2. The class GRAPH of (undirected and simple) graphs.
3. The class ORD of linear orderings.
4. The classes LOP of linear orderings with a distinguished point and LOU of linear orderings with a unary relation.

Let $C$ and $D$ be classes. We say that $C$ is *strongly isomorphism reducible to* $D$ and write $C \leq_{\text{iso}} D$, if there is a function $f: C \to D$ computable in polynomial time such that for all $\mathcal{A}, \mathcal{B} \in C$, $\mathcal{A} \simeq \mathcal{B}$ iff $f(\mathcal{A}) \simeq f(\mathcal{B})$. We then say that $f$ is a *strong isomorphism reduction* from $C$ to $D$ and write $f: C \leq_{\text{iso}} D$. If $C \leq_{\text{iso}} D$ and $D \leq_{\text{iso}} C$, denoted by $C \equiv_{\text{iso}} D$, then $C$ and $D$ have the same *strong isomorphism degree*.

Examples:
(a) The map sending a field to its multiplicative group shows that FIELD $\leq_{\text{iso}}$ CYCLIC.
(b) CYCLIC $\leq_{\text{iso}}$ ABELIAN $\leq_{\text{iso}}$ GROUP; more generally, if $C \subseteq D$, then $C \leq_{\text{iso}} D$ via the identity.
(c) SET $\equiv_{\text{iso}}$ FIELD $\equiv_{\text{iso}}$ ABELIAN $\equiv_{\text{iso}}$ CYCLIC $\equiv_{\text{iso}}$ ORD $\equiv_{\text{iso}}$ LOP. (For the proof see [2].)

PROPOSITION 19. *$C \leq_{\text{iso}}$ GRAPH for all classes $C$.*

The structure of $\leq_{\text{iso}}$ between LOU and GRAPH is linked with central open problems of descriptive complexity. Before turning to that I'll first consider the structure below LOU. That structure, even below LOP, is quite rich.

THEOREM 20. *The partial ordering of the countable atomless Boolean algebra is embeddable into the partial ordering induced by $\leq_{\text{iso}}$ on the degrees of strong isomorphism reducibility below LOP. More precisely, let $\mathcal{B}$ be the countable atomless Boolean algebra. Then there is a one-to-one function $b \mapsto C_b$ defined on $\mathcal{B}$ such that for all $b, b' \in \mathcal{B}$:*
*(i) $C_b$ is a subclass of LOP*
*(ii) $b \leq b'$ iff $C_b \leq_{\text{iso}} C_{b'}$.*

This result is obtained by comparing the number of isomorphism types of structures with universe of bounded cardinality in different classes. For a class $C$ we let $C(n)$ be the subclass consisting of all structures in $C$ with universe of cardinality $\leq n$ and we let $\#C(n)$ be the number of isomorphism types of structures in $C(n)$. Examples:
$\#\text{BOOLE}(n) = [\log n]$, $\#\text{CYCLIC}(n) = n$, $\#\text{SET}(n) = \#\text{ORD}(n) = n+1$.
$\#\text{LOP}(n) = \sum_{i=1}^{n} i = (n+1) \cdot n/2$ and $\#\text{LOU}(n) = \sum_{i=0}^{n} 2^i = 2^{n+1} - 1$.
$\#\text{GROUP}(n)$ is superpolynomial but subexponential (more precisely, it is bounded by $n^{O(\log^2 n)}$). See [1].

A class $C$ is *potentially reducible* to a class $D$, written $C \leq_{\text{pot}} D$, iff there is some polynomial $p$ such that $\#C(n) \leq \#D(p(n))$ for all $n \in \mathbb{N}$. Of course, by $C \equiv_{\text{pot}} D$ we mean $C \leq_{\text{pot}} D$ and $D \leq_{\text{pot}} C$.

LEMMA 21. *If $C \leq_{\text{iso}} D$, then $C \leq_{\text{pot}} D$.*

*Proof.* Let $f : C \leq_{\text{iso}} D$. As $f$ is computable in polynomial time, there is a polynomial $p$ such that for all $\mathcal{A} \in C$ we have $|f(\mathcal{A})| \leq p(|\mathcal{A}|)$, where $f(\mathcal{A})$ denotes the universe of $f(\mathcal{A})$. As $f$ strongly preserves isomorphisms, it therefore induces a one-to-one map from $\{\mathcal{A} \in C \mid |\mathcal{A}| \leq n\}/_\simeq$ to $\{\mathcal{B} \in D \mid |\mathcal{B}| \leq p(n)\}/_\simeq$. □

We state some consequences of this simple observation:

PROPOSITION 22. 1. *CYCLIC $\not\leq_{\text{iso}}$ BOOLE and LOU $\not\leq_{\text{iso}}$ LOP.*
2. *$C \leq_{\text{pot}} \text{LOU}$ for all classes $C$ and $\text{LOU} \equiv_{\text{pot}} \text{GRAPH}$.*
3. *The strong isomorphism degree of GROUP is strictly between that of LOP and GRAPH.*
4. *The potential reducibility degree of GROUP is strictly between that of LOP and LOU.*

The following concepts are used in the proof of Theorem 20. We call a function $f : \mathbb{N} \to \mathbb{N}$ *value-polynomial* iff it is increasing and $f(n)$ can be computed in time $f(n)^{O(1)}$. Let VP be the class of all value-polynomial functions. For $f \in \text{VP}$ the set $C_f = \{\mathcal{A} \in \text{LOP} \mid |\mathcal{A}| \in \text{im}(f)\}$ is in polynomial time and is closed under isomorphism. As there are exactly $f(k)$ pairwise non-isomorphic structures of cardinality $f(k)$ in LOP, we get

$$\#C_f(n) = \sum_{k \in \mathbb{N} \text{ with } f(k) \leq n} f(k).$$

The following proposition contains the essential idea underlying the proof of Theorem 20. Loosely speaking, it says that if the gaps between consecutive values of $f \in \text{VP}$ "kill" every polynomial, then there are classes $C$ and $D$ with $C \not\leq_{\text{pot}} D$.

PROPOSITION 23. *Let $f \in \text{VP}$ and assume that for every polynomial*

$p \in \mathbb{N}[X]$ there is an $n \in \mathbb{N}$ such that

$$\sum_{k \in \mathbb{N} \text{ with } f(2k) \leq n} f(2k) > \sum_{k \in \mathbb{N} \text{ with } f(2k+1) \leq p(n)} f(2k+1).$$

Then $\mathcal{C}_{g_0}$ is not potentially reducible to $\mathcal{C}_{g_1}$, where $g_0, g_1 : \mathbb{N} \to \mathbb{N}$ are defined by $g_0(n) := f(2n)$ and $g_1(n) := f(2n+1)$.

*Proof.* For contradiction assume that there is some polynomial $p$ such that $\#\mathcal{C}_{g_0}(n) \leq \#\mathcal{C}_{g_1}(p(n))$ for all $n \in \mathbb{N}$. Choose $n$ to satisfy the hypothesis. Then

$$\#\mathcal{C}_{g_0}(n) = \sum_{f(2k) \leq n} f(2k) > \sum_{f(2k+1) \leq p(n)} f(2k+1) = \#\mathcal{C}_{g_1}(p(n)),$$

a contradiction. □

The other needed ingredient for the proof of Theorem 20 is:

LEMMA 24. *The images of the functions in VP together with the finite subsets of $\mathbb{N}$ are the elements of a countable Boolean algebra $\mathcal{V}$ (under the usual set-theoretic operations). The factor algebra $\mathcal{V}/\equiv_{\text{pot}}$, where for $b, b' \in \mathcal{V}$*

$$b \equiv b' \iff (b \setminus b') \cup (b' \setminus b) \text{ is finite,}$$

*is a countable atomless Boolean algebra.*

This lemma shows that the set of images of functions in VP has a rich structure. To complete the proof of Theorem 20, the functions in VP are composed with a "stretching" function $h$, which guarantees that the gaps between consecutive values "kill" every polynomial. Then we can apply the idea of the proof of Proposition 23 to show that the set of the $\leq_{\text{pot}}$-degrees has a rich structure too. For the details see [2].

So far, in all concrete examples of classes $C$ and $D$ for which we know the status of $C \leq_{\text{iso}} D$ and of $C \leq_{\text{pot}} D$, we have $C \leq_{\text{iso}} D$ iff $C \leq_{\text{pot}} D$. So the question arises whether the relations of strong isomorphism reducibility and potential reducibility coincide. We believe that they are distinct but have only the following partial result:

THEOREM 25. *If $UEEXP \cap coUEEXP \neq EEXP$, then the relations of strong isomorphism reducibility and that of potential reducibility are distinct.*

Recall that $EEXP = DTIME\left(2^{2^{n^{O(1)}}}\right)$ and $NEEXP := NTIME\left(2^{2^{n^{O(1)}}}\right)$.

The complexity class $UEEXP$ consists of those $Q \in NEEXP$ for which there is a non-deterministic Turing machine of type NEEXP that for every $x \in Q$ has exactly one accepting run. Finally, $coUEEXP := \{\sim Q \mid Q \in UEEXP\}$.

Here is the idea of the proof: Assume $Q \in \text{UEEXP} \cap \text{coUEEXP}$. We construct classes $C$ and $D$ which contain structures in the same cardinalities and which contain exactly two non-isomorphic structures in these cardinalities. Therefore they are potentially reducible to each other. While it is trivial to exhibit two non-isomorphic structures in $C$ of the same cardinality, from any two non-isomorphic structures in $D$ we obtain information on membership in $Q$ for all strings of a certain length. If $C \leq_{\text{iso}} D$ held, then we would get non-isomorphic structures in $D$ (in time allowed by EEXP) by applying the strong isomorphism reduction to two non-isomorphic structures in $C$ and therefore obtain $Q \in \text{EEXP}$.

In the other direction we have:

THEOREM 26. *If strong isomorphism reducibility and potential reducibility are distinct, then $P \neq \#P$.*

Recall that $P = \#P$ means that for every polynomial time non-deterministic Turing machine $\mathbb{M}$ the function $f_{\mathbb{M}}$ such that $f_{\mathbb{M}}(x)$ is the number of accepting runs of $\mathbb{M}$ on $x \in \Sigma^*$ is computable in polynomial time. The class $\#P$ consists of all the functions $f_{\mathbb{M}}$.

Until now we have focused exclusively on isomorphism relations on invariant polynomial time classes of finite structures. But this theory can be put into the broader context of $NP$ equivalence relations in general. If $E$ and $E'$ are $NP$ equivalence relations, then we say that $E$ is *strongly equivalence reducible to $E'$*, and write $E \leq_{\text{eq}} E'$, iff there is a function $f$ computable in polynomial time such that for all strings $x, y$: $xEy$ iff $f(x)E'f(y)$. We then say that $f$ is a *strong equivalence reduction* from $E$ to $E'$ and write $f : E \leq_{\text{eq}} E'$. The following natural question then arises: Is there a *maximal $NP$ equivalence relation* under the reducibility $\leq_{\text{eq}}$? The final section of [2] relates this question to enumerations of clocked Turing machines, to $p$-optimal proof systems as well as to other central questions in complexity theory.

Another natural question is whether, in analogy to the computability theory context, every $NP$ equivalence relation is reducible to an isomorphism relation on a polynomial time invariant class of finite structures, or equivalenty, whether graph isomorphism is $\leq_{\text{eq}}$ complete among $NP$ equivalence relations. For this we have the following partial result:

PROPOSITION 27. *([2]) Assume that the polynomial time hierarchy does not collapse. Then* not *every NP equivalence relation reduces to graph isomorphism.*

Indeed there are many worthy open questions in this area waiting to be explored.

*In conclusion*

After decades of work focusing on the "unary" case, definability theory has been dramatically deepened by the study of binary relations, most importantly equivalence relations. An important step in this process was taken in Harvey's fundamental paper with Lee Stanley [8]. The extent to which the different areas of logic have been enriched through the study of analogues of Harvey's idea is only now being understood, and I look forward to seeing much exciting work in this direction during the coming years.

# BIBLIOGRAPHY

[1] H.U. Besche, B. Eick and E.A. O'Brien, The groups of order at most 2000, Electronic Research announcements of the American Mathematical Society, 7:1-4,2001.

[2] S. Buss, Y. Chen, J. Flum, S. Friedman and M. Müller, Strong isomorphism reductions in complexity theory, Journal of Symbolic Logic, December 2011.

[3] E. Fokina, S. Friedman, *Equivalence relations on classes of computable structures*, Proceedings of "Computability in Europe 2009", Heidelberg, Germany, Lecture Notes in Computer Science 5635, 198–207, 2009.

[4] E. Fokina, S. Friedman, *On $\Sigma_1^1$ equivalence relations over the natural numbers*, to appear, Mathematical Logic Quarterly.

[5] E. Fokina, S. Friedman, V. Harizanov, J. Knight, C. McCoy and A. Montalban, Isomorphism relations on computable structures, Journal of Symbolic Logic, March 2012.

[6] E. Fokina, S. Friedman, A. Törnquist, *The effective theory of Borel equivalence relations*, Annals of Pure and Applied Logic 161, pp. 837–850, 2010.

[7] E. Fokina, J. Knight, C. Maher, A. Melnikov, S. Quinn, *Classes of Ulm type, and relations between the class of rank-homogeneous trees and other classes*, submitted.

[8] Friedman, H., and L. Stanley, *A Borel reducibility theory for classes of countable structures*, J. Symb. Logic, 54(1989), 894–914.

[9] S. Friedman, Descriptive set theory for finite structures, Lecture at the Kurt Gödel Research Center, 2009. Available at http://www.logic.univie.ac.at/ sdf/papers/wien-spb.pdf.

[10] S. Friedman, T. Hyttinen and V. Kulikov, Generalized Descriptive Set Theory and Classification Theory, submitted, see http://www.logic.univie.ac.at/ sdf/papers/joint.tapani.vadim.pdf.

[11] S. D. Friedman, L. Motto Ros, *Analytic equivalence relations and bi-embeddability*, Journal of Symbolic Logic, Vol.76, No.1, pp. 1581–1587, 2011.

[12] S. Gao, Invariant Descriptive Set Theory, Pure and Applied mathematics, CRC Press/Chapman & Hall, 2009.

[13] S. Gao and P.M. Gerdes, Computably enumerable equivalence relations, Studia Logica, vol. 67, pp. 27-59, Feb. 2001.

[14] L. Harrington, McLaughlin's Conjecture, Handwritten notes, 1976.

[15] L. Harrington, Arithmetically Incomparable Arithmetical Singletons, Handwritten notes, 1975.

[16] L. Harrington, A. Kechris and A. Louveau, Glimm-Efros dichotomy for Borel equivalence relations, Journal of the American Mathematical Society, volume 3, number 4, pages 903–928, 1990.

[17] Hjorth, G., *The isomorphism relation on countable torsion-free Abelian groups*, Fund. Math., 175(2002), 241–257.

[18] G. Hjorth, *Classification and orbit equivalence relations*, Mathematical surveys and monographs 75, American Mathematical Society, 2000.

[19] G. Hjorth and A. Kechris, Recent developments in the theory of Borel reducibility, Fundamenta Mathematicae, volume 170, number 1–2, pages 21–52, 2001.

[20] T. Hyttinen and S. Shelah, Constructing strongly equivalent nonisomorphic models for unsuperstable theories, Part C, Journal of Symbolic Logic, Vol.64, No.2, 1999.

[21] T. Hyttinen and H. Tuuri, Constructing strongly equivalent nonisomorphic models, Annals of Pure and Applied Logic, Vol.52, Issue 3, 1991.
[22] S. Jackson, A. Kechris and A. Louveau, Countable Borel Equivalence Relations, Journal of Mathematical Logic, pages 1–80, volume 2, number 1, 2002.
[23] V. Kanovei, Borel equivalence relations. Structure and classification, Unviersity Lecture Series, 44, American Mathematical Society, 2008.
[24] A. Kechris, Measure and category in effective descriptive set theory, Annals of Pure and Applied Logic, 5:337–384, 1973.
[25] A. Kechris, *Classical Descriptive Set Theory*, Graduate Texts in Mathematics, Springer-Verlag, 1995.
[26] A. Kechris, *New directions in descriptive set theory*, Bull. Symbolic Logic, 5 (1999), 2, 161–174.
[27] A. Kechris, A. Louveau *The classification of hypersmooth Borel equivalence relations*, J. of the American Math. Society 10 (1997), 1, 215–242.
[28] M. Koerwien, A complicated $\omega$-stable depth 2 theory, to appear, Journal of Symbolic Logic.
[29] C. Laskowski, An old friend revisited: Countable models of omega-stable theories, Proceedings of the Vaught's Conjecture conference, Notre Dame Journal of Formal Logic 48 (2007) 133–141.
[30] A. Louveau, C. Rosendal, *Complete analytic equivalence relations*, Trans. Amer. Math. Soc. **357** (2005), 12, 4839–4866.
[31] D. Marker, The Borel Complexity of Isomorphism for Theories with Many Types, preprint.
[32] A. Montalbán, *On the equimorphism types of linear orderings*, Bulletin of Symbolic Logic, 13 (2007), 71-99.
[33] P. G. Odifreddi, *Classical Recursion Theory*, volume II, North-Holland Publishing Co., 1999.
[34] H. Rogers, Theory of recursive functions and effective computability, McGraw-Hill, 1967.
[35] G. Sacks, *Higher Recursion Theory*, Springer-Verlag, 1989.
[36] S. Shelah, *Classification theory, revised edition*, North Holland, 1990.
[37] J. H. Silver, Counting the number of equivalence classes of Borel and coanalytic equivalence relations, Annals of Mathematical Logic, volume 18, pages 1–18, 1980.

# An Application of Proof-Theory in Answer Set Programming
V.W. Marek and J.B. Remmel

*Dedicated to Harvey Friedman whose fundamental work in foundations has inspired both of us.*

ABSTRACT. We show that the stable models of a logic program $P$ are the satisfying valuations of a suitably chosen propositional theory, called the set of *reduced defining equations*, $r\Phi_P$, of $P$. In general, this propositional theory may contain infinite sentences. We show that $r\Phi_P$ contains only finite sentences if and only if the Gelfond-Lifschitz operator $GL_P$ associated with the program $P$ is an upper-half continuous antimonotone operator. We also discuss possible extensions of our results to programs with cardinality constraints.

## 1 Introduction

The use of proof theory in logic based formalisms for constraint solving is pervasive. For example, in Satisfiability (SAT), proof theoretic methods are used to find lower bounds on complexity of various SAT algorithms. However, proof-theoretic methods have not played as prominent role in Answer Set Programming (ASP) formalisms. This is not to say that there were no attempts to apply proof-theoretic methods in ASP. To give a few examples, Marek and Truszczynski in [29] used proof-theoretic methods to characterize Reiter's extensions in Default Logic and, hence, by implication, the stable semantics of logic programs. Bonatti [1] and separately Milnikel [30] devised non-monotonic proof systems to study skeptical consequences of programs and default theories. Lifschitz [15] used proof-theoretic methods to approximate the well-founded semantics of a logic program. Bondarenko et.al. [2] studied an approach to stable semantics using methods with a clear proof-theoretic flavor. Marek, Nerode, and Remmel in a series of papers, [20, 21, 22, 23, 24], developed proof theoretic methods to study what they termed *non-monotonic rule systems* which have as special cases almost all ASP formalisms that have been seriously studied in the literature. Recently the area of proof systems for ASP, and more generally, nonmonotonic logics, has received a lot of attention, see [10, 13]. It is clear that the ASP community feels that an additional research of this area is necessary. Nevertheless, there is no clear classification of proof systems for nonmonotonic reasoning analogous to that present in classical logic, and SAT in particular.

In this paper, we define a notion of $P$-proof schemes, which is a kind of a proof system that was previously used by Marek, Nerode, and Remmel to study complexity issues for the stable semantics of logic programs [23]. This proof system abstracts the $M$-proofs of [29] and produces Hilbert-style proofs. The nonmonotonic character of our $P$-proofs is provided by the presence of guards, called the *support* of the proof scheme, which insure context-dependence. A different but equivalent presentation of proof schemes using a variation of the unit resolution rule can be found in [28].

In particular, we use $P$-proof schemes to show that the stable models of a logic program $P$ are the set of satisfying valuations of a suitably chosen propositional theory called the set of *reduced defining equations*, $r\Phi_P$, of $P$. In general, these defining equations may be infinite. The main focus of this paper is to study the class of programs for which all these equations are finite which we call FSP-programs. FSP-programs turn out to be characterized by a form of continuity of the Gelfond-Lifschitz operator. That is, let $At$ denote the set of underlying atoms of the program $P$ and $\mathcal{P}(At)$ denote the set of all subsets of $At$. We say that any function $O : \mathcal{P}(At) \to \mathcal{P}(At)$ is an operator on the set $At$ of propositional atoms. An operator $O$ is *monotone* if for all sets $X, Y \subseteq At$, $X \subseteq Y$ implies $O(X) \subseteq O(Y)$. Likewise an operator $O$ is *antimonotone* if for all sets $X, Y \subseteq At$, $X \subseteq Y$ implies $O(Y) \subseteq O(X)$. Let $N$ denote the set of natural numbers. For a sequence $\langle X_n \rangle_{n \in N}$ of sets of atoms, we say that $\langle X_n \rangle_{n \in N}$ is *monotonically increasing* if for all $i, j \in N$, $i \leq j$ implies $X_i \subseteq X_j$ and we say that $\langle X_n \rangle_{n \in N}$ is *monotonically decreasing* if for all $i, j \in N$, $i \leq j$ implies $X_j \subseteq X_i$. We shall consider four types of operators in this paper, namely, upper-half and lower-half continuous monotone operators and and upper-half and lower-half continuous antimonotone operators. Here we say

1. a monotone operator $O$ is *upper-half continuous* if for every monotonically increasing sequence $\langle X_n \rangle_{n \in N}$, $O(\bigcup_{n \in N} X_n) = \bigcup_{n \in N} O(X_n)$,

2. a monotone operator $O$ is *lower-half continuous* if for every monotonically decreasing sequence $\langle X_n \rangle_{n \in N}$, $O(\bigcap_{n \in N} X_n) = \bigcap_{n \in N} O(X_n)$,

3. an antimonotone operator $O$ is *upper-half* continuous if for every monotonically increasing sequence $\langle X_n \rangle_{n \in N}$, $O(\bigcup_{n \in N} X_n) = \bigcap_{n \in N} O(X_n)$, and

4. an antimonotone operator $O$ is *lower-half* continuous if for every monotonically decreasing sequence $\langle X_n \rangle_{n \in N}$, $O(\bigcap_{n \in N} X_n) = \bigcup_{n \in N} O(X_n)$.

It is a classic result of van Emden and Kowalski [7] that the one-step provability operator $T_P$ for a Horn program $P$ is upper-half continuous monotone operator. In the logic programming literature, upper-half continuous operators are called just continuous operators. It is also well known that $T_P$ is not lower-half continuous in general. For normal logic programs, the Gelfond-Lifschitz operator is a lower-half continuous antimonotone operator. The main results of this paper are to show that

1. if $At$ is countable, then every lower-half continuous antimonotone operator $f$ on $At$ is of the form $GL_P$ for some normal logic program $P$ and

2. all equations in the set of reduced defining equations $r\Phi_P$ of a normal logic program $P$ are finite if and only if $GL_P$ is an upper-half continuous operator.

The outline of this paper is as follows. In section 2, we shall establish preliminary notation. In section 3, we shall define $P$-proofs schemes and the set of reduced defining equations for $P$. We then use the set reduced defining equations for $P$ to characterize the stable models and give a proof theoretic characterization of the Gelfond-Lifschitz operator $GL_P$. In section 4, we shall prove our main results on FSP programs and the continuity properties of the Gelfond-Lifschitz operator of programs. In section 5, we shall discuss possible extensions of our results to cardinality constraint logic programs. Finally, in section 6, we give some conclusions.

## 2 Preliminaries

Let $At$ be a countably infinite set of atoms. We will study programs consisting of clauses built of the atoms from $At$. A *program clause* $C$ is a string of the form

$$p \leftarrow q_1, \ldots, q_m, \neg r_1, \ldots, \neg r_n \qquad (1)$$

The integers $m$ or $n$ or both can be 0. The atom $p$ will be called the head of $C$ and denoted $head(C)$. We let $posBody(C)$ denote the set $\{q_1, \ldots, q_m\}$ and $negBody(C)$ denote the set $\{r_1, \ldots, r_n\}$. For any set of atoms $X$, we let $\neg X$ denote the conjunction of negations of atoms from $X$. Thus, we can write clause (1) as

$$head(C) \leftarrow posBody(C), \neg negBody(C).$$

Let us stress that the set $negBody(C)$ is a set of atoms, not a set of negated atoms as is sometimes used in the literature. A normal propositional program is a set $P$ of such clauses. For any $M \subseteq At$, we say that $M$ is a model of $C$ of the form (1) if whenever $q_1, \ldots, q_m \in M$ and $\{r_1, \ldots, r_n\} \cap M = \emptyset$, then $p \in M$. We say that $M$ is a model of a program $P$ if $M$ is a model of each clause $C \in P$. Horn clauses are clauses with no negated literals, i.e. clauses of the form (1) where $n = 0$. We will denote by $Horn(P)$ the part of the program $P$ consisting of its Horn clauses. Horn programs are logic programs $P$ consisting entirely of Horn clauses. Thus for a Horn program $P$, $P = Horn(P)$.

For any set $I \subseteq At$, we let $T_P(I)$ equal the set of all $p \in At$ such that there is a clause $C = p \leftarrow q_1, \ldots, q_m$ in $P$ and $q_1, \ldots, q_m \in I$. Then $T_P$ has a least fixed point $F_P$ which is obtained by iterating $T_P$ starting at the empty set for $\omega$ steps, i.e., $F_P = \bigcup_{n \in N} T_P^n(\emptyset)$ where for any $I \subseteq At$, $T_P^0(I) = I$ and $T_P^{n+1}(I) = T_P(T_P^n(I))$. Then $F_P$ is the least model of $P$. The classic result due to van Emden and Kowalski, see [19], is the following.

PROPOSITION 1. *For every Horn program $P$, the operator $T_P$ is an upper-half continuous monotone operator.*

The semantics of interest for us is the *stable semantics* of normal programs, although we will discuss some extensions in Section 5. The stable models of a program $P$ are defined as fixed points of the operator $T_{P,M}$. This operator is defined on the set of all subsets of $At$, $\mathcal{P}(At)$. If $P$ is a program and $M \subseteq At$ is a subset of the Herbrand base, define operator $T_{P,M} : \mathcal{P}(At) \to \mathcal{P}(At)$ as follows:

$$T_{P,M}(I) = \{p: \text{there exist a clause } C = p \leftarrow q_1, \ldots, q_m, \neg r_1, \ldots, \neg r_n$$
$$\text{in } P \text{ such that } q_1 \in I, \ldots, q_m \in I, r_1 \notin M, \ldots, r_n \notin M\}$$

We observe the following well-known fact.

PROPOSITION 2. *For every program $P$ and every set $M$ of atoms the operator $T_{P,M}$ is an upper-half continuous monotone operator.*

Thus the operator $T_{P,M}$ like all monotonic continuous operators, possesses a least fixed point $F_{P,M}$.

Given program $P$ and $M \subseteq At$, we define the *Gelfond-Lifschitz reduct* of $P$, $P_M$, as follows. For every clause $C = p \leftarrow q_1, \ldots, q_m, \neg r_1, \ldots, \neg r_n$ of $P$,

execute the following operations.
(1) If some atom $r_i$, $1 \leq i \leq n$, belongs to $M$, then eliminate $C$ altogether.
(2) In the remaining clauses that have not been eliminated by operation (1), eliminate all the negated atoms.
The resulting program $P_M$ is a Horn propositional program. The program $P_M$ possesses a least Herbrand model. If the least model of $P_M$ coincides with $M$, then $M$ is called a *stable model* for $P$. This gives rise to an operator $GL_P$ which associates to each $M \subseteq At$, the least fixed point of $T_{P,M}$. We will discuss the operator $GL_P$ and its proof-theoretic connections in section 4.1.

## 3 Proof schemes and reduced defining equations

In this section we recall the notion of a *proof scheme* as defined in [20, 29] and introduce a related notion of *defining equations*.

Given a propositional logic program $P$, a proof scheme is defined by induction on its length. Specifically, a proof scheme with respect to $P$ or a $P$-proof scheme, for short, is a sequence $S = \langle \langle C_1, p_1 \rangle, \ldots, \langle C_n, p_n \rangle, U \rangle$ such that either

**I.** $n = 1$ in which case $\langle \langle C_1, p_1 \rangle, U \rangle$ is a $P$-proof scheme if $C_1 \in P$, $p_1 = head(C_1)$, $posBody(C_1) = \emptyset$, and $U = negBody(C_1)$ or

**II.** $n > 1$ in which case

$$\langle \langle C_1, p_1 \rangle, \ldots, \langle C_n, p_n \rangle, \langle C, p \rangle, U \cup negBody(C) \rangle$$

is a $P$-proof scheme if and only if $\langle \langle C_1, p_1 \rangle, \ldots, \langle C_n, p_n \rangle, U \rangle$ is a $P$-proof scheme, $C = p \leftarrow posBody(C), \neg negBody(C)$ is a clause in the program $P$, and $posBody(C) \subseteq \{p_1, \ldots, p_n\}$.

If $S = \langle \langle C_1, p_1 \rangle, \ldots, \langle C_n, p_n \rangle, U \rangle$ is a $P$-proof scheme, then we call (i) the integer $n$ – the *length* of $S$, (ii) the set $U$ – the *support* of $S$, and (iii) the atom $p_n$ – the *conclusion* of $S$. We shall denote $U$ by $supp(S)$.

EXAMPLE 3. Let $P$ be a program consisting of four clauses: $C_1 = p \leftarrow$, $C_2 = q \leftarrow p, \neg r$, $C_3 = r \leftarrow \neg q$, and $C_4 = s \leftarrow \neg t$. Then we have the following examples of $P$-proof schemes:

(a) $\langle \langle C_1, p \rangle, \emptyset \rangle$ is a $P$-proof scheme of length 1 with conclusion $p$ and empty support.

(b) $\langle \langle C_1, p \rangle, \langle C_2, q \rangle, \{r\} \rangle$ is a $P$-proof scheme of length 2 with conclusion $q$ and support $\{r\}$.

(c) $\langle\langle C_1, p\rangle, \langle C_3, r\rangle, \{q\}\rangle$ is a $P$-proof scheme of length 2 with conclusion $r$ and support $\{q\}$.

(d) $\langle\langle C_1, p\rangle, \langle C_2, q\rangle, \langle C_3, r\rangle, \{q, r\}\rangle$ is a $P$-proof scheme of length 3 with conclusion $r$ and support $\{q, r\}$.

Proof scheme in (c) is an example of a proof scheme with unnecessary items (the first term). Proof scheme (d) is an example of a proof scheme which is not internally consistent in that $r$ is in the support of its proof scheme and is also its conclusion. □

A $P$-proof scheme carries within itself its own applicability condition. In effect, a $P$-proof scheme is a *conditional* proof of its conclusion. It becomes applicable when all the constraints collected in the support are satisfied. Formally, for any set of atoms $M$, we say that a $P$-proof scheme $S$ is $M$-*applicable* if $M \cap supp(S) = \emptyset$. We also say that $M$ admits $S$ if $S$ is $M$-applicable.

The fundamental connection between proof schemes and stable models [20, 29] is given by the following proposition.

PROPOSITION 4. *For every normal propositional program $P$ and every set $M$ of atoms, $M$ is a stable model of $P$ if and only if the following conditions hold.*

(i) *For every $p \in M$, there is a $P$-proof scheme $S$ with conclusion $p$ such that $M$ admits $S$.*

(ii) *For every $p \notin M$, there is no $P$-proof scheme $S$ with conclusion $p$ such that $M$ admits $S$.*

Proposition 4 says that the presence and absence of the atom $p$ in a stable model depends *only* on the supports of proof schemes. This fact naturally leads to a characterization of stable models in terms of propositional satisfiability. Given $p \in At$, the *defining equation* for $p$ with respect to $P$ is the following propositional formula:

$$p \Leftrightarrow (\neg U_1 \vee \neg U_2 \vee \ldots) \qquad (2)$$

where $\langle U_1, U_2, \ldots \rangle$ is the list of all supports of $P$-proof schemes with the conclusion $p$. If $p$ is not the conclusion of any proof scheme, then we set the defining equation of $p$ to be $p \Leftrightarrow \bot$. In the case, where all the supports of proof schemes of $p$ are empty, we set the defining equation of $p$ to be $p \Leftrightarrow \top$. Up to a total ordering of the finite sets of atoms such a formula is unique. For example, suppose we fix a total order on $At$, $p_1 < p_2 < \cdots$. Then given two sets of atoms, $U = \{u_1 < \cdots < u_m\}$ and $V = \{v_1 < \cdots < v_n\}$, we say that $U \prec V$, if either (i) $u_m < v_n$, (ii) $u_m = v_n$ and $m < n$, or (iii) $u_m = v_n$, $n = m$, and $(u_1, \ldots, u_n)$ is lexicographically less than $(v_1, \ldots, v_n)$. We say that (2) is the *defining equation* for $p$ relative to $P$ if $U_1 \prec U_2 \prec \cdots$. We will denote the defining equation for $p$ with respect to $P$ by $Eq_p^P$.

For example, if $P$ is a Horn program, then for every atom $p$, either the support of all its proof schemes are empty or $p$ is not the conclusion of any proof scheme. The first of these alternatives occurs when $p$ belongs to the least model of $P$, $lm(P)$. The second alternative occurs when $p \notin lm(P)$. The defining equations are $p \Leftrightarrow \top$ (that is $p$) when $p \in lm(P)$ and $p \Leftrightarrow \bot$ (that is $\neg p$) when $p \notin lm(P)$. When $P$ is a stratified program the defining equations are more complex, but the resulting theory is logically equivalent to

$$\{p : p \in \mathit{Perf}_P\} \cup \{\neg p : p \notin \mathit{Perf}_P\}$$

where $\mathit{Perf}_P$ is the unique stable model of $P$.

Let $\Phi_P$ be the set $\{Eq_p^P : p \in At\}$. We then have the following consequence of Proposition 4.

PROPOSITION 5. *Let $P$ be a normal propositional program. Then stable models of $P$ are precisely the propositional models of the theory $\Phi_P$.*

When $P$ is *purely negative*, i.e. all clauses $C$ of $P$ have $\mathit{PosBody}(C) = \emptyset$, the stable and supported models of $P$ coincide [5] and the defining equations reduce to Clark's completion [3] of $P$.

Let us observe that, in general, the propositional formulas on the right-hand-side of the defining equations may be infinite.

EXAMPLE 6. Let $P$ be an infinite program consisting of clauses $p \leftarrow \neg p_i$, for all $i \in N$. In this case, the defining equation for $p$ in $P$ is infinite. That is, it is

$$p \Leftrightarrow (\neg p_1 \vee \neg p_2 \vee \neg p_3 \vee \ldots)$$

□

The following observation is quite useful. If $U_1, U_2$ are two finite sets of propositional atoms, then

$$U_1 \subseteq U_2 \text{ if and only if } \neg U_2 \models \neg U_1$$

Here $\models$ is the propositional consequence relation. The effect of this observation is that not all the supports of proof schemes are important, only the inclusion-minimal ones.

EXAMPLE 7. Let $P$ be an infinite program consisting of clauses $p \leftarrow \neg p_1, \ldots, \neg p_i$, for all $i \in N$. The defining equation for $p$ in $P$ is

$$p \Leftrightarrow [\neg p_1 \vee (\neg p_1 \wedge \neg p_2) \vee (\neg p_1 \wedge \neg p_2 \wedge \neg p_3) \vee \ldots]$$

which is infinite. But our observation above implies that this formula is *equivalent* to the formula

$$p \Leftrightarrow \neg p_1$$

□

Motivated by the Example 7, we define the *reduced defining equation* for $p$ relative to $P$ to be the formula

$$p \Leftrightarrow (\neg U_1 \vee \neg U_2 \vee \ldots) \qquad (3)$$

where $U_i$ range over *inclusion-minimal* supports of $P$-proof schemes for the atom $p$ and $U_1 \prec U_2 \prec \cdots$. Again, if $p$ is not the conclusion of any proof scheme, then we set the defining equation of $p$ to be $p \Leftrightarrow \bot$. In the case, where there is a proof scheme of $p$ with empty support, then we set the defining equation of $p$ to be $p \Leftrightarrow \top$. We denote this formula as $rEq_p^P$, and define $r\Phi_P$ to be the theory consisting of $rEq_p^P$ for all $p \in At$. We then have the following strengthening of Proposition 5.

PROPOSITION 8. *Let $P$ be a normal propositional program. Then stable models of $P$ are precisely the propositional models of the theory $r\Phi_P$.*

In our example 7, the theory $\Phi_P$ involved formulas with infinite disjunctions, but the theory $r\Phi_P$ contains only normal finite propositions.

Given a normal propositional program $P$, we say that $P$ is a *finite support program* (FSP-program) if all the reduced defining equations for atoms with respect to $P$ are finite propositional formulas. Equivalently, a program $P$ is an *FSP*-program if for every atom $p$ there is only finitely many inclusion-minimal supports of $P$-proof schemes for $p$.

## 4 Continuity properties of operators and proof schemes

In this section we investigate continuity properties of operators and we will see that one of those properties characterizes the class of FSP programs. In particular, we shall look at the continuity properties of the operators $T_P$ and $GL_P$ described in the previous section.

As we noted above, $T_P$ is always an upper-half continuous monotone operator. In general, the operator $T_P$ for Horn programs is *not* lower-half continuous. For example, let $P$ be the program consisting of the clauses $p \leftarrow p_i$ for $i \in N$. Then the operator $T_P$ is not lower-half continuous. That is, if $X_i = \{p_i, p_{i+1}, \ldots\}$, then clearly $p \in T_P(X_i)$ for all $i$. However, $\bigcap_i X_i = \emptyset$ and $p \notin T_P(\emptyset)$.

Lower-half continuous monotone operators have appeared in the Logic Programming literature [6]. Even more generally, for a monotone operator $O$, let us define its *dual* operator $O^d$ as follows:

$$O^d(X) = At \setminus O(At \setminus X).$$

Then an operator $O$ is upper-half continuous if and only if $O^d$ is lower-half continuous [12]. Therefore, for any Horn program $P$, the operator $T_P^d$ is lower-half continuous.

## 4.1 Gelfond-Lifschitz operator $GL_P$ and proof-schemes

Recall that the Gelfond-Lifschitz operator for a program $P$ which we denote $GL_P$, assigns to a set of atoms $M$ the least fixpoint of the operator $T_{P,M}$ or, equivalently, the least model $N_M$ of the program $P_M$ which is the Gelfond-Lifschitz reduct of $P$ via $M$ [11]. The following fact is crucial.

PROPOSITION 9 ([11]). *The operator $GL$ is antimonotone.*

Here is a useful proof-theoretic characterization of the operator $GL_P$.

PROPOSITION 10. *Let $P$ be a normal propositional program and $M$ be a set of atoms. Then*

$$GL_P(M) = \{p : \text{there exists a P-proof scheme } S \text{ such that } M \text{ admits } S,$$
$$\text{and } p \text{ is the conclusion of } S\}$$

Proof: Let us assume that $p \in GL_P(M)$, that is, assume $p \in N_M$. As $N_M$ is the least model of the Horn program $P_M$, $N_M = \bigcup_{n \in N} T_{P_M}^n(\emptyset)$. It is then easy to prove by induction on $n$, that if $p \in T_{P_M}^n(\emptyset)$, then there is a $P$-proof scheme $S_p$ such that $p$ is the conclusion of $S_p$ and $S_p$ is admitted by $M$. Conversely, we can show, by induction on the length of the $P$-proof schemes, that whenever such $P$-proof scheme $S$ is admitted by $M$, then $p$ belongs to $GL_P(M)$. □

## 4.2 Continuity properties of the operator $GL_P$

This section will be devoted to proving results on the continuity properties of the operator $GL_P$. First, we prove that for every program $P$, the operator $GL_P$ is lower-half continuous. We then show that if $f$ is a lower-half continuous antimonotone operator, then $f = GL_P$ for a suitably chosen program $P$. Finally, we show that the operator $GL_P$ is upper-half continuous if and only if $P$ is an *FSP*-program. That is, $GL_P$ is upper-half continuous if for all atoms $p$ the reduced defining equation for any $p$ (w.r.t. $P$) is finite.

PROPOSITION 11. *For every normal program $P$, the operator $GL_P$ is lower-half continuous.*

Proof: We need to prove that for every program $P$ and every monotonically decreasing sequence $\langle X_n \rangle_{n \in N}$,

$$GL_P(\bigcap_{n \in N} X_n) = \bigcup_{n \in N} GL_P(X_n).$$

Our goal is to prove two inclusions: $\subseteq$, and $\supseteq$.
We first show $\supseteq$. Since

$$\bigcap_{j \in N} X_j \subseteq X_n$$

for every $n \in N$, by antimonotonicity of $GL_P$ we have

$$GL_P(X_n) \subseteq GL_P(\bigcap_{j \in N} X_j).$$

As $n$ is arbitrary,

$$\bigcup_{n \in N} GL_P(X_n) \subseteq GL_P(\bigcap_{j \in N} X_j).$$

Thus the inclusion $\supseteq$ holds.
Conversely, let $p \in GL_P(\bigcap_{n \in N} X_n)$. Then, by Proposition 10, there must be a proof scheme $S$ with support $U$ and conclusion $p$ such that

$$U \cap \bigcap_{n \in N} X_n = \emptyset.$$

But the family $\langle X_n \rangle_{n \in n}$ is monotonically descending and the set $U$ is finite. Thus there is an integer $n_0$ so that

$$U \cap X_{n_0} = \emptyset.$$

This, however, implies that $p \in GL_P(X_{n_0})$, and thus

$$p \in \bigcup_{n \in N} GL_P(X_n).$$

As $p$ is arbitrary, the inclusion $\subseteq$ holds.  Thus $GL_P(\bigcap_{n \in N} X_n) = \bigcup_{n \in N} GL_P(X_n)$. $\square$

The lower-half continuity of antimonotone operators is closely related to programs, as shown by the following result.

PROPOSITION 12. *Let $At$ be a denumerable set of atoms. Let $f$ be an antimonotone and lower-half continuous operator on $\mathcal{P}(At)$. Then there exists a normal logic program $P$ such that $f = GL_P$.*

**Proof.**
We define the program $P = P_f$ as follows:
$$P = \{p \leftarrow \neg q_1, \ldots, \neg q_i : p \in f(At \setminus \{q_1, \ldots, q_i\})\}.$$
We claim that $f = GL_P$, that is, for all $X$, $f(X) = GL_P(X)$.
Let $X \subseteq At$ be given. We consider two cases.
**Case 1:** $X$ is cofinite, say $X = At \setminus \{q_1, \ldots, q_i\}$. We need to prove two inclusions, (a) $f(X) \subseteq GL_P(X)$ and (b) $GL_P(X) \subseteq f(X)$.
For (a), note that if $p \in f(X)$, then the clause $p \leftarrow \neg q_1, \ldots, \neg q_i$ belongs to $P$. Hence $p \leftarrow$ belongs to $P_X$ and $p \in GL_P(X)$.
For (b), note that if $p \in GL_P(X)$, then given the form of the clauses in $P$, there must be some clause $p \leftarrow \neg q_{i_1}, \ldots, \neg q_{i_j}$ in $P$ where $\{q_{i_1}, \ldots, q_{i_j}\} \subseteq \{q_1, \ldots, q_i\}$. But this means that $p \in f(At \setminus \{q_{i_1}, \ldots, q_{i_j}\})$. Since $f$ is antimonotone and $At \setminus \{q_1, \ldots, q_i\} \subseteq At \setminus \{q_{i_1}, \ldots, q_{i_j}\}$, we must have
$$f(At \setminus \{q_{i_1}, \ldots, q_{i_j}\}) \subseteq f(At \setminus \{q_1, \ldots, q_i\}) = f(X)$$
and, hence, $p \in f(X)$. Thus $GL_P(X) \subseteq f(X)$.

**Case 2:** $X$ is not cofinite. Let $\{q_0, q_1, \ldots\}$ be an enumeration of $At \setminus X$. Let $Y_i = At \setminus \{q_0, \ldots, q_i\}$. Then, clearly, $X \subseteq Y_i$ for all $i \in N$. Moreover the sequence $\langle Y_i \rangle_{i \in N}$ is monotonically decreasing and $\bigcap_{i \in N} Y_i = X$. Therefore, by our assumptions on the operator $f$,
$$f(X) = \bigcup_{i \in N} f(Y_i).$$
Again, we need to prove two inclusions, (a) $f(X) \subseteq GL_P(X)$ and (b) $GL_P(X) \subseteq f(X)$. For (a), note that if $p \in f(X)$, then for some $i \in N$, $p \in F(Y_i)$. Therefore, for that $i$, $p \leftarrow \neg q_0, \ldots, \neg q_i$ is a clause in $P$. But then $X \cap \{q_0, \ldots, q_i\} = \emptyset$ so that the clause $p \leftarrow$ is in $P_X$ and $p \in GL_P(X)$.

For the proof of (b), note that if $p \in GL_P(X)$, then because of the syntactic form of the clauses in our program there are atoms $r_0, \ldots, r_k$ so that the clause $p \leftarrow \neg r_0, \ldots, \neg r_k$ belongs to the program $P$, and $r_0, \ldots, r_k \notin X$. Thus $\{r_0, \ldots, r_k\} \subseteq \{q_0, q_1, \ldots\}$ and, hence, for some $i \in N$, $\{r_0, \ldots, r_k\} \subseteq \{q_0, \ldots, q_i\}$. Now, consider such a $Y_i$. Since $Y_i$ is cofinite, it follows from Case 1 that $f(Y_i) = GL_P(Y_i)$. Since $X \subseteq Y_i$, $f(Y_i) \subseteq f(X)$ by the antimonotonicity of $f$. But $p \in GL_P(Y_i)$ because $r_0, \ldots, r_k \notin Y_i$ and, hence, $p \in f(Y_i)$. Now, since $f(Y_i) \subseteq f(X)$, $p \in f(X)$, as desired. □
We are now ready to prove the main result of this paper.

**PROPOSITION 13.** *Let $P$ be a normal propositional program. The following are equivalent:*

(a) *P is an* FSP-*program.*

(b) *The operator* $GL_P$ *is upper-half continuous, i.e.*
$$GL_P(\bigcup_{n \in N} X_n) = \bigcap_{n \in N} GL_P(X_n)$$
*for every monotonically increasing sequence* $\langle X_n \rangle_{n \in N}$.

Proof: Two implications need to be proved: $(a) \Rightarrow (b)$, and $(b) \Rightarrow (a)$.
Proof of the implication $(a) \Rightarrow (b)$. Here, assuming $(a)$, we need to prove two inclusions:
(i) $GL_P(\bigcup_{n \in N} X_n) \subseteq \bigcap_{n \in N} GL_P(X_n)$, and
(ii) $\bigcap_{n \in N} GL_P(X_n) \subseteq GL_P(\bigcup_{n \in N} X_n)$.
To prove (i), note that since $X_n \subseteq \bigcup_{j \in N} X_j$, we have
$$GL_P(\bigcup_{j \in N} X_j) \subseteq GL_P(X_n).$$

As $n$ is arbitrary,
$$GL_P(\bigcup_{j \in N} X_j) \subseteq \bigcap_{n \in N} GL_P(X_n).$$

This proves (i).
To prove (ii), let $p \in \bigcap_{n \in N} GL_P(X_n)$. Then, for every $n \in N$, $p \in GL_P(X_n)$ and so, for every $n \in N$, there is an inclusion-minimal support $U$ for $p$ such that
$$U \cap X_n = \emptyset.$$
But by (a) there are only finitely many inclusion-minimal supports for $P$-proof schemes for $p$. Therefore there is a support of an inclusion minimal support of a proof scheme of $p$, $U_0$, such that for infinitely many $n$'s
$$U_0 \cap X_n = \emptyset.$$
But the sequence $\langle X_n \rangle_{n \in N}$ is monotonically increasing. Therefore for all $n \in N$, $U_0 \cap X_n = \emptyset$. But then
$$U_0 \cap \bigcup_{n \in N} X_n = \emptyset,$$
so that $p \in GL_P(\bigcup_{n \in N} X_n)$. Thus (ii) holds and the implication $(a) \Rightarrow (b)$ follows.

To prove that $(b) \Rightarrow (a)$, assume that the operator $GL_P$ is upper-half continuous. We need to show that for every $p$, the reduced defining equation for $p$ is finite. So let us assume that $rEq_p^P$ is not finite. This means that there is an infinite set $\mathcal{X} = \{U_1, U_2, \ldots\}$, where $U_1 \prec U_2 \prec \cdots$, such that

1. each $U_i$ is finite,

2. the elements of $\mathcal{X}$ are pairwise inclusion-incompatible, and

3. for every set of atoms $M$, $p \in GL_P(M)$ if and only if for some $U_i \in \mathcal{X}$, $U_i \cap M = \emptyset$.

We will now define two sequences:

1. a sequence $\langle K_n \rangle_{n \in N}$ of infinite sets of integers and

2. a sequence $\langle p_n \rangle_{n \in N \setminus \{0\}}$ of atoms.

We define $K_0 = N$, and we define $p_1$ as the first element $q \in U_1$ such that

$$\{j : q \notin U_j\}$$

is infinite. Clearly, $K_0$ is well-defined. We need to show that $p_1$ is well-defined. If $p_1$ is not well-defined, then for every $q \in U_1$ there is an integer $i_q$ such that for all $m > i_q$, $q \in U_m$. But $U_1$ is finite so taking $n = \max_{q \in U_1} i_q$, we find that for *all* $m > n$, $U_1 \subseteq U_m$ - which contradicts the fact that the sets in $\mathcal{X}$ are pairwise inclusion-incompatible. Thus $p_1$ is well-defined. We now set

$$K_1 = \{n \in K_0 : p_1 \notin U_n\} = \{n \in K_0 : \{p_1\} \cap U_n = \emptyset\}.$$

Clearly. $K_1$ is infinite.
Now, let us assume that we already defined $p_l$ and $K_l$ so that $K_l = \{n : U_n \cap \{p_1, \ldots, p_l\} = \emptyset\}$ is an infinite subset of $N$. We select $p_{l+1}$ as the first element $q \in U_{l+1}$ so that

$$\{j : j \in K_l \text{ and } q \notin U_j\}$$

is infinite. Clearly, by an argument as above, there is such $q$, and so $p_{l+1}$ is well-defined. We then set

$$K_{l+1} = \{j \in K_l : p_{l+1} \notin U_j\}.$$

Since $\{p_1, \ldots, p_l\} \cap U_j = \emptyset$ for all $j \in K_l$, $\{p_1, \ldots, p_{l+1}\} \cap U_j = \emptyset$ for all $j \in K_{l+1}$. By construction, the set $K_{l+1}$ is infinite.

Now, we complete the argument as follows. We set $X_n = \{p_1, \ldots, p_n\}$. The sequence $\langle X_n \rangle_{n \in N}$ is monotonically increasing. For each $n$, there is $j$, in fact infinitely many $j$'s, so that $X_n \cap U_j = \emptyset$. Therefore, for each $n$, $p \in GL_P(X_n)$. Hence $p \in \bigcap_{n \in N} GL_P(X_n)$.
On the other hand, let $X = \bigcup_{n \in N} X_n$. Then

$$X = \{p_1, p_2, \ldots\}.$$

By our construction, $p_n \in U_n$, and so $U_n \cap X \neq \emptyset$. Therefore $X$ does not admit *any* $P$-proof scheme for $p$. Thus $p \notin GL_P(X) = GL_P(\bigcup_{n \in N} X_n)$. But this would contradict our assumption that $GL_P$ is upper-half continuous. Thus there can be no such $p$ and hence $P$ must be a $FSP$-program. □

## 5 Extensions to $CC$-programs

In [31], Niemelä and coauthors defined a significant extension of logic programming with stable semantics which allows for programming with cardinality constraints, and, more generally, with weight constraints. This extension has been further studied in [26, 25]. To keep things simple, we will limit our discussion to cardinality constraints only, although it is possible to extend our arguments to any class of convex constraints [14]. *Cardinality constraints* are expressions of the form $lXu$, where $l, u \in N$, $l \leq u$ and $X$ is a finite set of atoms. The semantics of an atom $lXu$ is that a set of atoms $M$ satisfies $lXu$ if and only if $l \leq |M \cap X| \leq u$. When $l = 0$, we do not write it, and, likewise, when $u \geq |X|$, we omit it, too. Thus an atom $p$ has the same meaning as $1\{p\}$ while $\neg p$ has the same meaning as $\{p\}0$.

The stable semantics for $CC$-programs is defined via fixpoints of an analogue of the Gelfond-Lifschitz operator $GL_P$; see the details in [31] and [26]. The operator in question is neither monotone nor antimonotone. But when we limit our attention to the programs $P$ where clauses have the property that the head consists of a single atom (i.e. are of the form $1\{p\}$), then one can define an operator $CCGL_P$ which is antimonotone and whose fixpoints are stable models of $P$. This is done as follows.

Given a clause $C$

$$p \leftarrow l_1 X_1 u_1, \ldots, l_m X_m u_m,$$

we transform it into the clause

$$p \leftarrow l_1 X_1, \ldots, l_m X_m, X_1 u_1, \ldots, X_m u_m \tag{4}$$

[25]. We say that a clause $C$ of the form (4) is a $CC$-Horn clause if it is of the form

$$p \leftarrow l_1 X_1, \ldots, l_m X_m. \tag{5}$$

A $CC$-Horn program is a $CC$-program all of whose clauses are of the form (5). If $P$ is a $CC$-Horn program, we can define the analogue of the one step provability operator $T_P$ by defining that for a set of atoms $M$,

$$T_P(M) = \{p : (\exists C = p \leftarrow l_1 X_1, \ldots, l_m X_m \in P)(\forall i \in [m])(|X_i \cap M| \geq l_i)\}$$

where $[m] = \{1, \ldots, m\}$. It is easy to see that $T_P$ is a monotone operator and that the least fixed point of $T_P$ is given by

$$lfp(T_P) = \bigcup_{n \in N} T_P^n(\emptyset). \tag{6}$$

We can define the analogue of the Gelfond-Lifschitz reduct of a $CC$-program, which we call the *NSS*-reduct of $P$, as follows. Let $\bar{P}$ denote the set of all transformed clauses derived from $P$. Given a set of atoms $M$, we eliminate from $\bar{P}$ those clauses where some upper-constraint $(X_i u_i)$ is not satisfied by $M$, i.e. $|M \cap X_i| > u_i$. In the remaining clauses, the constraints of the form $X_i u_i$ are eliminated altogether. This leaves us with a $CC$-Horn program $P_M$. We then define $CCGL_P(M)$ to be the least fixed point of $T_{P_M}$ and say that $M$ is a $CC$-stable model if $M$ is a model of $P$ and $M = CCGL_P(M)$. The equivalence of this construction and the original construction in [31] for normal $CC$-programs is shown in [25].

Next we define the analogues of $P$-proof schemes for normal $CC$-programs, i.e. programs which consists entirely of clauses of the form (4). This is done by induction as follows. When

$$C = p \leftarrow X_1 u_1, \ldots, X_k u_k$$

is a normal $CC$-clause without the cardinality-constraints of the form $l_i X_i$ then

$$\langle\langle C, p\rangle, \{X_1 u_1, \ldots, X_k u_k\}\rangle$$

is a $P$-$CC$-proof scheme with support $\{X_1 u_1, \ldots, X_k u_k\}$. Likewise, when

$$S = \langle\langle C_1, p_1\rangle, \ldots, \langle C_n, p_n\rangle, U\rangle$$

is a $P$-$CC$-proof scheme,

$$p \leftarrow l_1 X_1, \ldots, l_m X_m, X_1 u_1, \ldots, X_m u_m$$

is a clause in $P$, and $|X_1 \cap \{p_1, \ldots, p_n\}| \geq l_1, \ldots, |X_m \cap \{p_1, \ldots, p_n\}| \geq l_m$, then

$$\langle\langle C_1, p_1\rangle, \ldots, \langle C_n, p_n\rangle, \langle C, p\rangle, U \cup \{X_1 u_1, \ldots, X_m u_m\}\rangle$$

is a $P$-$CC$-proof scheme with support $U \cup \{X_1 u_1, \ldots X_m u_m\}$. The notion of admittance of a $P$-$CC$-proof scheme is similar to the notion of admittance of $P$-proof scheme for normal programs $P$. That is, if $\mathcal{S} = \langle\langle C_1, p_1\rangle, \ldots, \langle C_n, p_n\rangle, \langle C, p\rangle, U\rangle$ is a $CC$-proof scheme with support $U = \{X_1 u_1, \ldots X_n u_n\}$, then $\mathcal{S}$ is admitted by $M$ if for every $X_i u_i \in U$, $M \models X_i u_i$, i.e. $|M \cap X_i| \leq u_i$.

Similarly, we can associate a propositional formula $\phi_U$ so that $M$ admits $\mathcal{S}$ if and only if $M \models \phi_U$ as follows:

$$\phi_U = \bigwedge_{i=1}^{n} \bigvee_{W \subseteq X_i, |W|=|X_i|-u_i} \neg W. \tag{7}$$

Then we can define a partial ordering on the set of possible supports of proof schemes by defining $U_1 \preceq U_2 \iff \phi_{U_2} \models \phi_{U_1}$. For example if $U_1 = \langle \{1, 2, 3\}2, \{4, 5, 6\}2\rangle$ and $U_2 = \langle\{1, 2, 3, 4, 5, 6\}, 4\rangle$, then

$$\phi_{U_1} = \bigvee_{1 \leq i < j \leq 6} (\neg i \wedge \neg j) \text{ and}$$

$$\phi_{U_2} = (\neg 1 \vee \neg 2 \vee \neg 3) \wedge (\neg 4 \vee \neg 5 \vee \neg 6).$$

Then clearly $\phi_{U_2} \models \phi_{U_1}$ so that $U_1 \preceq U_2$. We then define a normal propositional $CC$-program to be *FPS CC-program* if for each $p \in At$, there are finitely many $\preceq$-minimal supports of $P$-$CC$-proof schemes with conclusion $p$.

We can also define analogue of the defining equation $CCEq_p^P$ of $p$ relative to a normal $CC$-program $P$ as

$$p \Leftrightarrow (\phi_{U_1} \vee \phi_{U_2} \vee \cdots) \tag{8}$$

where $\langle U_1, U_2, \ldots\rangle$ is a list of supports of all $P$-$CC$-proofs schemes with conclusion $p$. Again up to a total ordering of possible finite supports, this formula is unique. Let $\Phi_P$ be the set $\{CCEq_p^P : p \in At\}$. Similarly, we define the *reduced defining equation* for $p$ relative to $P$ to be the formula

$$p \Leftrightarrow (\neg \phi_{U_1} \vee \neg \phi_{U_2} \vee \ldots) \tag{9}$$

where $U_i$ range over $\preceq$-*minimal* supports of $P$-$CC$-proof schemes for the atom $p$.

Then we have the following analogues of Propositions 4 and 5.

PROPOSITION 14. *For every normal propositional $CC$-program $P$ and every set $M$ of atoms, $M$ is a $CC$-stable model of $P$ if and only if the following two conditions hold:*

*(i) for every $p \in M$, there is a $P$-$CC$-proof scheme $S$ with conclusion $p$ such that $M$ admits $S$ and*

*(ii) for every $p \notin M$, there is no P-CC-proof scheme S with conclusion p such that M admits S.*

**PROPOSITION 15.** *Let P be a normal propositional CC-program. Then CC-stable models of P are precisely the propositional models of the theory $\Phi_P$.*

We also can prove the analogues of Propositions 9 and 10.

**PROPOSITION 16.** *For any CC-program P, the operator $CCGL_P$ is antimonotone.*

Proof: It is easy to see that if $M_1 \subseteq M_2$, then for any clause

$$C = p \to l_1 X_1, \ldots, l_m X_m, X_1 u_1, \ldots X_m l_m,$$

$M_2 \models X_i u_i$ implies $M_1 \models X_i u_i$. Thus it follows that $P_{M_2} \subseteq P_{M_1}$ and hence $lfp(T_{P_{M_2}}) \subseteq lfp(T_{P_{M_1}})$. □

**PROPOSITION 17.** *Let P be a normal propositional CC-program and M be a set of atoms. Then*

$$CCGL_P(M) = \{p: \text{ there exists a P-proof scheme } S \text{ such that } M \text{ admits } S, \text{ and } p \text{ is the conclusion of } S\}$$

Proof: Let us assume that $p \in CCGL_P(M)$, i.e. $p \in lfp(T_{P_M})$. Since $lfp(T_{P_M}) = \bigcup_{n \geq 1} T_{P_M}^n(\emptyset)$, we can easily show by induction on $n$ that if $p \in T_{P_M}^n(\emptyset)$, then there is a P-CC-proof scheme $S_p$ such $p$ is the conclusion of $S_p$ and $S_p$ is admitted by $M$.

Conversely, we can show, by induction on the length of the P-CC-proof schemes, that whenever there is P-CC-proof scheme $S$ admitted by $M$, then $p$ belongs to $lfp(T_{P_M})$. □

Next we prove that analogue of Proposition 11.

**PROPOSITION 18.** *For every normal CC-program P, the operator $CCGL_P$ is lower-half continuous.*

Proof: We need to prove that for every normal CC-program $P$ and every monotonically decreasing sequence $\langle X_n \rangle_{n \in N}$

$$CCGL_P(\bigcap_{n \in N} X_n) = \bigcup_{n \in N} CCGL_P(X_n).$$

We need to prove two inclusions: $\subseteq$, and $\supseteq$.
We first show $\supseteq$. Since

$$\bigcap_{j \in N} X_j \subseteq X_n$$

for every $n \in N$, it follows from the antimonotonicity of $CCGL_P$ that we have
$$CCGL_P(X_n) \subseteq GL_P(\bigcap_{j \in N} X_j).$$
As $n$ is arbitrary,
$$\bigcup_{n \in N} CCGL_P(X_n) \subseteq CCGL_P(\bigcap_{j \in N} X_j).$$
Thus the inclusion $\supseteq$ holds.

Conversely, let $p \in CCGL_P(\bigcap_{n \in N} X_n)$. Then, by Proposition 17, there must be a $CC$-proof scheme $S$ with support support $U = \{Y_1 u_1, \ldots, Y_n u_n\}$ and conclusion $p$ such that
$$|Y_i \cap \bigcap_{n \in N} X_n| \leq u_i \text{ for } i = 1, \ldots, n.$$
Since the family $\langle X_n \rangle_{n \in n}$ is monotonically descending, it follows that
$$Y_i \cap X_1 \supseteq Y_i \cap X_2 \supseteq \cdots.$$
Since $Y_i$ is finite, it is the case that if $|Y_i \cap \bigcap_{n \in N} X_n| \leq u_i$, then there is some $m_i$ such that $|Y_i \cap X_{m_i}| \leq u_i$. Hence if $m = \max(m_1, \ldots, m_n)$, then
$$|Y_i \cap X_m| \leq u_i \text{ for } i = 1, \ldots, n.$$
This, however, implies that $p \in CCGL_P(X_m)$, and thus
$$p \in \bigcup_{n \in N} CCGL_P(X_n).$$
As $p$ is arbitrary, the inclusion $\subseteq$ holds. Thus $CCGL_P(\bigcap_{n \in N} X_n) = \bigcup_{n \in N} CCGL_P(X_n)$. □

Next we can prove the analogue of the first half of Proposition 13.

PROPOSITION 19. *Let $P$ be a normal propositional $CC$-program. Then if $P$ is an* FSP-*program, the operator $CCGL_P$ is upper-half continuous, i.e.*
$$CCGL_P(\bigcup_{n \in N} X_n) = \bigcap_{n \in N} CCGL_P(X_n)$$
*for every monotonically increasing sequence $\langle X_n \rangle_{n \in N}$.*

Proof: We need to prove two inclusions:
(i) $CCGL_P(\bigcup_{n \in N} X_n) \subseteq \bigcap_{n \in N} CCGL_P(X_n)$, and

(ii) $\bigcap_{n \in N} CCGL_P(X_n) \subseteq CCGL_P(\bigcup_{n \in N} X_n)$.

To prove (i), note that since $X_n \subseteq \bigcup_{j \in N} X_j$, we have

$$CCGL_P(\bigcup_{j \in N} X_j) \subseteq CCGL_P(X_n).$$

As $n$ is arbitrary,

$$CCGL_P(\bigcup_{j \in N} X_j) \subseteq \bigcap_{n \in N} CCGL_P(X_n).$$

This proves (i).

To prove (ii), let $p \in \bigcap_{n \in N} CCGL_P(X_n)$. Then, for every $n \in N$, $p \in CCGL_P(X_n)$ and so, for every $n \in N$, there is a minimal support $U_n = \{Y_1^{(n)} u_1^{(n)}, \ldots, Y_{m_n}^{(n)} u_{m_n}^{(n)}\}$ for $p$ such that

$$|Y_i^{(n)} \cap X_n| \leq u_i^{(n)} \text{ for } i = 1, \ldots, m_n.$$

But there are only finitely many $\preceq$-minimal supports for $P$-$CC$-proof schemes for $p$. Therefore there is a support $U_0 = \{Z_1 w_1, \ldots, Z_t w_t\}$ for a $P$-$CC$-proof scheme with conclusion $p$ such that for infinitely many $n$'s

$$|Z_i \cap X_n| \leq w_i \text{ for } i = 1, \ldots, t.$$

But the sequence $\langle X_n \rangle_{n \in N}$ is monotonically increasing. Therefore for *all* $n \in N$,

$$|Z_i \cap X_n| \leq w_i \text{ for } i = 1, \ldots, t.$$

But since each $Z_i$ is finite, then it must be the case that

$$|Z_i \cap \bigcup_{n in N} X_n| \leq w_i \text{ for } i = 1, \ldots, t.$$

so that $p \in CCGL_P(\bigcup_{n \in N} X_n)$. □

We leave the question of whether the converse of Proposition 19 holds as an open problem. We note that, alternatively, one can give an alternative proof of Proposition 19 by giving a direct reduction of our $CC$-programs to normal logic programs using the methods of [8] and the distributivity result of [17]. However, such a reduction leads to an exponential blow up in the size of the representation.

## 6 Conclusions

We note that investigations of proof systems in a related area, SAT, play a key role in establishing lower bounds on the complexity of algorithms for finding the models. We wonder if there are analogous results in ASP. For achieving such a goal, we need to find and investigate proof systems for ASP. One candidate for such a proof system is provided in this paper by using $P$-proof schemes.

As noted in the introduction several proof systems have been developed for Answer Set Programming. However, all such proof systems must use *some* mechanism to block conclusions if they become inapplicable since we are dealing with a nonmonotonic logic. This creates significant technical problems which have been controlled either with semantics or with some intentional techniques such as modal logic (or both). The situation gets even worse when an additional layer of indeterminism is added as in the case of cardinality constraints or set constraints [31, 26]. To the best of our knowledge, the proof system considered in Section 5 is the first attempt to provide proof-theoretic foundations of Answer Set Programming with cardinality constraints. A similar system can be built for general set constraints. Hopefully, better and more tranparent proof systems of this sort will be discovered. Finally, we raise the question of whether such proof systems can be used to develop a deeper understanding of the complexity issues related to finding stable models.

## Acknowledgments

The research of the first author was supported by the National Science Foundation under Grant IIS-0325063. The research of the second author was supported by the National Science Foundation under Grant DMS 0654060. The authors express gratitude to anonymous reviewers for suggestions and corrections.

## BIBLIOGRAPHY

[1] Bonatti, P.A., Reasoning with infinite stable models. *Artificial Intelligence 156*:75–111, 2004.
[2] Bondarenko, A., Toni, F. and Kowalski, R.A., An Assumption-Based Framework for Non-Monotonic Reasoning. In *Logic Programming and Non-monotonic Reasoning*, LPNMR-93, L.M. Pereira and A. Nerode, eds., pages 171–189, MIT Press, 1993.
[3] Clark, K., Negation as failure. In *Logic and data bases*, H. Gallaire and J. Minker, Eds. Plenum Press, pages 293–322, 1978.
[4] Davey, B.A., and Priestley, H.A., *Introduction to Lattices and Order*, Cambridge University Press, 1992.
[5] Dung, P.M. and Kanchanasut, K., On the generalized predicate completion of non-Horn programs, *Logic Programming. Proceedings of the North American Conference*, pages 587–603, MIT Press, 1989.
[6] Doets, K., *From Logic to Logic Programming*, MIT Press, 1994.
[7] M.H. van Emden and R.A. Kowalski. The semantics of predicate logic as a programming language. *Journal of the ACM* 23:733–742, 1976.
[8] Ferraris, P., and Lifschitz, V., Weight constraints as nested expressions, *Theory and Practice of Logic Programming* 5:45-74, 2005.

[9] Ferraris, P., Lee, J. and Lifschitz, V., A generalization of Lin-Zhao theorem. *Annals of Mathematics and Artificial Intelligence* 47:79–101, 2006.

[10] Gebser, M. and Schaub, T., Generic Tableaux for Answer Set Programming, *Proceedings of 23$^{rd}$ International Conference on Logic Programming*, V. Dahl and I. Niemelä, eds. Springer Lecture Notes in Computer Science 4670, pages 119–133, 2007.

[11] Gelfond, M. and Lifschitz, V., The stable model semantics for logic programming. In *Proceedings of the International Joint Conference and Symposium on Logic Programming*, pages 1070–1080, 1988.

[12] Jonsson, B. and Tarski, A., Boolean Algebras with Operators. *American Journal of Mathematics* 73:891–939, 1951.

[13] Järvisalo, M. and Oikarinen, E., Extended ASP Tableaux and Rule Redundancy in Normal Logic Programs, *Proceedings of 23$^{rd}$ International Conference on Logic Programming*, V. Dahl and I. Niemelä, eds. Springer Lecture Notes in Computer Science 4670, pages 134–148, 2007.

[14] Liu, L. and Truszczyński, M., Properties of programs with monotone and convex constraints. In *Proceedings of the 20$^{th}$ National Conference on Artificial Intelligence*, M.N. Veloso amd S. Kambaheti, eds., pages 701-706, AAAI Press, 2005.

[15] Lifschitz, V., Foundations of logic programming, in *Principles of Knowledge Representation*, CSLI Publications, pages 69-127, 1996.

[16] Lifschitz, V. and Razborov, A., Why are there so many loop formulas. *Annals of Mathematics and Artificial Intelligence* 7:261–268, 2006.

[17] Lifschitz, V., Tang, L.R. and Turner, H., Nested expressions in logic programs, *Annals of Mathematics and Artificial Intelligence*, 25:369-389, 1999.

[18] Lin, F. and Zhao Y., ASSAT: Computing answer sets of a logic program by SAT solvers. *Proceedings of the 18$^{th}$ National Conference on Artificial Intelligence*, pages 112–117, 2002.

[19] Lloyd, J., *Foundations of Logic Programming*, Springer-Verlag, 1989.

[20] Marek, W., Nerode, A. and Remmel, J.B., Nonmonotonic Rule Systems I. *Annals of Mathematics and Artificial Intelligence*, 1:241–273, 1990.

[21] Marek, W., Nerode, A. and Remmel, J.B., Nonmonotonic Rule Systems II. *Annals of Mathematics and Artificial Intelligence*, 5:229-264, 1992.

[22] Marek, W., Nerode, A. and Remmel, J.B., How Complicated is the Set of Stable Models of a Logic Program? *Annals of Pure and Applied Logic*, 56:119-136, 1992.

[23] Marek, W., Nerode, A. and Remmel, J.B., The stable models of predicate logic programs. *Journal of Logic Programming* 21:129-154, 1994.

[24] Marek, W., Nerode, A. and Remmel, J.B., Context for belief revision: Forward chaining-normal nonmonotonic rule systems, *Annals of Pure and Applied Logic* 67:269-324, 1994.

[25] Marek, V.W., Niemelä, I. and Truszczynski, M. Logic programs with monotone abstract constraint atoms, *Theory and Practice of Logic Programming* 8:167–199, 2008.

[26] Marek, V.W. and Remmel, J.B. Set Constraints in Logic Programming. In *Logic Programming and Nonmonotonic Reasoning, Proceedings of the 7th International Conference* (LPNMR-04). *Springer Lecture Notes in Artificial Intelligence 2923*, pages 154–167, Springer-Verlag, 2004.

[27] Marek, V.W. and Remmel, J.B. On the Continuity of Gelfond-Lifschitz operator and Other Applications of Proof-theory in Answer Set Programming. In *International Conference on Logic Programming*, Springer Lecture Notes in Computer Science 5366, pages 223–237, Springer-Verlag, 2008.

[28] Marek, V.W. and Remmel, J.B. Guarded resolution and Answer Set Programming, *Theory and Practice of Logic Programming*, 11:111–123, 2010.

[29] Marek, W. and Truszczyński, M. *Nonmonotonic Logic*, Springer-Verlag, Berlin, 1993.

[30] Milnikel, R.S., Sequent Calculi for Skeptical Reasoning in Predicate Default Logic and Other Nonmonotonic Systems, *Annals of Mathematics and Artificial Intelligence* 44:1-34, 2005.

[31] Simons, P., Niemelä, I., and Soininen, T. 2002. Extending and implementing the stable model semantics. *Artificial Intelligence* 138:181–234, 2002.

www.ingramcontent.com/pod-product-compliance
Lightning Source LLC
Chambersburg PA
CBHW051038160426
43193CB00010B/988